Bryozoan Paleobiology

Bryozoan Paleobiology

Paul D. Taylor
Natural History Museum
London, UK

WILEY Blackwell

This edition first published 2020
© 2020 Natural History Museum

The right of Paul D. Taylor to be identified as the author of this work has been asserted in accordance with law.

Registered Office
John Wiley & Sons, Inc., 111 River Street, Hoboken, NJ 07030, USA
John Wiley & Sons Ltd, The Atrium, Southern Gate, Chichester, West Sussex, PO19 8SQ, UK

Editorial Office
9600 Garsington Road, Oxford, OX4 2DQ, UK

For details of our global editorial offices, customer services, and more information about Wiley products visit us at www.wiley.com.

Wiley also publishes its books in a variety of electronic formats and by print-on-demand. Some content that appears in standard print versions of this book may not be available in other formats.

Library of Congress Cataloging-in-Publication Data
Names: Taylor, Paul D., author.
Title: Bryozoan paleobiology / Paul D. Taylor, Natural History Museum, London, UK.
Description: Hoboken, NJ : Wiley-Blackwell, 2020. | Series: Topa topics in paleobiology | Includes bibliographical references and index.
Identifiers: LCCN 2020012771 (print) | LCCN 2020012772 (ebook) | ISBN 9781118455005 (paperback) | ISBN 9781118454985 (adobe pdf) | ISBN 9781118454992 (epub)
Subjects: LCSH: Bryozoa, Fossil. | Bryozoa–Biology. | Bryozoa–Ecology.
Classification: LCC QE798 .T39 2020 (print) | LCC QE798 (ebook) | DDC 564/.67–dc23
LC record available at https://lccn.loc.gov/2020012771
LC ebook record available at https://lccn.loc.gov/2020012772

Cover Design: Wiley
Cover Image: © Screw axes of the bryozoan *Archimedes* from the Carboniferous of Alabama
© Trustees of the Natural History Museum, London

Set in 9/11.5pt Trump Mediaeval LT Std by SPi Global, Pondicherry, India

V4556ADDC-7A51-4930-9561-8E99F779FEDF_070820

Contents

Preface

Until the early nineteenth century, natural historians were puzzled by organisms at the time known as zoophytes: were they animals (zoo-), plants (-phyte), or something in between? Perhaps they were even the common ancestors of animals and plants? Zoophytes as then conceived included sponges, corals, and coralline algae, as well bryozoans, the subject of this book. The so-called 'zoophyte problem' greatly engaged Charles Darwin when he set sail from Plymouth Sound on board HMS *Beagle* in December 1831. Indeed, Darwin's first scientific paper, which was read by his mentor Robert Grant before both the Wernerian and Plinian societies when Darwin was a medical student at the University of Edinburgh, had concerned species of zoophytes we now know to be the bryozoans *Flustra* and *Carbasea*. And he made detailed observations of the intriguing behaviour of the peculiar 'bird-head' structures in bryozoans dredged off Patagonia during the *Beagle* voyage (Keynes 2003).

Zoophyta has long been abandoned as a taxonomic group and we now know much more about the biology of the diverse animals formerly grouped together under this name. Nevertheless, our anthropocentric view of life still makes it difficult to comprehend these peculiar plant-like, colony-forming animals that are so very different from the dogs, spiders, and other animals we encounter daily. In their immobility and growth,

colonial animals resemble plants but they are not autotrophs that photosynthesize but are instead heterotrophs that must obtain their nutrition by consuming other plants or animals. The resemblance in form between benthic colonial animals and higher terrestrial plants reflects not only their sessile lifestyles but also a shared modular construction (Hughes 2005).

Major advances have been made in recent years in our knowledge of some types of colonial animals, especially corals reflecting their importance in reef ecosystems. But bryozoans have been far less intensively studied and remain poorly understood, particularly by non-specialists. This belies the fact that the Bryozoa are a diverse phylum, with more than 6000 named species living today and a predicted 5000 more yet to be described (Gordon and Costello 2016), and have ecological importance in many marine and some freshwater habitats at the present day (Plate 1). Their rich fossil record – comprising more than 1300 genera – gives them considerable geological significance as well: bryozoans are common fossils in Ordovician–Holocene rocks deposited in shallow marine environments (Plates 2–3). Sometimes bryozoan colonies inhabiting the seafloor have supplied sufficient carbonate skeletal material to the sediment to form bryozoan limestones.

While the utility of bryozoans in applied geology as zonal fossils or palaeoenvironmental indicators has been limited, which is one of the reasons they have attracted too little attention from geologists, Bryozoa are the best phylum in which to study the evolution of coloniality. Furthermore, their skeletons can preserve key aspects of their life histories such as the timing of sexual reproduction, and overgrowths between encrusting bryozoans provide rare instances in which competition is 'frozen' in the fossil record. Bryozoans living today and preserved as fossils are also intriguing and frequently enigmatic creatures that offer great opportunities for making new discoveries.

This book brings together information from the scattered literature on living and fossil bryozoans with the intention of providing a broad overview of the palaeobiology (and biology) of these fascinating animals. It updates the standard general text on bryozoans written by Ryland (1970), and the excellent book of McKinney and Jackson (1989) focusing on the adaptive morphology of bryozoans. Beginning with an introduction covering the basics of bryozoan morphology, ecology, and systematics, *Bryozoan Paleobiology* progresses from the smallest to the largest scale: from the skeleton and its microstructure, via zooid- and colony-level features and functions to biotic interactions, ecology, biogeography, phylogeny, and evolution. Important topics are highlighted – such as zooidal polymorphism – and critical areas for future research are identified.

Our understanding of bryozoan palaeobiology inevitably depends on making comparisons with the biology of living bryozoans, hence the large number of references to neontological studies. In common with other synthesis of this kind, personal experience has played a leading role on what is included and conversely what is excluded. While this undoubtedly colours many of the interpretations presented, I hope to have avoided unjustifiably strong biases.

Paul D. Taylor
London
November 2019

Acknowledgements

Serendipity invariably has a role in the career pathways of scientists and academics. I was fortunate in finding a Jurassic bryozoan during my independent geological mapping project while an undergraduate at the University of Durham in the early 1970s, and equally fortunate when, unbeknownst to me at the time, my main lecturer in palaeontology, Gilbert Larwood, just happened to be a bryozoan specialist. Gilbert took me under his wing, became my first mentor, and went on to supervise my doctoral research. I then enjoyed two years as a postdoc at the University College of Swansea where I was attracted by the presence of John Ryland's group studying living bryozoans, including Peter Hayward and John Thorpe to whom I owe a major debt for teaching me so much about bryozoan biology. Another influence in Swansea was Derek Ager. Derek was a highly original thinker about the fossil record and palaeoecology who did not receive the credit he deserved.

After Swansea I was appointed to a research post on bryozoans in the Department of Palaeontology at the then British Museum (Natural History). Pat Cook, my counterpart in the Department of Zoology, taught me never to be too dogmatic when making statements about bryozoans – there are many surprises waiting around the corner to embarrass the unwary in the study of these diverse, complex, and often enigmatic animals.

I will forever be grateful to Daphne Lee at the University of Otago who encouraged me to apply for a William Evans Fellowship there in the late 1980s. This proved to be my personal *Beagle* voyage. Keith Probert at the Portobello Marine Laboratory introduced me to the wonderful diversity of animals on the Otago Shelf that interact with bryozoans, not just the hermit crabs I was there to study, while Doug Campbell and Dave MacKinnon (Canterbury University) took me in the field to see the rich Cenozoic bryozoan faunas of New Zealand (Doug's son Hamish was later to guide me around the glorious Chatham Islands on two memorable geological fieldtrips). My collaborations with Dennis Gordon (NIWA, Wellington), the global authority on cheilostome bryozoans, describing the taxonomy of fossil and Recent bryozoans of New Zealand, began at this time. While at Portobello I was contacted by Richard Boardman (Smithsonian Institution) and F. Ken McKinney (Appalachian State University) who were partway through a study of the peculiar cyclostome bryozoan *Cinctipora*, one of the bioconstructional species on the Otago Shelf. Their invitation to join them in this research marked the beginning of my interest in bryozoan biomineralization, the skeleton of *Cinctipora* showing a striking ultrastructure (Boardman, McKinney, and Taylor 1992).

Before migrating to the world of ornithology, Mike Weedon undertook two postdocs with me at the NHM on bryozoan skeletal ultrastructures. Other scientists to whom I am indebted for various collaborative biomineralization projects include Chiara Lombardi (ENEA, La Spezia), Piotr Kuklinski (Polish Institute of Oceanology, Sopot), Noel James (Kingston University, Ontario), and Bill Schopf (UCLA).

The late F. Ken McKinney, who has already been mentioned, deserves my special thanks. He offered so many fresh insights and was a wonderful person to work with, as well as a great friend. To Dennis Gordon and Ken McKinney, I must add a third long-standing collaborator, Mark Wilson (College of Wooster, Ohio). It has been a delight working with Mark over the years and benefitting from his knowledge of hard substrate palaeoecology combined with a clarity of thought and expression that makes him such a superb teacher.

More recently, I have been fortunate to collaborate with two of the rising stars of bryozoology. Andrea Waeschenbach (Life Sciences, NHM) undertook a postdoc with Tim Littlewood and me on bryozoan molecular phylogeny, ferreting out the false data in GenBank and adding a lot of new data of her own to show the power of molecular sequences in understanding bryozoan evolution. Lee Hsiang Liow (University of Oslo) has been employing her analytical and modelling prowess to investigate biotic interactions through geological time, in particular competition for substrate space, a field where bryozoans have the potential to make a wider impact in macroecology and evolutionary ecology.

Supervision of doctoral students has served to broaden my perspectives. In chronological order, I thank them all: Richard Carthew, Julian Hammond, Caroline Buttler, Jon Todd, Kevin Tilbrook, Sian Evans, Phil Watts, Jo Snell, Lais Ramalho, Tanya Knowles, Scott Tompsett, Caroline Sogot, Emanuela Di Martino, and Peter Batson.

Others with whom I have been privileged to collaborate and learn from include Ehrhard Voigt, Eckart Håkansson, Andrei Grischenko, Andrej Ernst, Roger Cuffey, Alan Cheetham, Aaron O'Dea, Eckart Håkansson, Andrew Ostrovsky, Al MacGowan, Ma Junye, Seo Ji Yun, Shun Mawatari, Kamil Zágoršek, Noel James, Jeremy Jackson, Antonietta Rosso, Joachim Scholz, Beth Okamura, Tim Palmer, Andrew Smith, Silviu Martha, Urszula Hara, Patrick Wyse Jackson, Hans Arne Nakrem, Matt Dick, Scott Lidgard, David Jablonski, Abby Smith, Seabourne Rust, Kjetil Voje, Loïc Villier, Leandro Vieira, Francoise Bigey, Helen Jenkins, Consuelo Sendino, and Mary Spencer Jones.

I am grateful to the photographers of the NHM, particularly Harry Taylor, Phil Crabb, and Phil Hurst, for their skilled macrophotography. Most of the SEM images were taken by me during countless productive hours spent in the NHM's imaging and analysis laboratories where Alex Ball, Chris Jones, and several others were always at hand when help was needed. Piotr Kuklinski, Andrej Ernst, Caroline Buttler, and Thomas Schwaha are thanked for generously providing additional images used in this book.

The study of bryozoans has never attracted the funding it deserves – as Sir David Attenborough once remarked, you can't care about something unless you know it exists. Alas, far too few people are aware of the existence of these perenielly 'unfashionable' animals, let alone have any inkling of their fascinating natural history. Nevertheless, I have benefitted over the years from grants awarded by the Natural Environment Research Council (NERC), Leverhulme Trust, European Union, Royal Society, British Council, and the Japan Society for the Promotion of Science (JSPS), all of which I gratefully acknowledge.

Finally, I would like to thank Caroline Buttler, Andrea Waeschenbach, Lee Hsiang Liow, and Emanuela Di Martino for their comments on the manuscript, and Louise Spencely for her skilled copy editing.

1 Introduction

Once known as 'polyzoans' or 'ectoprocts', bryozoans are unique in being the only animal phylum in which the great majority, if not all, species form colonies. But what exactly constitutes a colony? In the context employed here it is an aggregate of genetically identical, conjoined modules, unlike a colony of sea-birds or of ants. Coloniality has evolved on numerous occasions among aquatic invertebrates. However, coloniality is not homologous between bryozoans and corals or hemichordates, although these independently evolved groups of colonial animals do exhibit several similarities. The most important feature of colonial invertebrates is their modular construction: new modules (zooids) are added during the growth of the colony and, with a few exceptions, remain physically attached to their neighbours throughout the life of the colony. The process of adding a zooid – termed budding – involves mitotic cell divisions only. Thus, all of the zooids in a bryozoan colony are genetically identical clones.

1.1 Zooids

Bryozoan zooids are small, typically measuring under a millimetre in length. However, because bryozoan colonies can contain a large number of zooids, individual colonies are typically a centimetre to a few decimetres in size. For example, seven-year-old colonies of *Flustra foliacea*, the so-called 'hornwrack', less than 10 cm tall have been estimated to contain more than 100 000 zooids (Stebbing 1971). Much larger colonies occur in a few bryozoan species: a 2-metre diameter colony of the Recent cheilostome *Pentapora foliacea* (the 'Ross coral') was recorded from British coastal waters in the early nineteenth century (Lombardi, Taylor, and Cocito 2010). In freshwater habitats, gelatinous masses formed by aggregations of colonies of the

Bryozoan Paleobiology, First Edition. Paul D. Taylor.
© 2020 Natural History Museum. Published 2020 by John Wiley & Sons Ltd.

phylactolaemate bryozoan *Pectinatella magnifica* can be as much as one metre in diameter (Dendy 1963; see also Cahuzac and d'Hondt 2017). And fossil trepostome bryozoan colonies up to 66 cm across have been recorded from the Late Ordovician of Kentucky (Cuffey and Fine 2005), and over a metre wide in Permian deposits of Tasmania (Reid 2003). At the other extreme are the tiny colonies of the ctenostome *Monobryozoon* and related genera, little more than a millimetre in length and consisting of a single feeding zooid plus stolons and detachable buds (Franzén 1960; Berge, Leinaas, and Sandøy 1985; Schwaha et al. 2019). Bryozoans are able to benefit from being simultaneously diminutive in terms of zooid size – bringing the advantages incurred by a large surface area relative to volume – but large with respect to colony size – enhancing their survival from various sources of mortality such as predation and physical disturbance.

In common with other colonial animals, bryozoan colonies are able to endure the death of one or more zooids in the colony. This phenomenon is known as **partial mortality**. Death of zooids can be due to their natural ageing, or external factors such as predation. In Mediterranean colonies of the palmate cheilostome *Pentapora fascialis* studied by Cocito, Sgorbini, and Bianchi (1998), local tissue necrosis caused by silt accumulation and algal overgrowth impacted mainly older colonies and caused collapse of their centres while younger zooids at the periphery of the colony continued to live.

As colonies grow, they accumulate an ever-increasing number of dead zooids in their older parts, forming a **necromass**. For example, the basal branches of bushy and foliaceous colonies may consist entirely of dead zooids, like the dead wood of trees (e.g. McKinney and Taylor 2006, fig. 3). The skeletons of zooids in the necromass remain functional in the sense of providing structural support for the feeding zooids closer to the branch tips. Describing these old zooids as dead is not strictly accurate. In at least some species, colonies may be reactivated if, for example, the colony is broken and reparative growth ensues.

One important implication of partial mortality is that bryozoan colonies may exhibit **negative growth** – with time, more zooids may be lost than new zooids budded, resulting in a net decrease in colony size. In a study of encrusting cheilostomes on settlement panels in Jamaica, Jackson and Winston (1981) found an increasing proportion of colonies showing negative growth through a period of two years. The existence of shrinking colonies means that colony age cannot always be estimated from size – colonial animals may 'lie about their age'.

The budding of new zooids in bryozoan colonies is usually restricted to specific growth zones, for example, around the perimeters of patch-like encrusting colonies, or at the branch tips of tree-like erect colonies. In some species, however, new zooids may be budded more widely across the entire outer surface of the colony.

The basic bryozoan zooid (Figure 1.1) consists of a body wall (**cystid**) containing a fluid-filled cavity (**coelom**) within which is suspended a **polypide** with a **lophophore** and **gut**. This arrangement has led to bryozoans occasionally being portrayed as 'animals-in-a-box', a concept that can be misleading because the box-like cystid, including the mineralized skeleton, if present, is actually an integral living part of the zooid rather than an inert container. Nevertheless, some degree of independence between the cystid and the polypide is evident in bryozoan zooids. This is most clearly manifested by the phenomenon of **polypide cycling** (Figure 1.2) in which the polypide periodically degenerates and is replaced by another within the same cystid, a process characterizing nearly all marine bryozoans (cf. the cheilostome *Epistomia bursaria* where the zooids have only one polypide generation: Dyrynda 1981). The lifespans of individual polypides range from a few days (Bayer and Todd 1997) to 9–10 months for some slow-growing species from the Antarctic (Barnes 2000). When placed under stress (e.g. low pH, simulating ocean acidification), polypide cycling is diminished in favour of the budding of new zooids (Lombardi et al. 2017; Swezey et al. 2017).

The remains of the degenerated polypide form a **brown body** (Gordon 1973, 1977). This is either retained within the coelom of the zooid, or defecated

Figure 1.1 Basic cheilostome bryozoan labelling some of the most important anatomical features of the zooids. Growth direction is from left (proximal) to right (distal). Only the zooid on the right has its lophophore expanded for feeding; in the other three, the lophophore is retracted into the security of the box-like skeleton (after Taylor, Lombardi, and Cocito 2015, fig. 1. © The Trustees of the Natural History Museum, London).

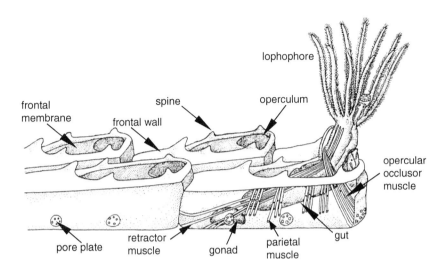

through the gut of the new replacement polypide. In the cheilostome *Steginoporella*, individual zooids can contain as many as 22 brown bodies (Palumbi and Jackson 1983), signifying an equivalent number of polypide cycles. As marine bryozoans lack a specialized excretory system, polypide cycling furnishes a means of removing waste material, crudely comparable with autumnal leaf fall in trees.

Although polypide cycling is a mechanism for prolonging the lifespans of zooids, they are not immortal and evidence for zooid senescence can be seen in the decreasing durations of successive polypides and increasing time required for a new polypide to be formed in zooids of some cheilostomes (Bayer and Todd 1997). In addition, older zooids tend to be more heavily fouled and exhibit slower regeneration of colony growth following damage (Palumbi and Jackson 1983).

Fossil evidence for the degeneration of polypides in bryozoans occurs in the form of diagenetically altered brown bodies – **brown deposits** – which are often seen in thin sections of Palaeozoic bryozoans

(e.g. Morrison and Anstey 1979; Ernst and Voigt 2002; Key et al. 2008; Plate 4A). Most brown deposits have a high iron content and some appear to have been pyritized. Calcified brown bodies are known in a few living cheilostomes (Cummings 1975; Gordon and Parker 1991a) but have yet to be reported in the fossil record. However, so-called phosphatic pearls or calculi, 30–600 µm in diameter, described from within the zooidal chambers of the Silurian cystoporate bryozoan *Favositella* by Oakley (1934), may perhaps represent encapsulated brown bodies (Lindskog et al. 2017, p. 36).

Polypides are the food gathering and digestive organs of bryozoans. They consist of an inverted cone-, bell- or horseshoe-shaped crown of tentacles (lophophore), with a mouth at the centre where the tentacles converge (Plate 5). The mouth leads into the U-shaped gut comprising a sucking pharynx (Nielsen 2013), stomach, pylorus, rectum and, in a few species, a gizzard (e.g. Gordon 1975). Bryozoan lophophores have between 8 and over 100 tentacles, depending on species. The largest lophophores are found in the freshwater phylactolaemates and

Figure 1.2 Schematic figure of polypide cycling in a bryozoan in which brown bodies are retained by the zooid. Starting from a single zooid (lower left), six zooids are budded in succession, making a total of seven zooids in the final growth stage depicted (vertical row on the far right). By the third growth stage, the polypide of zooid 1 has degenerated, leaving a brown body in the cystid, a fate repeated by the other, successively younger zooids as they age. However, through time, new polypides form within the old cystids, giving a second polypide generation. This in turn is followed by a second degeneration, as indicated by the presence of two brown bodies in the oldest zooid (lower right).

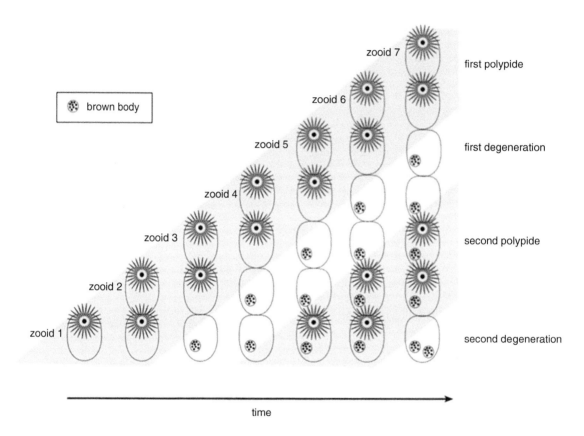

are horseshoe-shaped in plan view (Wood 1983; Plate 5E). All bryozoans are active suspension feeders. Marine bryozoans mainly consume phytoplankton (Chapter 3.1), but recent research by Wood (2019) on phylactolaemates has identified ciliate protists and rotifers as major food sources, which may explain the generally larger size of their zooids. The tentacles of the lophophore bear several lines of tiny, hair-like **cilia**, including a paired series of lateral cilia that beat in a metachronal wave to force water with entrained plankton through the open end of the lophophore and towards the mouth.

A key characteristic of bryozoans is the ability of zooids to withdraw the lophophore into the sanctuary of the cystid, expanding again into the water column to resume feeding. Consequently, feeding zooids in bryozoans always have an opening in the cystid for passage of the lophophore. Such openings in the zooidal skeleton are conventionally referred to as **apertures** in stenolaemates, and **orifices** in gymnolaemates (Figure 1.3).

Figure 1.3 Comparative external skeletal morphology of the feeding zooids in a cyclostome stenolaemate (A) and a cheilostome gymnolaemate (B). Note the simple, rounded aperture at the distal end of the cyclostome but the more complex orifice with lateral processes (condyles) at the distal end of the cheilostome zooid. Both zooids in these examples have calcified exterior frontal walls, that of the cyclostome pierced by tiny pseudopores, while that of the cheilostome comprises nine overarching spines fused laterally and along the midline of the zooid, each spine containing a few large pores (pelmata). A, *Reptomultisparsa incrustans* (Jurassic, Bathonian, Caillasse de la Basse-Ecarde; Ranville, Calvados, France). B, *Hayamiellina* aff. *constans* (Pleistocene, Setana Fm.; Kuromatsunai, Hokkaido, Japan). Scale bars = 200 μm.

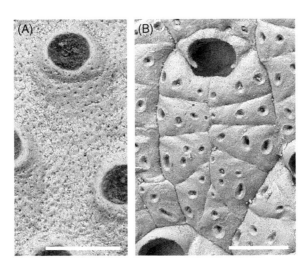

Withdrawal of the lophophore is achieved by contraction of retractor muscles that are attached to the base of the lophophore. Protraction is a more complicated process that relies on the contraction of muscles pulling on the cystid walls or a membrane, the precise mechanism varying in different major taxonomic group. Muscular contractions squeeze the coelom and cause the **tentacle sheath** – a sock-like structure enclosing the retracted lophophore – to turn inside out. This eversion of the tentacle sheath pushes the lophophore out through the orifice or aperture (Chapter 3.1). In some bryozoans, including most species belonging to the class Stenolaemata, the zooids have the shape of straight or gently curved tubes (Figure 1.4A), typically becoming progressively wider in diameter distally towards the skeletal opening for the lophophore. In other bryozoans, such as most species of the order Cheilostomata, the zooids are approximately box-shaped (Figure 1.4B) and

have a distinct frontal surface containing a distal opening through which the lophophore extrudes.

The outermost component layer of the cystid, at the interface with the external environment, is a thin organic coating called a **cuticle**. Like the periostraca of molluscs and brachiopods, this is secreted by an underlying epithelium. Those parts of the cystid wall that are deformed by the muscles involved in tentacle sheath eversion must be flexible and therefore cannot develop a rigid biomineralized layer beneath the cuticle. However, parts of the cystid not deformed by parietal muscles can potentially develop a biomineralized skeleton between the secretory epithelium and the outer cuticle (Chapter 2.1). Bryozoan skeletons are composed of the calcium carbonate biominerals calcite and aragonite. Both of these biominerals, calcite in particular, are relatively stable and have a high fossilization potential. The palaeontological literature often

Figure 1.4 Typically tubular zooidal skeletons of a stenolaemate compared with the box-shaped zooidal skeletons of a cheilostome: (A) fractured colony of the trepostome *Stenopora crinita* with long prismatic zooids intersected by prominent growth bands at the bottom and top of the image (Permian; Illawara, New South Wales, Australia); (B) broken colony of the cheilostome *Schizoporella errata* showing five frontally budded layers of box-like zooids (Recent; Haifa Bay. Israel). Scale bars: A = 1 mm; B = 500 μm.

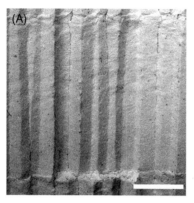
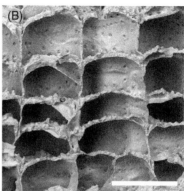

employs the terms **zooecium** (pl. **zooecia**) and **zoarium** (pl. **zoaria**), which are the skeletal parts of the zooid and colony, respectively. These terms are largely superfluous and are not used in this book.

1.2 Colonies

Based on zooids that range from tubular to box-like in shape, bryozoans have evolved a myriad of **colony-forms** (Plates 6–8, 10–12). Bryozoan zooids may be likened to building bricks that can be arranged in different ways to generate structures diverse in shape and varied in function. The inherent plasticity endowed by the modularity that characterizes bryozoans accounts for the large disparity in colony-forms (Chapter 4.1). Some colony-forms are common, particularly sheet-like encrustations, whereas others are found in only a handful of taxa. Identical colony-forms have evolved repeatedly in different bryozoan clades through the long geological history of the phylum (Chapter 9.11). Evolutionary convergence in colony-form can be viewed in terms of fitness landscape theory: within the limits imposed by various developmental, structural,

and phylogenetic constraints, particular colony-forms represent adaptive peaks – they are optimal morphological strategies for coping successfully with the challenges of living as a sessile benthic suspension-feeder.

An important consequence of the rampant convergent evolution of colony-forms in bryozoans is that the overall shape of a colony is seldom a reliable indicator of its taxonomic identity. Therefore, bryozoan taxonomy has been founded mainly on the morphological characters of the zooids. For bryozoan groups having mineralized skeletons, most of the characters used in taxonomy are skeletal, both in living and fossil bryozoans. Indeed, it is standard practise for living bryozoans to be bleached prior to identification, a process mimicking taphonomic loss of soft tissues during fossilization. This means that the same morphological characters are usually employed in the identification and classification of fossil bryozoans as those used for living taxa. Unlike some phyla, a parataxonomy for fossils has been unnecessary in bryozoans. Notwithstanding the identical taxonomic procedures and classifications employed for living and fossil bryozoans, it is becoming apparent from molecular studies that traditional skeletal characters are not always closely correlated with phylogeny (Chapter 8). Bryozoan classifications will need to be comprehensively overhauled if they are to reflect phylogeny more clearly.

Discordance in taxonomy can also result from the different methods employed to study the skeletons of fossil bryozoans. Palaeozoic bryozoans are most often studied in thin sections, which are seldom used for post-Palaeozoic bryozoans where surface features are employed. There are two main reasons for this contrast in study methods. The first is that Palaeozoic bryozoans more often occur in hard, strongly lithified rocks from which it is difficult to extract specimens showing reasonably well-preserved colony surfaces. Secondly, the dominant bryozoans of the Palaeozoic belonging to the superorder Palaeostomata have fewer external characters than the cyclostome and cheilostome bryozoans of the post-Palaeozoic. Thin sections of palaeostomate bryozoans are cut in three standard orientations – transverse, longitudinal, and tangential – to reveal internal skeletal characters and build up an understanding of the three-dimensional morphology of the skeleton (Figure 1.5).

1.2.1 How has coloniality arisen?

Coloniality has evolved multiple times in metazoans (e.g. Ryland 1981). It is easy to envisage how a colony may have developed from a unitary ancestor by reference to the well-known freshwater cnidarian *Hydra*. This animal reproduces asexually, budding clonal daughter individuals on the side of the body. These grow to a certain size and then drop off. If the buds had instead remained attached to the parent, then a colony would result. Retention of asexual buds to initiate coloniality has· been termed **clonoteny** (Rosen 1986). In the case of bryozoans, the unitary ancestor would have resembled a modern 'worm' of the lophophorate phylum Phoronida. The most reliable recent molecular phylogenetic analyses (Nesnidal et al. 2013; Laumer et al. 2019) have corroborated earlier anatomical suggestions (Hyman 1959) that phoronids are the sister-group of bryozoans. Whereas most phoronids are unitary animals, one

Figure 1.5 Cutaway diagram of a branch from a ramose trepostome colony (after Madsen and Håkansson 1989), showing the three standard sections employed by bryozoologists, as well as the division of the branch into an inner endozone, in which the zooids have thin walls and parallel the length of the branch, and an outer exozone in which zooids have thick walls and are oriented at right angles to branch length.

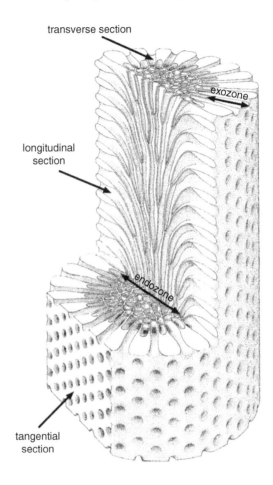

species (*Phoronis ovalis*) forms true colonies by budding, while several others are able to multiply asexually. The model of Farmer, Valentine, and Cowen (1973; see also Farmer 1977) for the origin of bryozoans starts from a unitary phoronid worm in which the asexual buds are retained. This is followed by size reduction of the zooids constituting the new colony. Because the genetic individual (genet) is the entire colony, the zooids can be small without sacrificing the advantages that large overall body size brings to competitive interactions, resistance to disturbance etc. The disproportionate relationship between body size and metabolic rate – metabolic allometry – acts as a constraint on the physiological rates of animals which can be ameliorated to some extent by having a modular colonial construction (see Hartikainen, Humphries, and Okamura 2014). Small zooid size brings with it the physiological advantages for metabolic exchange and efficiency of having a relatively large surface area/volume ratio (Ryland and Warner 1986). Indeed, bryozoan zooids dispense with the sophisticated circulatory and excretory systems found in larger unitary animals.

1.3 Colony Propagation

Sexual reproduction in bryozoans is important not only as a means of generating genetic variation, but also because the larvae produced by sexual reproduction are the principal agents of dispersal. Most bryozoan colonies – and often their constituent zooids – are hermaphroditic and produce both testes and ovaries (see Reed 1991 and Ostrovsky 2013 for reviews of bryozoan reproduction). In contrast, a few species of crisiid cyclostomes seem to be dioecious with colonies that are either male or female (Robertson 1903; Jenkins, Bishop, and Hughes 2015). Individual zooids are typically sequential hermaphrodites, functioning first as males and then as females. In some species, however, the sexes are separate and male and female zooids are distinct polymorphs that differ morphologically and, in some instances, lack the ability to feed (Chapter 3.6).

While self-fertilization between zooids belonging to the same colony is in theory possible, cross-fertilization between colonies is believed to be the norm. For example, colonies of the cheilostome

Celleporella kept in isolation fail to produce larvae (Cancino, Castañeda, and Orellana 1991) and invest less in producing female zooids (Hughes, Manriquez, and Bishop 2002). In another cheilostome, *Bugulina stolonifera*, selfing can occur if colonies are kept in isolation but results in fewer viable larvae and these are less likely to complete metamorphosis after settling (Johnson 2010).

The production of female polymorphs (gonozooids) in cyclostomes can be considerably reduced in the absence of sperm from other colonies (Jenkins, Bishop, and Hughes 2015). Sperm are released into the water through pores in the tips of the tentacles and must be acquired by female zooids from another colony, a process which is aided by the feeding currents of the recipient zooids drawing water towards themselves. The period of viability of water-borne sperm is limited, with an estimated half-life of just 1.2 hours in the cheilostome *Celleporella hyalina* (Manríquez, Hughes, and Bishop 2001). Depending on species concerned, fertilization can occur either externally or internally in gymnolaemate bryozoans (Temkin 1996), while in stenolaemates the location is unknown but is almost certain to be internal, with the sperm being gathered by the transient polypide present during the early development of the gonozooid. After fertilization of the egg, the embryo in most bryozoan species is brooded by the parent colony before being released as a swimming larva when sufficiently developed. The diverse styles of embryonic brooding or incubation found among bryozoans (e.g. Ström 1977) are discussed in Chapter 3.

Brooded bryozoan larvae are short-lived, incapable of feeding, and must settle quickly to establish a new colony before their provision of energy resources from the parent becomes exhausted. Therefore, settlement commonly occurs in close proximity to the parent colony (e.g. Mariani 2003), a phenomenon called **philopatry**. In the cheilostome *Bugula neritina* larvae also tend to settle close to sibling larvae (Keough 1984). Settlement is followed by metamorphosis, entailing radical changes in tissue organization and resulting in the founding individual of the new colony called the **ancestrula**. The ancestrula initially adheres to the substrate using a sticky acid mucopolysaccharide secretion from the pyriform organ of the larva (Loeb and Walker 1977). The first asexually budded zooids originate directly from the ancestrula and these in turn bud further zooids to continue colony growth. Rates of budding in new colonies vary greatly: fast-growing encrusting bryozoans such as *Conopeum tenuissimum* are capable of producing as many as 150 budded zooids in the first week of their life (Dudley 1973), while the bushy fouling cheilostome *Bugula neritina* may grow to a height of 20 mm in the first month of life (Mawatari 1951).

Although sexual reproduction is the main mechanism for colony multiplication in bryozoans, fragmentation of colonies provides an additional means of propagation in some species, especially in free-living species (see Chapter 4.4). Barnes, Webb, and Linse (2006b) found that nearly half of the colonies of the erect palmate cheilostome *Cellarinella nutti* they collected from the Weddell Sea, Antarctica, had grown from fragments of pre-existing colonies, a similar proportion to that of Neogene fossils of another palmate cheilostome, *Metrarabdotos*, from Venezuela (Cheetham et al. 2001).

A study of Late Cretaceous and Paleocene palmate colonies of the cheilostome family Coscinopleuridae (Håkansson and Thomsen 2001) revealed an increase in asexually propagated colonies through geological time at the expense of colonies formed by larvae. The same authors (Thomsen and Håkansson 1995) had earlier found that most of the erect species in their samples reproduced predominantly through fragmentation. In one of these – *Columnotheca cribrosa* – there was a strong positive correlation between the proportion of ovicellate brooding zooids in the colony and the proportion of sexually recruited colonies when samples from different habitats were compared (Figure 1.6). Colonies formed through fragmentation dominated in reef mounds, larvally recruited colonies in most other habitats.

Very few bases of attachment indicative of larvally recruited colonies are known in the

Figure 1.6 Correlation between the proportion of brooding zooids within colonies and that of sexually produced colonies in the cheilostome *Columnotheca cribrosa* from 10 samples collected in the Danian of Denmark (based on Thomsen and Håkansson 1995, fig. 6).

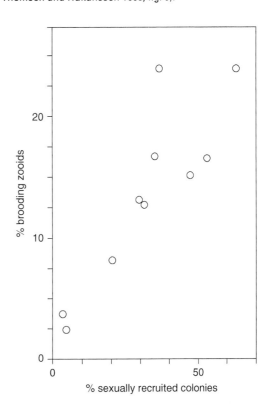

common Carboniferous fenestrate *Archimedes* (McKinney and Burdick 2001). Taken together with the abundant evidence of axial screws originating from the edges of meshworks, this led McKinney (1983) to propose that colony multiplication by fragmentation was a dominant process in *Archimedes* (Figure 1.7). Many of the larger colonies toppled and came to rest horizontally on the seabed. New screws grew from the edges of the prostrate colonies before eventually breaking off and forming the nuclei for a new generation of screws.

Fragments derived from a single colony are, of course, genetically identical. Borrowing from

Figure 1.7 Life history of the fenestrate bryozoan *Archimedes* (based on McKinney 1983, fig. 2). A colony recruited from a larva (left) is shown attached to a cylindrical substrate (as in McKinney and Burdick 2001). After toppling of the colony onto the sea-bed, a secondary screw axis grows vertically. Subsequent detachment of this screw axis initiates an asexual cycle of propagation.

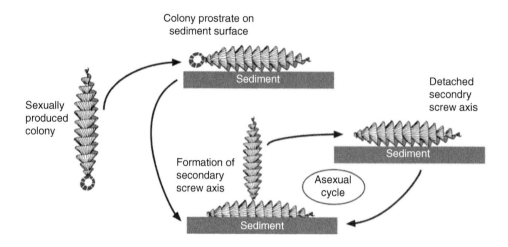

nomenclature devised for clonal plants, each fragment is called a **ramet**, whereas the entire complement of ramets is known as the **genet**. Occasionally, the ramets of bryozoans come back into contact during growth, in which case they are able to fuse. Fusion between parts of the same genet, including not only separated ramets but also separate lobes of a single integral colony, has been termed **autosyndrome**. Although less common than autosyndrome, fusion between two genetically distinct colonies of the same species can also occur (e.g. Craig and Wasson 2000; Hughes et al. 2004). This is known as **homosyndrome**. 'Colonies' produced by homosyndrome are **chimaeras** as they consist of zooids of more than one genotype. In other colonial animals (corals and tunicates), chimaeras show reduced fitness (Rinkevich and Weissman 1987), implying that homosyndrome should be selected against. Ishii and Saito (1995) described four possible outcomes when colonies of *Watersipora* (referred to by them as *Dakaria*) came into contact: (i) overgrowth of one by the other; (ii) back-to-back growth of the two colonies to produce a bilamellar erect structure; (iii) non-fusion stand-off (see Chapter 5.1);

and (iv) fusion between the colonies and subsequent growth as a continuous sheet. Fusion between colonies of another cheilostome, *Membranipora membranacea*, results in temporary neural integration, which enables behavioural coordination of lophophore retraction following disturbance, but there is no exchange of metabolites between the fused colonies (Shapiro 1996).

Production of asexual reproductive propagules called **statoblasts** (Figure 1.8) is ubiquitous in the freshwater phylactolaemate bryozoans. These resistant seed-like bodies have a chitinous covering and are used in dispersal, sometimes by becoming attached to the plumage of waterfowl, and also in overwintering. Statoblasts are able to survive freezing and desiccation (e.g. Hengherr and Schill 2011). The fossil record of statoblasts is poorly known: most examples are from the Holocene or Pleistocene (e.g. Francis 1997; Courtney Mustaphi et al. 2016), but they have been recorded back to the Late Triassic (Kohring and Pint 2005) and possibly even the Permian (Vinogradov 1996).

Hibernacula are incipient zooids with thickened cuticles found in some ctenostomes, such as *Victorella*, which function as asexual overwintering

Figure 1.8 Statoblasts produced by freshwater phylactolaemate bryozoans for asexual reproductive, dispersal and overwintering: (A) statoblasts of the living phylactolaemate *Cristatella mucedo* with hook-like spines around the annulus (Recent; Luxembourg); (B) putative fossil statoblast showing cracked convex capsule and marginal annulus (Lower Cretaceous, Korumburra Gp; South Gippsland, Victoria, Australia). Scale bars: A = 1 mm; B = 200 μm.

 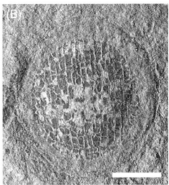

propagules (e.g. Carter et al. 2010) or as resting stages at times of food shortage (e.g. Jebram 1975).

'Overwintering' may also be achieved among some marine cheilostome bryozoans. For example, *Aetea* is capable of producing a special type of zooid called a **saccule** that persists on the substrate after loss of the other zooids in the colony and is capable of reinitiating colony growth (Balduzzi, Barbieri, and Gristina 1991).

A recent study of bryozoans from the Cretaceous Chalk (Taylor, Di Martino, and Martha 2018) found that some species encrusting echinoid tests apparently underwent periods of dormancy and regeneration, comparable to the overwintering described above for the living bryozoans. In a few species of cheilostomes, sealed, non-feeding zooids were budded around the margins of the sheet-like colonies and growth evidently ceased, later resuming with the budding of new feeding zooids. The best example is found in '*Micropora*' *eleanorae* where the sealed zooids survived on the substrate, despite the destruction of the autozooids which had more thinly calcified skeletons, to give rise to new generations of feeding zooids (Figure 1.9).

In *Hislopia*, specialized zooids break away from the parent colony and swim for a time before settling to found a new clonal colony (Wood, Anurakpongsatorn, and Mahujchariyawong 2006), a remarkable mode of asexual propagation known only from this single genus of freshwater ctenostomes.

1.4 Ecology

Bryozoans are widely distributed today across the globe, from polar regions to the equator. Diversities are high in some parts of the world. For example, no fewer than 556 species have been recorded living in the Mediterranean Sea (Rosso and Di Martino 2016).

Although most marine bryozoans inhabit subtidal environments on the continental shelf, they are known from the intertidal down to abyssal depths: the greatest recorded depth for a bryozoan is 8300 metres (Hayward 1981). Freshwater bryozoans are locally abundant in rivers and lakes. While most marine bryozoans can be categorized as stenohaline (i.e. intolerant of salinities differing significantly from normal marine), some species prosper in lower salinities (Winston 1977), including the brackish waters of estuaries and the inner parts of the Baltic Sea, as well as the fluctuating

Figure 1.9 Dormancy and regrowth in the Cretaceous cheilostome '*Micropora*' *eleanorae* (Campanian, Chalk Gp.; Norwich, Norfolk, UK): (A) autozooids (bottom) transitioning to closed kenozooids (top) believed to be formed during periods of colony dormancy; (B) regrowth from the edge of a band of kenozooids (bottom) through the formation of a new subcolony commencing with a pseudoancestrula (arrowed). Scale bars = 500 μm.

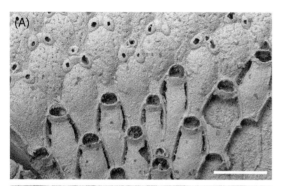

salinities found in coastal lagoons (see Chapter 6.1). Calcification tends to be relatively weak in these bryozoans. Menon and Nair (1974) remarked that during low salinity monsoonal times the euryhaline cheilostome *Einhornia crustulenta* in the Cochin backwaters of India showed little calcification and the normally calcified opercula consisted only of unmineralized cuticle.

Of fundamental importance for both marine and freshwater bryozoans are hard or firm substrates onto which the larvae can settle to establish new colonies (e.g. Eggleston 1972a). Commonly used substrates are rocks, shells, and plants, the latter including coastal algae and seagrasses in marine environments, and submerged branches of trees in freshwater environments. A few bryozoan species, most belonging to the cheilostome genus *Jellyella*, are specialized to grow on floating *Sargassum* or on the buoyant dead shells of the squid *Spirula* and consequently have a pelagic, pseudoplanktonic ecology (Taylor and Monks 1997). Another floating bryozoan is the Antarctic ctenostome *Alcyonidium pelagosphaerum* which forms hollow, ball-like colonies up to 23 mm in diameter and lacking an obvious substrate (Peck, Hayward, and Spencer Jones 1995; Porter and Hayward 2004).

All bryozoans are suspension feeders, consuming predominantly dinoflagellates. Some other potential sources of nutrition have been suggested. Exudates from seaweeds may be used as an alternative trophic resource by epiphytic bryozoans (De Burgh and Fankboner 1978; Manríquez and Cancino 1996), dissolved organic matter has also been proposed as a source of nutrition (Best and Thorpe 1991), while uptake of amino acids directly from seawater by the cheilostome *Bugula neritina* has been demonstrated (Stephens and Schinske 1961). However, there is as yet no indication that these contribute significantly to the diet of bryozoans, at least not for species inhabiting shallow-water environments.

Bacteria living within bryozoan tissues have also been hypothesized as nutritionally beneficial to the host (e.g. Karagodina et al. 2018). Perhaps of more interest in a palaeobiological context is the notion of bryozoan symbiosis with nutrition-providing zooxanthellae, given the effects of such symbioses on host skeletons, including hypercalcification, and reef-building capacities of other invertebrates through geological time. Zahl and McLaughlin (1957) mentioned the occurrence of zooxanthellae in the tissues of bryozoans but this has not been confirmed. Good candidates for bryozoans hosting symbiotic zooxanthellae are giant colonies of the trepostome *Tabulipora* sp. from the Early Permian of North Greenland with branches up to 7 cm thick. However, analysis of the stable light isotopes in skeletons of these bryozoans failed to find the expected higher $\partial^{13}C$ values, leading to rejection of the zooxanthellae hypothesis (Key et al. 2005). Symbiosis was also dismissed as a cause of gigantism in some Moroccan Late Ordovician bryozoans by Jiménez-Sánchez, Vennin, and Villas (2015) because the trepostomes concerned lived in low mesophotic to oligophotic conditions and at a high palaeolatitude.

Bryozoans themselves feature in the diets of a wide range of animals, notably pycnogonids (sea spiders) and nudibranchs (sea slugs) (see Chapter 5.2). Some bryozoan predators take single zooids at a time, while others consume the entire colony or large parts of the colony.

1.4.1 How long do bryozoan colonies live and how fast do they grow?

The lifespan of individual colonies varies according to species. Many species living in latitudes experiencing strong seasonality survive for a year or less (e.g. Eggleston 1972b). Colonies of some such species die after sexual reproduction. Their demise in the winter can be due to the deterioration or destruction of their substrates. Bryozoans such as *Membranipora membranacea* living as epiphytes of macroalgae may perish when the fronds they colonize become detached or decay, which occurs annually in some algae (e.g. Seed

et al. 1981; Cook, Bock, and Gordon 2018, p. 73). In polar environments winter ice-scour can totally obliterate benthic communities annually, including any bryozoans (Conlan et al. 1988). At the other end of the spectrum are perennial colonies. Conspicuous annual growth bands are evident in the skeleton of some long-lived cheilostomes (Plate 7A). For example, *Pentapora* has dense but narrow winter bands that alternate with broad summer bands (Lombardi et al. 2008). Similar bands have allowed colony ages to be estimated in some living species (Stebbing 1971; Brey et al. 1998; Barnes, Webb, and Linse 2006a, b). In the Antarctic cheilostome, *Melicerita obliqua*, growth bands reveal a maximum age of 45 years (Bader and Schäfer 2004). The congeneric *Melicerita chathamensis* from New Zealand has smaller colonies that live for about nine years, growing at a linear rate of about 5.3 mm per annum and adding approximately 110 zooids each year to the single palmate branch of the colony (Smith and Lawton 2010; Key et al. 2018).

Bryozoan growth rates, which have been quantified in several ways, vary enormously between colonies of the same species from different localities, and even more so between different species. For example, Smith (2014) compiled published data on linear growth rates which are summarized here in Figure 1.10. Most of the studied colonies

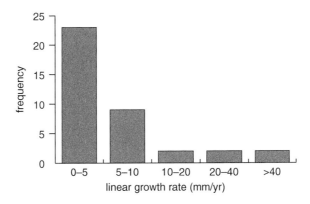

Figure 1.10 Linear growth rates of bryozoan species using data compiled from the literature by Smith (2014, supplementary material).

grew at less than 5 mm per annum but some grew much more quickly. Possibly the highest linear growth rate recorded is from the runner-like encrusting cheilostome *Aetea* in which branch tips can advance across the substrate at 72 cm per annum (Jackson and Coates 1986). Kuklinski et al. (2012) reported growth rates of sheet-like encrusting bryozoans on settlement panels across a latitudinal gradient of 44–78°N in Europe. They found growth rates in terms of colony area to be lower at higher latitudes, both overall and for particular taxa, e.g. colonies of the cyclostome *Diplosolen* from the Adriatic Sea grew at a rate of about 180 mm^2 per annum compared to 4 mm^2 per annum in Spitsbergen.

Sokolover, Ostrovsky, and Ilan (2018) found the growth rate of *Schizoporella errata* to be much greater in high ambient flow velocities than in low flow, echoing findings from other taxa (e.g. Hughes and Hughes 1986). Both field- and laboratory-reared colonies of the cheilostome *Membranipora membranacea* showed significant increases in growth rate with increasing temperature, with a roughly fivefold increase in mean growth rate between colonies in the laboratory kept at 6°C and 14°C (Saunders and Metaxas 2009). A prodigious rate of growth was recorded by Cocito et al. (2006) in colonies of the erect cheilostome *Pentapora fascialis* living close to submarine freshwater springs in the Adriatic Sea. Here colonies grew at almost 10 cm per year, compared to a 'normal' rate of growth of about 3 cm per year elsewhere in the Mediterranean. High concentrations of bicarbonate in the freshwater of the springs may be a key factor in the higher growth rate in this case.

Growth bands interpreted as annual increments have been noted in a few fossil bryozoans. Using this evidence, a large cyclostome colony from the Late Cretaceous was estimated as having lived for more than 35 years on the basis of its growth bands (Taylor and Voigt 1999), while some trepostome colonies from the Permian of Tasmania may have reached 75 years in age (Reid 2014). Cyclical variations in zooidal aperture spacing in a colony of the cryptostome *Rhombopora* from the Carboniferous of Ireland (Hageman et al. 2011) could be detected at average frequencies of every 5.3, 9.4, and 23.3 zooids. On the basis of growth rates in modern bryozoans, the 23.3 zooid cycle was interpreted as annual and the 5.3 cycle as lunar/tidal, allowing the inference to be made that the 20 cm-tall colony lived for about 20 years. There remains considerable scope for studies of geochemical signatures to estimate growth rates and colony ages in fossil bryozoans, extending research using isotopic profiles in some Recent species (e.g. Pätzold, Ristedt, and Wefer 1997).

1.5 Taxonomy

Three bryozoan classes – Phylactolaemata, Gymnolaemata, and Stenolaemata – have long been recognized on the grounds of major anatomical differences. Their status as monophyletic clades has been confirmed recently by molecular phylogenetic studies (Waeschenbach, Taylor, and Littlewood 2012). Phylactolaemates form the sister-group of gymnolaemates + stenolaemates. Whereas all phylactolaemates living today are found in freshwater environments, the other two classes have an overwhelmingly marine distribution. Key morphological features of the three bryozoan classes are summarized in Table 1.1.

Order-level subdivisions are employed in gymnolaemates and stenolaemates but not in the low diversity phylactolaemates in which there are fewer than 100 described species. The two gymnolaemate orders are: ctenostomes, which lack mineralized hard parts, and cheilostomes, which are characterized by biomineralized skeletons as well as zooidal orifices closed by a lid-like operculum. Elevation of Ctenostomata and Cheilostomata to subclass level has been proposed recently by d'Hondt (2016), together with numerous other categorical rank adjustments. However, large-scale changes to the classifications used for gymnolaemates (and other bryozoans) must await a better understanding of bryozoan phylogeny that will emerge as molecular sequence data becomes available in a greater number of taxa.

Table 1.1 Morphological characteristics of the three classes of bryozoans.

	Phylactolaemata	Gymnolaemata	Stenolaemata
Mineralized skeleton	absent	present or absent	present
Zooid shape	tubular	tubular or box-like	tubular
Polymorphism	lacking	present	present
Lophophore	large and usually horseshoe-shaped	small to moderate in size and circular	small and circular
Musculature	intrinsic body wall muscles	parietal muscles	parietal muscles
Epistome[*]	present	absent	absent
Lophophore frontal cilia	present	present	absent
Habitat	freshwater	marine, rarely brackish or freshwater	marine

[*] the **epistome** is a flap above the mouth, shared with Phoronida and believed by some to represent a separate body cavity – the protocoel – an interpretation contested by Gruhl, Grobe, and Bartolomaeus (2005) for phoronids, and later by Gruhl, Wegener, and Bartolomaeus (2009) for phylactolaemates.

1.5.1 Ctenostomata

Most modern ctenostome species are 'weedy' and inconspicuous, although a few develop bushy or gelatinous colonies of considerable size. For an impression of colony-form disparity in recent ctenostomes, see d'Hondt (1983) and Hayward (1985), while the paper on Brazilian ctenostomes by Vieira, Migotto, and Winston (201a) contains excellent photographs of living colonies.

Cryptopolyzoon is unusual in agglutinating grains of sand which stick to the zooids, in some instances almost entirely covering the colony (Dendy 1889). The small pedunculate ctenostome *Clavopora* has a stalk of muscular heterozooids which can contract to alter the position and orientation of the head on which the feeding zooids are situated (Mawatari 1968).

Aside from their lack of a mineralized skeleton, a characteristic structure found in ctenotomes, plus a few cheilostomes (e.g. *Cellaria*: Perez and Banta 1996), is the setigerous or pleated collar. This is a thin membrane with a cuff- or collar-like form that emerges just before the lophophore is protruded and can be reinforced by rods (M.J. McKinney and Dewel 2002). Its function may be to push aside debris and generally prevent the lophophore from being fouled in the muddy habitats colonized by many ctenostomes.

Despite their soft-bodied nature, ctenostome bryozoans can be found in the fossil record, either as borings in calcareous substrates (Chapter 4.8), or as **bioimmurations** (Figure 1.11) resulting from overgrowth by other organisms with hard skeletons (e.g. Voigt 1979; Taylor 1990a, b; Todd 1994; Todd, Taylor, and Favorskaya 1997). Bioimmurations can take the form of natural moulds on the underside of the overgrowing organism (Figure 1.11A, G), or casts on the substrate surface when the spaces left by decay of the bryozoan zooids are filled by diagenetic minerals such as calcite or pyrite (Figure 1.11B–D, F). A related, non-biogenic process – **lithoimmuration** – may have been responsible for the preservation of the ctenostome *Pierrella larsoni* encrusting the insides of the body chambers of baculite ammonites in the Late Cretaceous Western Interior Seaway of the United States (Wilson and Taylor 2012; Figure 1.11E). Early diagenetic growth of authigenic calcite associated with formation of the concretions containing the ammonites seems to have moulded the bryozoan zooids before their decomposition. In addition, outlines of encrusting ctenostome zooids

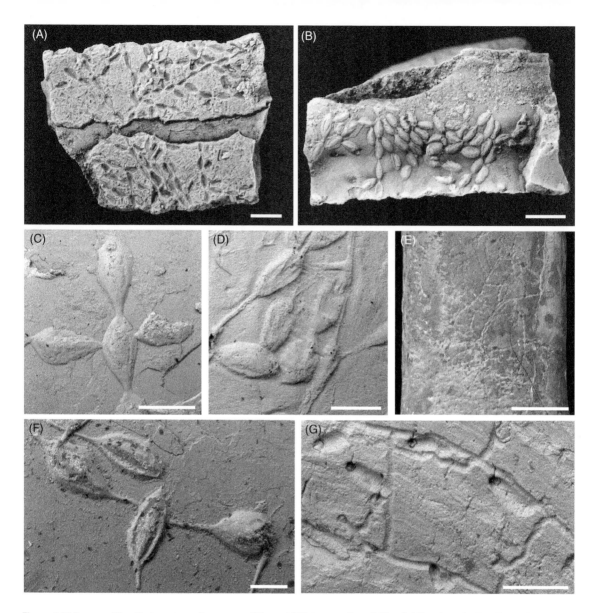

Figure 1.11 Immured fossil ctenostome bryozoans: (A) mould bioimmuration of *Simplicidium brandesi* on the underside of a serpulid annelid tube (Cretaceous, Berriasian; near Balki, Crimea); (B) natural cast biommuration of another colony on the outside of the same serpulid; (C) cruciate branching pattern in a natural cast bioimmuration of *Simplicidium smithii* (Jurassic, Kimmeridgian; South Ferriby, Lincolnshire, UK); (D) artificial cast of a bioimmured zooids of an un-named arachnidiid (Triassic, Anisian, Upper Muschelkalk; Künzelsau-Garnberg, Baden-Würtemberg, Germany); (E) steinkern of the body chamber of a baculite ammonite with impressions of zooids of the ?lithoimmured *Pierrella larsoni* (Cretaceous, Campanian or Maastrichtian, Pierre Shale; Red Bird, Montana, USA); (F) natural cast bioimmuration of an un-named arachnidiid, each zooid with a pair of longitudinal grooves possibly marking the edge of the frontal membrane (Jurassic, Upper Callovian or Early Oxfordian, Oxford Clay Fm.; Stanton Harcourt, Oxfordshire, UK); (G) mould bioimmuration of *Buskia waiinuensis* with a stolonal network to which were attached formerly erect autozooids flattened against the substrate during overgrowth by a cyclostome bryozoan from bottom right towards top left (Pleistocene, Nukumaru Limestone; Waiinu Beach, Whanganui, New Zealand). Scale bars: A, B = 2 mm; C, D, G = 500 μm; E = 10 mm; F = 200 μm.

can be preserved on shell substrates through a process termed **epibiont shadowing** (Palmer, Taylor, and Todd 1993). Described from the Jurassic, these traces are formed when the surface of the shell not covered by the bryozoan was bored by microendoliths, creating a pale background to the dark, unbored patches where the bryozoan zooids were situated.

Putative ctenostome body fossils include *Demafinga pennsylvanica* (Plate 7E) from the Carboniferous of Pennsylvania which was described as an *Alcyonidium*-like ctenostome by Cuffey, Dodge, and Skema (2014), and *Syringothenia bystrowi* from the Middle Ordovician of the St Petersburg region redescribed by Viskova and Ivatsov (1999) as an alcyonidiid. As it is unclear how either of these soft-bodied, possible ctenostomes came to be preserved, their affinities require further research for confirmation.

Other Palaeozoic fossils (notably *Ascodictyon* and *Allonema*) with calcareous skeletons previously identified as ctenostomes (e.g. Ulrich and Bassler 1904; d'Hondt and Horowitz 2007) are now regarded as Problematica (Olempska and Rakowicz 2014; Wilson and Taylor 2014).

1.5.2 Cheilostomata

More than 80% of bryozoan species living at the present-day belong to the Cheilostomata (Bock and Gordon 2013), yet this order first appeared in the fossil record as recently as the Late Jurassic and remained uncommon until the mid-Cretaceous (Taylor 1988b). The classification of cheilostomes is in a state of flux. The two traditional suborders – Anasca and Ascophora – are clearly morphological grades, anascans being paraphyletic and ascophorans polyphyletic. Attempts to replace these with suborders more reflective of phylogeny have begun in recent years. However, a robust new classification must await the molecular sequencing of a far greater number of taxa and the construction of a more complete molecular tree than currently available. The rich fossil record and importance of cheilostomes means that, unlike ctenostomes but like stenolaemates, they recur throughout this book and are not discussed further at this point.

1.5.3 Stenolaemata

All stenolaemates have calcareous skeletons and the prolific fossil record of the class begins in the Early Ordovician (Ma *et al.* 2015). At the time of the first bryozoan part of the *Treatise on Invertebrate Paleontology* (Bassler 1953), only three stenolaemate orders were recognized: Cyclostomata, Trepostomata, and Cryptostomata. However, the new orders Cystoporata, Esthonioporata, and Fenestrata have since been separated off, respectively from Cyclostomata, Trepostomata, and Cryptostomata. An additional order – Timanodictyina (Morozova 1966) – has also been recognized, making a total of seven stenolaemate orders in all (Figure 1.12). Cryptostomata is occasionally subdivided into two orders – Cryptostomida and Rhabdomesida – while to avoid homonymy of jawless fishes with the same name, Cyclostomata is sometimes referred to as Tubuliporida (Gorjunova 1996). Both the phylogenetic relationships between stenolaemate orders, and indeed the distinctiveness and validity of these orders, remain to be reliably resolved.

The dominantly Palaeozoic stenolaemate orders (i.e. all apart from the extant cyclostomes) are now placed in the subclass Palaeostomata, erected recently by Ma, Buttler, and Taylor (2014). Palaeostomates have a 'free-walled' skeletal organization lacking calcified frontal exterior walls and most possess styles embedded within their interior walls. Table 1.2 summarizes the morphological features, and geological ranges, of the seven stenolaemate orders.

1.6 Collecting and Studying Fossil Bryozoans

Fossil bryozoans are collected in the field in two main ways: as individually picked specimens or in bulk samples. The first of these methods is most useful for obtaining larger erect colonies and their fragments. Encrusting colonies can be found by seeking potentially encrusted substrates, such as brachiopods in the Palaeozoic

Figure 1.12 Examples of fossil bryozoans belonging to the eight orders possessing mineralized skeletons: (A) Cheilostomata – *Copidozoum planum* (Plio-Pleistocene, Burica Fm.; Burica Peninsula, Panama); (B) Cyclostomata – *Hyporosopora rugosa* (Jurassic, Kimmeridgian; Ponte du Chay, Charente Maritime, France); (C) Esthonioporina – *Nekhorosheviella* sp. (Ordovician, Upper Tremadocian, Fenhsiang Fm.; Hubei, China); (D) Cystoporata – *Fistuliphragma* sp. (Devonian, Eifelian, Upper Junckenberg Fm.; Gondelsheim, Prüm syncline, Germany); (E) Trepostomata – *Monotrypa pulchella* (Silurian, Wenlock, Much Wenlock Limestone Fm.; Dudley, West Midlands, UK); (F) Cryptostomata – *Nematopora lineata* (Ordovician. Ashgill, Ojl Myr flint; Gotland, Sweden); (G) Fenestrata – silicified phylloporinid ?*Moorephylloporina typica* (Upper Ordovician, Edinburg Fm.; Strasburg, Virginia, USA); (H) Timanodictyina – tangential thin section of *Timanodictya* sp. (Permian; USA). Scale bars: A, F, H = 200 µm; B, C, D = 500 µm; E = 2 mm; G = 1 mm.

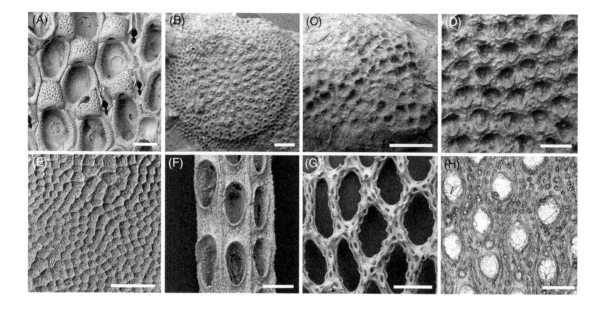

and oysters in the Mesozoic and Cenozoic. Fortunately, localities depleted in other macrofossils favoured by fossil enthusiasts may remain rich in bryozoans, and bryozoan-encrusted shells are often discarded by collectors in search of more pristine shells. Fragments of erect bryozoans, free-living cheilostomes and encrusters detached from their substrates are best recovered from bulk samples after washing and sieving. The majority of fossil bryozoans occur in shallow-water carbonates, and interbedded limestones and shales can be particularly bryozoan-rich in the Palaeozoic (Plate 2A).

Robust fossil colonies can be prepared by cleaning uncemented sediment from the colony surfaces using a soft brush under running water, while all but the most fragile of colonies benefit from treatment in an ultrasonic bath. Silicified bryozoans can be etched from limestones using acids (e.g. Elias and Condra 1957; Tavener-Smith 1973a; Gautier, Wyse Jackson, and McKinney 2013). However, silicification is often surficial and internal walls are not always preserved (Taylor and Curry 1985), while in some cases the fidelity of silicification is too coarse for satisfactory preservation of zooid-level morphology (e.g. Schneider et al. 2013). Phosphatic steinkerns (internal moulds), which have been very little studied, may also be recovered using acid dissolution (e.g. Datillo et al. 2016). Steinkerns of bryozoan zooids

Table 1.2 Main morphological features and geological ranges of the orders of stenolaemate bryozoans. Order names were derived as follows: Cyclostomata = round mouthed, referring to the shape of the skeletal aperture; Esthonioporata = porous bryozoan from Estonia; Cystoporata = referring to the vesicular cystopores found in many species; Trepostomata = derived from 'trepos' meaning to turn, apparently referring to the zooids bending from the endozone into the exozone; Cryptostomata = hidden mouthed, referring to the notion that the mouth is located beneath the skeletal aperture, at the position of the hemisepta; Fenestrata = derived from fenestra, or window, in reference to the openings (fenestrules) in the often reticulate colonies; Timanodictyina = bryozoan from Timan, Russia.

	Colony-form	Calcified exterior frontal walls	Pores in skeletal walls	Zooidal polymorphism	Range
Cyclostomata	Varied	yes/no	yes	common: gonozooids, kenozooids, nanozooids	Ordovician–Recent
Esthonioporata	Dome-shaped or ramose	no	no	generally lacking	Ordovician–Devonian
Cystoporata	Varied	no	rarely	rare	Ordovician–?Triassic
Trepostomata	Varied	no	no	common: mesozooids and exilazooids	Ordovician–Triassic
Cryptostomata	Ramose, bifoliate or cribrate	no	no	occasional exilazooids	Ordovician–Triassic[**]
Fenestrata	Mostly reticulate or pinnate[***]	no	no	uncommon[*]	Ordovician–Permian
Timanodictyina	Ramose or bifoliate	no	no	rare	Devonian–Permian

[*] a large number of polymorphs have been named but their validity it is uncertain as it is unclear whether they represent functionally distinct types of zooids; [**] *Tebitopora* is a cryptostome from the Triassic of Tibet (Hu 1984); [***] *Schischatella* is a rare example of an encrusting fenestrate bryozoan (Ernst and Bohaty 2009).

are very occasionally found in the fine fraction of bulk sediment samples (Hara 2015). Decalcified bryozoans preserved as moulds are difficult or impossible to identify, especially those of palaeostomates in which the interiors of colonies are lacking because the presence of basal diaphragms prevents penetration of sediment.

As already mentioned, the study of most Palaeozoic bryozoans relies on the preparation of thin sections, preferably cut in three standard orientations (Figure 1.5): longitudinal, tangential, and transverse (Wyse Jackson and Buttler 2015). These can be prepared using rock samples containing bryozoans, in which case obtaining standard orientations may depend on chance

intersections, or from loose branches that may first need to be embedded in resin to avoid loss of the outer parts of the specimen and to aid manipulation. As an alternative to thin sectioning, cellulose acetate peels can be prepared from polished and etched surfaces flooded with acetone followed by the application of acetate sheet that is lifted off when dry to reveal an impression of the cut surface. Serial peels have been employed to reconstruct three-dimensional morphology and budding patterns (e.g. Boardman and McKinney 1976).

Post-Palaeozoic bryozoans are studied predominantly using surface morphology. Scanning electron microscopy (SEM) has become an essential tool for observing small-scale morphological features as well as producing clear images at the level of the zooid. The advent of low-vacuum and environmental chamber instruments that do not require coating of specimens with a conducting metal has revolutionized bryozoan SEM, allowing type and other valuable specimens to be studied unaltered (Taylor and Jones 1996). Coupled with voluminous specimen chambers, this has also permitted large colonies or colonies encrusting large substrates to be examined intact.

X-ray CT- and synchrotron-tomography are being used increasingly for fossil bryozoans and offer immense potential for future advances (e.g. Viskova and Pakhnevich 2009; Koromyslova and Pakhnevich 2016; Federov, Koromyslova, and Martha 2017; Koromyslova, Martha, and Pakhnevich 2018). However, aside from the large amount of time needed to process and clean image stacks, these techniques suffer from the problem of the minor density contrast between the calcareous bryozoan skeleton and the carbonate cement or sediment that normally infills the chambers of the zooids.

2 Biomineralization and Geochemistry

All stenolaemate bryozoans possess biomineralized skeletons, as do the cheilostome gymnolaemates, which are the overwhelmingly dominant order of bryozoans living today. The fine-scale structure of these skeletons are the smallest morphological features pertinent to our understanding of bryozoan palaeobiology.

Biomineralized skeletons are composite materials combining inorganic and organic phases (Dove, De Yoreo, and Weiner 2003). In the case of bryozoans, the inorganic phase consists of one or both of the calcium carbonate minerals calcite and aragonite. These minerals take the form of numerous tiny, micron-sized crystallites arranged in distinctive patterns. The organic phase comprises an outer covering called the *cuticle*, which is the equivalent of the periostra-cum of brachiopod and mollusc shells, as well as thin organic sheets between the crystallites. Both the cuticle and intercrystalline organic sheets in bryozo-ans have a predominantly proteinaceous composition but also contain chitin.

Research on other biomineralizing groups has shown that biogenic crystallites are composed of numerous nanogranules, each submicron-sized (see De Yoreo et al. 2015). The nanogranules are arranged with their crystallographic axes in the same orientation. Hence the component crystallites of the skeleton resemble single inorganic crystals. A substructure evident in some SEM images suggests that nanogranules may also be the building blocks of bryozoan crystallites.

Specialized epithelial cells are responsible for the secretion of the bryozoan skeleton. At the growing edges of the cheilostome *Membranipora*, the secretory epithelium initially consists of columnar palisade cells responsible for formation of

Bryozoan Paleobiology, First Edition. Paul D. Taylor.
© 2020 Natural History Museum. Published 2020 by John Wiley & Sons Ltd.

the cuticle (Tavener-Smith and Williams 1972). New cells are formed at the growing edge in a conveyor belt-like manner, pushing back the older palisade cells, which become flattened and commence secreting the mineralized part of the skeleton. Thus, mature skeletal walls have an outer cuticle underlain successively by mineralized skeleton and an epithelial layer.

2.1 Skeletal Wall Types

Skeletal walls of the type described above are called **exterior walls** in bryozoans. Their basic structure resembles mollusc and brachiopod shells and, as in these related lophotrochozoans, they are developed at the boundary between the animal and the external environment. Secretory epithelia are present only on one side of exterior walls. However, bryozoans also possess **interior walls** formed through folding and invagination of secretory epithelia. Interior walls very often separate the zooids in a colony but may also form the surface calcification of the colony, in which case the skeleton is overlain by an investment of soft tissue including a coelom or a pseudocoel. Unlike exterior walls, interior walls do not have a cuticular layer and are calcified from an epithelium that is present on both sides of the walls, passing over the distal edges.

Distinguishing between exterior and interior walls in fossil bryozoans is crucial if the disposition of soft tissues in the living animal is to be inferred, which in turn allows various aspects of their functional morphology to be interpreted. The appearance of these two basic wall types in thin section can usually enable their distinction: unlike exterior walls, which are accreted from one side only, interior walls generally have a mirror image structure – the medial part of the wall is the first to be formed and is overlain by subsequently formed calcification on both sides. The surface textures of interior and exterior walls are also different. Under an optical microscope, transverse folds and wrinkles ('growth checks') often embellish the outer surfaces of exterior walls, whereas interior walls tend to be smooth-surfaced or are covered by pustules, nodes, spines, or reticulations (Figure 2.1A). Scanning electron microscopy

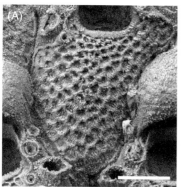 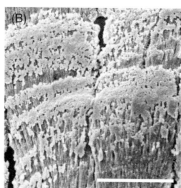 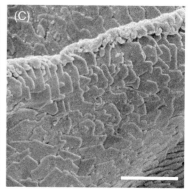

Figure 2.1 Surface textures of skeletal walls in bryozoans: (A) interior wall (cryptocyst) of the anascan cheilostome *Doryporella alcicornis* with reticulations (Pleistocene, Setana Fm.; Kuromatsunai, Hokkaido, Japan); (B) planar spherulitic fabric of the frontal exterior wall in the cyclostome *Crisia sigmoidea* (Recent; Mediterranean Sea off Marseille, France); (C) imbricated crystallites at the edge of an interior wall of the cyclostome *Entalophoroecia deflexa* (Recent; Mediterranean Sea off Marseille, France). Scale bars: A = 100 μm; B, C = 10 μm.

of pristine fossil bryozoans reveals a clear contrast in the microtexture of wall surfaces when not obscured by cement or sediment. Outer surfaces of exterior walls are smooth at high magnifications, although slight corrosion often reveals a fabric termed **planar spherulitic**, comprising fans of acicular crystallites lying parallel to the wall surface (Figure 2.1B). In contrast, because both sides of interior walls are surfaces of skeletal accretion, they expose the faces of growing crystallites often arranged in imbricated patterns (Figure 2.1C).

Also of importance in understanding the functional morphology of bryozoan skeletons is the distinction between three topological wall types – **basal**, **vertical**, and **frontal** (Figure 2.2). Basal walls form the floors of the zooids, vertical walls their sides, and frontal walls their outer-facing surfaces. Each of these topological wall types can be developed as an interior or an exterior wall, depending on the species concerned and the exact location within the colony. Sometimes, the vertical walls between zooids are exterior walls, in which case the two juxtaposed exterior walls together constitute a **compound wall** that has a cuticular layer in its centre. The suture between the exterior walls may be corru-

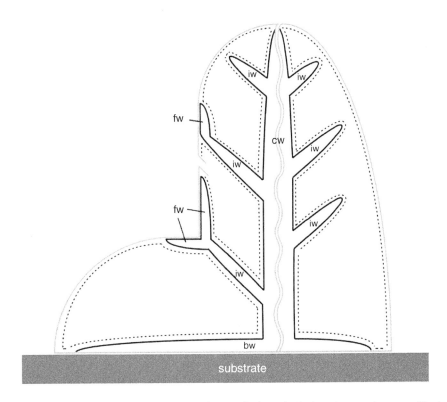

Figure 2.2 Stylized vertical section through the base and stem of a hypothetical cyclostome bryozoan illustrating some of the main types of skeletal walls. Cuticle is in grey. Epithelia, which are potentially capable of secreting mineralized skeleton, are indicated by the dashed lines. The colony has an exterior basal wall (bw) from which arises a compound wall (cw) with autozooids having exterior frontal walls (fw) budding on the left side, and kenozooids lacking such walls on the right side. Vertical interior walls (iw) occur between the zooids (based on Taylor, Lombardi, and Cocito 2015, fig. 2. © The Trustees of the Natural History Museum, London).

Figure 2.3 Transversely broken branch of the cheilostome *Chiplonkarina campbelli* showing corrugated compound walls fractured along the cuticular layer between zooids (Cretaceous, Campanian or Maastrichtian, Kahuitara Tuff; Pitt Island, Chatham Islands, New Zealand). Scale bar = 500 μm.

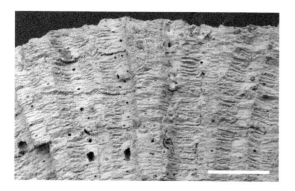

gated (Figure 2.3). For example, compound walls are commonly present between rows of zooids in cheilostomes. The existence of a perishable cuticle at the centres of compound walls in these cheilostomes is the reason why dried recent specimens, as well as some minimally cemented fossils, can disaggregate into zooidal rows. Compound walls are less common in stenolaemates, a notable exception being the Pliocene cyclostome *Blumenbachium globosum* in which a digitate compound wall forms the boundary between adjacent subcolonies (Tavener-Smith 1975, text-fig. 1; Balson and Taylor 1982).

2.2 Pores and Pseudopores

Pores can be developed in all types of skeletal walls, providing vital conduits for soft tissue linkage between zooids and between the visceral and hypostegal body cavities of zooids. Those penetrating exterior walls pass through the bimineralized layer only and are sealed by cuticle. In cyclostomes such pores in exterior walls are termed **pseudopores** (Figure 2.4A, B). Unfortunately, however, the term pseudopore has been applied differently in cheilostomes where it is generally used for pores in frontal interior walls. Pores in interior walls of stenolaemates are referred to as interzooidal pores, mural pores, or communication pores. Interzooidal pores are ubiquitous in post-Palaeozoic cyclostome stenolaemates (Figure 2.4D–F). However, with the exception of a few genera of cystoporates (Figure 2.4C), pores are lacking in the walls of palaeostomes. Combined with the typical presence of basal diaphragms to seal-off the proximal parts of the zooids, the absence of interzooidal pores in palaeostomes means that the deep interiors of colonies would have been sealed-off from the living tissues that formed an outer rind on the colony surface (Chapter 9.12). The porous-walled cyclostomes, by contrast, can potentially maintain living tissue deep into the interior of the colony.

Cyclostome interzooidal pores are <10 μm in diameter and are commonly ringed by tiny inward-growing spines that limit the opening (Nielsen and Pedersen 1979). The body cavities of adjacent zooids are continuous through the interzooidal pores, at least in the living genus *Crisia*. Pseudopores in cyclostomes are the exterior-wall homologues of interzooidal pores. They are particularly well developed in frontal walls but are usually lacking in basal walls. In living cyclostomes, a pore cell fills each pseudopore but its function, and that of the pseudopore as a whole, has yet to be established. Possibilities include maintenance of the exterior cuticle, secretion of chemicals to deter would-be fouling organisms, and respiration at times that the lophophore is retracted and cannot be used for this purpose. With respect to the proposed respiratory function, it may be significant that reproductive polymorphs (gonozooids) in cyclostomes always have a significantly higher density of pseudopores than do feeding zooids (autozooids): not only might the developing embryos that are being brooded inside gonozooids require a plentiful supply of oxygen (Ryland 1970), but gonozooids lack the lophophores which can be used as organs of respiration in autozooids. Calcified exterior basal

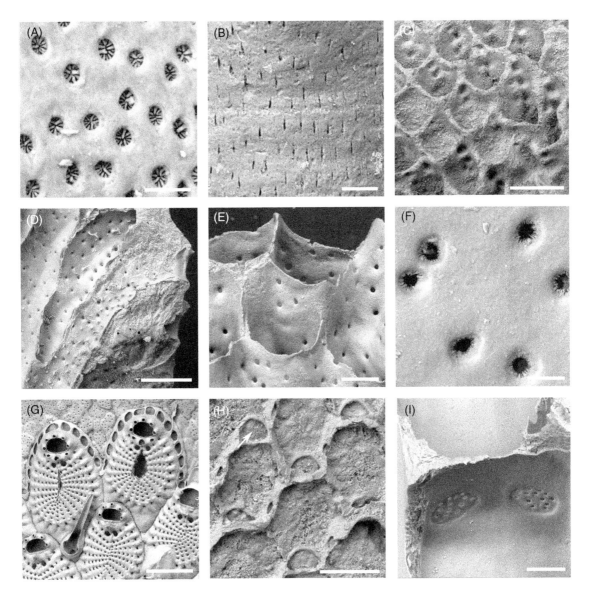

Figure 2.4 Pores and pseudopores in bryozoan skeletal walls: (A) round pseudopores partly closed by radial spines in the exterior wall of a gonozooid of the cyclostome *Liripora pseudosarniensis* (Recent; southern New Zealand); (B) slit-like pseudopores in the cyclostome *Stomatopora* cf. *dichotomoides* (Jurassic, Middle Bathonian, Fullers Earth Rock; East Cranmore, Somerset, UK); (C) ceramoporid cystoporate with large pores (or pits?) in the walls (Silurian, Llandovery; Lickershamn, Gotland, Sweden); (D) fractured branch of the cyclostome '*Heteropora*' *neozelanica* showing porous walls (Recent; Otago Shelf, New Zealand); (E) pores in interior walls of the cyclostome '*Telopora*' cf. *flabellaris* (Recent; New Zealand); (F) pores ringed by radial spines in the cyclostome *Hornera* (Recent; Otago Shelf, New Zealand); (G) growing edge of the cheilostome *Puellina* with multiple pore chambers along the distolateral borders of two autozooids (Recent; Puerto Rico); (H) abraded colony of the cheilostome *Wilbertopora* exposing large semicircular, distal pore chambers, one of which is arrowed (Cretaceous, Upper Albian, Upper Greensand; Salcombe Regis, Devon, UK); (I) broken zooid of the cheilostome *Pentapora foliacea* showing two multiporous septulae in the distal wall of a zooid (Recent; Plymouth, Devon, UK). Scale bars: A, F = 20 μm; B, I = 50 μm; C = 500 μm; D, G, H = 200 μm; E = 100 μm.

walls in cyclostomes typically only contain pseudopores in parts of colonies that grow off the substratum but not in basal exterior walls adpressed to the substrate.

Palaeostomates usually lack calcified exterior frontal walls, with the possible exception of occasional terminal diaphragms in some trepostomes. They do, however, possess calcified exterior basal walls on the undersides of encrusting colonies and occasionally the obverse surfaces of erect colonies such as *Rhombocladia* (Ernst et al. 2016). These seem always to lack pseudopores, in contrast to equivalent walls in cyclostomes.

Cheilostome pores typically have a distinct polarity defined by the **rosette cells** that occupy them and are often reflected in the skeleton. Some are associated with small chambers called **dietellae** or **pore chambers** (e.g. Gordon and Hastings 1979) (Figure 2.4G, H). Ovoidal patches of thinned skeleton in cheilostomes may be pierced by a single pore or a cluster of pores, forming uniporous and multiporous **septulae** (Figure 2.4I), respectively.

There appears to be a greater degree of control of passage of substances through the pores of cheilostomes than cyclostomes. Strands of the **funiculus**, a blood vessel-like circulatory system (Carle and Ruppert 1983) that allows rapid transport of metabolites from the stomach to other parts of the zooid (Lutaud 1985), link directly to the rosette cells occupying pores in cheilostome skeletal walls, whereas this appears not to be the case in cyclostomes where the pores appear to be open coelomic conduits in at least some species (Nielsen and Pedersen 1979).

Clearly, pores in skeletal walls play an important role in allowing soft tissue connectivity between the zooids of bryozoan colonies, and hence in colonial integration (Chapter 4.7). When pores are lacking, as in the great majority of palaeostomates, the only soft tissue linkage is over the outer, distal ends of the interzooidal walls separating the zooids via the hypostegal body cavity and associated tissues that cover the outer surface of the colony.

2.3 Skeletal Growth

Bryozoan skeletal walls grow in length by distal extension while simultaneously becoming thicker. Both processes involve the formation of new crystallites as well as the growth of already existing crystallites. The cuticle of exterior walls originates from undifferentiated epithelial or palisade cells at the growing edge of the colony (Tavener-Smith and Williams 1972). Epithelial cells in cheilostomes are characterized by having a nucleus with a prominent nucleolus, a small amount of heterochromatin, densely distributed granular endoplasmic reticulum, and scattered mitochondria. The epithelium secreting the mineralized component of the skeleton originates from a ring of generative cuboidal cells proximal to the cap of palisade cells. The new epithelium initially exudes cuticle and then an irregularly developed electron-dense sheet onto which the first crystallites of the biomineralized part of the skeleton are seeded.

Very little is known about calcification processes in bryozoans but mineral deposition is generally assumed to occur outside the cells, i.e. extracellularly. The often-precise arrangements of crystallites in bryozoan skeletons point to a degree of biological control. Indeed, bryozoans are better characterized as 'biologically controlled biomineralizers' than 'biologically induced biomineralizers' in the spectrum introduced by Lowenstam (1981).

2.4 Skeletal Ultrastructure

The crystallites making up the skeletal walls of bryozoans vary markedly in their morphology and configuration, defining a series of ultrastructural types or fabrics (e.g. Sandberg 1977). Variations in ultrastructural fabric (Figure 2.5) are apparent not only among taxa, but also between different types of walls within the same colony, and even within individual walls according to their stage of development. Fabrics are best observed on the growing surfaces of the walls, particularly close to their distal edges. This is easily done following bleaching of recent bryozoans

Figure 2.5 Skeletal wall ultrastructural fabrics in some recent bryozoans: (A) platy crystallites distally imbricated and growing towards the leading edge of an interior wall in the cyclostome *Fasciculipora ramosa* (Antarctica); (B) imbricated, lath-like crystallites accreting at right angles to the overall growth direction of an interior wall in the cyclostome *Cinctipora elegans* (Otago Shelf, New Zealand); (C) hexagonal semi-nacre in the cyclostome *Platonea stoechas* (Mediterranean Sea); (D) rhombic semi-nacre, with spiral overgrowths (screw dislocations) in the cheilostome *Acanthodesia savartii* (India); (E) fractured wall of the cheilostome *Bugula neritina* with fibres oriented almost perpendicular to the wall surface at the left (South Africa); (F) spindle-shaped crystallites disaggregating on the wall surface of the cheilostome *Jellyella eburnea* (Western Australia). Scale bars: A = 50 μm; B = 10 μm; C, D, E = 5 μm; F = 2 μm.

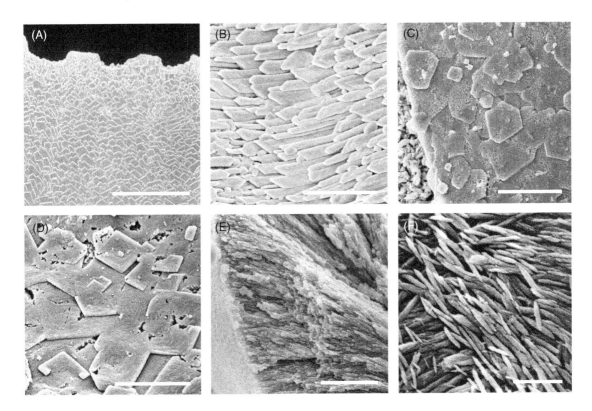

to remove organic material. However, observation of fabrics on wall surfaces of fossils is seldom possible as adherent sediment and/or epitaxial cement invariably obscures these surfaces. Examples of exceptional preservation revealing surface crystallites do exist, notably the bryozoans from the Carboniferous Buckhorn asphalt quarry in Oklahoma (Ernst et al. 2016; Figure 2.6), where growth of diagenetic cements was retarded by an asphalt coating, and also the mid-Cretaceous Upper Greensand of Devon in which fragile colonies are occasionally found that lack significant epitaxial cements (Taylor and Weedon 1996).

The first biomineralized part of bryozoan skeletal walls typically comprises minute crystallites that have a granular appearance in both thin section and when viewed using SEM. More or less equidimensional and lacking obvious patterning, these tiny crystallites are less than a micron in diameter and presumably grew rapidly. The presence of amorphous calcium carbonate (ACC) in the first-formed parts of the skeletons of other animals (Addadi, Raz, and Weiner 2003) raises the possibility that such granular crystallites in bryozoans are formed through early crystallization of ACC.

Aside from granular fabrics, there are two main types of ultrastructural wall fabric in bryozoans, which can be called wall-perpendicular and wall-parallel. Wall-perpendicular fabrics are known only among cheilostomes, whereas wall-parallel fabrics occur in both cheilostomes and stenolaemates.

Crystallites in wall-perpendicular fabrics are oriented approximately at 90° to the surface of the wall and are finely prismatic or acicular in morphology (Figure 2.5E), sometimes forming botryoidal fans in fractured walls. Surfaces of walls with wall-perpendicular fabrics are smooth, the boundaries of the component crystallites not usually being apparent. Wall thickening occurs by incremental growth of the crystallites along their long axes.

Wall-parallel fabrics result in skeletal walls that are laminated when observed in thin sections. The crystallites are most often arranged in a stepped, imbricated pattern on wall surfaces, resembling miniscule roofing tiles in scanning electron micrographs (Figure 2.5A, B). Only a small part of each imbricated crystallite is normally visible, most of the crystallite being hidden beneath other, overgrowing crystallites. The orientation of imbricated crystallites varies among species and even within individual walls depending on the age of the wall. Crystallites can be directed proximally towards the base of the wall, distally towards the growing edge of the wall (Figure 2.5A), or transversely at right angles to wall growth direction (Figure 2.5B). The shape of the crystallites varies from fibre- to lath- to plate-like. They occasionally have a spindle shape (Figure 2.5F). Crystallites are much longer than wide, whereas their width is typically greater than their depth. Lateral edges of crystallites abut edges of adjacent crystallites. The free, distal growing edges of the crystallites may be formed of well-defined crystallite faces meeting at a geometrically consistent interfacial angle, or can be rounded. Continuous accretion at the growing edges of the imbricated crystallites leads to a progressive thickening of the wall. Crystallites are often observed to split, sometimes into fan-like arrays, and what appear to be distinct crystallites on wall surfaces may in fact represent separate lobes of larger crystallites.

The extreme lengths of individual crystallites in the imbricated wall-parallel fabrics described above contrasts with a different type of wall-parallel fabric called semi-nacre. Here, numerous small, four- or six-sided crystallites are visible on wall surfaces viewed using SEM (Figure 2.5C, D). These crystallites grow radially outwards until they meet adjacent crystallites. New crystallites then originate as small 'seeds' developing on the surfaces of the older crystallites (Figure 2.5C). Alternatively, screw dislocations may initiate a pattern of continuous spiral overgrowth (Figure 2.5D), leading to a thickening of such semi-nacreous skeletal walls.

As already remarked, fossil bryozoans are seldom preserved well enough to reveal clearly skeletal fabrics on wall surfaces, but fractured walls may expose underlying fabrics if diagenesis has not been too severe. For example, a distinctive fabric of transverse fibres, which is widely distributed in modern cyclostomes, is occasionally also seen in fossil cyclostomes and may even have constituted the 'primary layer' or 'colonial plexus' identified in thin sections of the skeletons of Palaeozoic Fenestrata (Ernst et al. 2016). More often, however, loss of the thin sheets of organic material that wrapped crystallites, accompanied by epitaxial growth of diagenetic cement and/or neomorphism of the skeleton, results in obliteration of crystallite boundaries, obscuring the original ultrastructural fabric even when there is no adherent sediment.

Individual walls are commonly built up of several different wall fabrics in succession, constituting a **fabric suite**. In cheilostomes, for example, a layer of wall-perpendicular fabric may be followed by a layer of wall-parallel fabric, or vice-versa. Transitions between semi-nacre and imbricated wall-parallel fabrics occur in the walls of some cyclostomes, with the small, six-sided semi-nacre crystallites close to the distal edges of walls becoming larger and proximally imbricated away from the growing edge. Pristine wall growing tips are needed to appreciate fully the fabric suite as the early-formed fabrics rapidly become overgrown. Crystallites are typically ill-defined on the surfaces of old wall surfaces. This may perhaps correlate with slowing and cessation of active crystallite accretion and possibly also localized resorption of skeleton.

The crystallography of bryozoan skeletons has received scant attention, which is probably in part due to the small size of the crystallites making them challenging to study using petrographical and other methods such as EBSD. Hall *et al.* (2002) used electron diffractometry to determine the crystallographic axes of six Recent bryozoan species. Measurements of interfacial angles of growing calcite crystallites on wall surfaces of recent cyclostomes also permits crystallographic orientations to be inferred based on the known crystal form of inorganic calcite (see Runnegar 1984).

2.5 Spines

Skeletal spines are widespread among bryozoans (Figure 2.7). Most of the large, hollow spines characteristic of cheilostomes are believed to be polymorphic zooids – spinozooids – and are described elsewhere (Chapter 3.8). However, solid spines or **spinules** also occur in some cheilostomes. Stenolaemates possess two main kinds of spines – **mural spines** and **styles** – the former oriented outwards approximately at right angles to the surfaces of skeletal walls, and the latter paralleling the distal growth direction of the wall within which they become embedded as the wall thickens during growth. Mural spines are found in a wide range of cyclostomes (e.g. Farmer 1979), and occasionally in palaeostomates such as the Ordovician trepostomes *Batostoma*? *cornula* (Corneliussen and Perry 1970) and *Eichwaldopora ovulum* (Pushkin and Popov 2005). The pristine surface preservation in bryozoans from the Carboniferous Buckhorn Asphalt quarry reveals the three-dimensional morphology of mural spines in several palaeostomate species (Figure 2.6B).

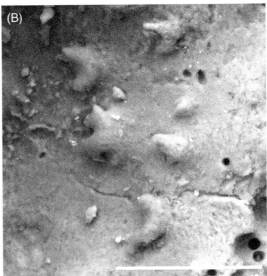

Figure 2.6 Unusually pristine preservation of Palaeozoic bryozoans occurs in the Carboniferous Buckhorn Asphalt (Carboniferous, Moscovian, Boggy Fm.; Sulphur, Oklahoma, USA): (A) longitudinally fractured interior wall of the trepostome *Tabulipora hispida* showing lamination; (B) surface of wall of the trepostome *Stenoporella* with proximally bent mural spines, some of which are bifid. Scale bars = 50 μm.

Figure 2.7 Cyclostome mural spines (A–D), and cheilostome spinules (E) and spicules (F): (A) proximally curved mural spine in *Crisulipora occidentalis* (Pacific Ocean, California); (B) apertures of '*Heteropora*' sp. packed with mural spines (Spirits Bay, New Zealand); (C) barbed head of a mural spine of '*Heteropora*' *neozelanica* (Otago Shelf, New Zealand); (D) pair of flattened, barbed mural spines in *Telopora* sp. (New Zealand); (E) delicately branched spinule of *Jellyella eburnea* (Western Australia); (F) three spicules of the caliper type in *Thalamoporella harmelini* (Mediterranean Sea, Israel). Scale bars: A, C = 5 μm; B = 100 μm; D, F = 20 μm; E = 50 μm.

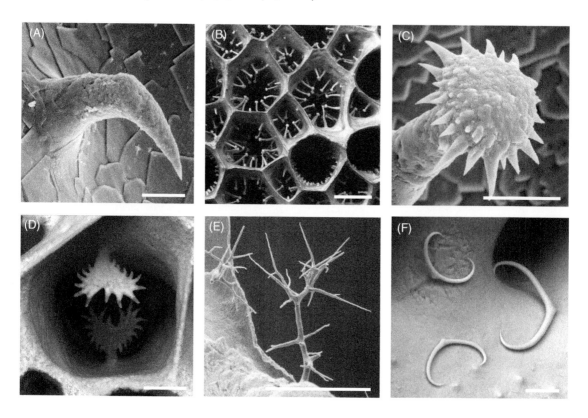

Mural spines and spinules show a wide range of morphologies. The simplest are short, unbranched, and tubercle-like. Longer spines may be straight or curved (Figure 2.7A), unbranched or branched, and sometimes terminate in an expanded and barbed head (Figure 2.7B–D). Spine type and density may differ according to polymorph type in stenolaemates, with kenozooids and gonozooids often having larger and more densely packed mural spines than autozooids (e.g. Taylor 1999, fig. 41.9C). Various spine- and node-like structures occur on the surfaces of cheilostome zooids, and elaborately branched, delicate spinules grow into the zooidal chambers of a few species (Figure 2.7E). It is assumed that they have a largely protective function, although the apparent fragility of spinules in cheilostomes such as *Jellyella eburnea* (Taylor and Monks 1997, fig. 16) casts doubt on this role. The anascan family Thalamoporellidae is unique in producing **spicules** that 'float' freely within the coeloms of the zooids (e.g. Soule, Soule, and Chaney 1992). Resembling miniature calipers (Figure 2.7F) or outstretched dividers, their functional significance has not been established.

Little is known about spinule and mural spine development and function in bryozoans. They clearly represent sites of enhanced and differentiated biomineralization but how this is controlled has yet to be determined. With regard to function, two main roles have been hypothesized: defence and soft tissue anchorage/positioning. Given the fragility of the spines, a defensive function would only be effective against very small predators (Chapter 5.2), and then only after the micropredators had first entered into the zooidal cavity and potentially caused damage. On the basis of histological sectioning of two cyclostomes, Farmer (1979) believed that mural spines initially served as attachment sites for vestibular extensor muscles before becoming points of anchorage for the muscles or ligaments of the membranous sac as the zooid lengthened and the spines occupied a position progressively more proximal of the aperture. Schäfer (1985) noted that mural spines in 'Cardioecia' watersi prevented the membranous sac from being pressed against the skeletal walls of the zooids. The presence in some cyclostomes (e.g. Disporella novaehollandiae: Gordon and Taylor 2001) of a greater number of mural spines in kenozooids than autozooids is not consistent with either the micropredator or tissue anchorage hypotheses, pointing to a different function. One hypothesis worth testing is that mural spines function to hamper the entry of parasites into zooidal chambers.

Short knob- or node-like tubercles in cheilostomes growing on algae were suggested by Voigt (1982) to function as spacers to protect the surface of the bryozoan colony from mechanical damage in strong currents. An experimental study of the epiphytic ctenostome Flustrellidra hispida showed that cuticular spines increased when colonies were cultured in higher water flow velocities (Whitehead, Seed, and Hughes 1996), for which several functional explanations were considered, including reduction in abrasion from adjacent algal fronds and particles in suspension. Other functional possibilities for surface spines could be to increase the thickness of the boundary layer above colonies, thereby permitting lophophores to extend when

flow was greater, or to reduce the impact pressure of breaking waves.

Styles are an apomorphic character of palaeostomates (Figure 2.8). Sometimes called acanthostyles or stylets, they typically have an unlaminated core surrounded by a laminated sheath, although some styles lack a core (Blake 1973). Three kinds of styles were recognized among rhabdomesine cryptostomes by Blake (1983): paurostyles with non-laminated continuous cores; heterostyles with discontinuous, lens-like non-laminated cores; and aktinostyles with laminated cores. The older literature often refers to styles as 'acanthopores' in the mistaken belief that the core represented a narrow tube that accommodated the soft tissues of a minute polymorphic zooid. In a few cases styles contain tiny spines within the sheath laminae oriented perpendicular to the core. Styles vary in size and morphology, with small and large size classes often occurring together in the same colony (Figure 2.8B). The largest styles equal the diameter of the autozooids, as in the Late Permian trepostome Dyscritellina (Figure 2.8C). Narrow styles may be contained within the thickness of interzooidal walls but thick styles are wider than the interzooidal walls and can indent the edges of the aperture giving it a petaloid appearance (Figure 2.8A; Plate 4N).

Styles of minute diameter occur in the laminated walls of many fenestrates (e.g. Gautier 1973; Gautier, Wyse Jackson, and McKinney 2013) and timanodictyines (e.g. Gilmour and Snyder 1986; Figure 1.12H). They are sometimes inappropriately termed 'capillaries', 'stenostyles' or, misleadingly, 'spicules' (Elias and Condra 1957). The thickened, 'capillary'-containing skeletons of these fenestrate bryozoans were once believed to have formed as a result of a symbiotic relationship between the bryozoan and an alga, the so-called 'algal consortium hypothesis' (Condra and Elias 1944; Rigby 1957), but it is now clear that the skeleton is entirely the construct of the bryozoan.

An unusual Devonian ptilodictyine cryptostome, Cryptostyloecia, possesses non-protruding

Figure 2.8 Palaeostomate styles: (A) thin section of the esthonioporine *Ibexella multiphragmata* with large styles indenting the zooidal chambers (Middle Ordovician, Kanosh Formation; Ibex, Utah, USA); (B) tangential thin section of the trepostome *Tabulipora chaoi* in which the thick walls contain styles of two sizes visible as small and large dots (Lower Carboniferous; Hunan, China); (C) arrows pointing to four huge styles in a thin section of the trepostome *Dyscritellina clivosa* (Permian, Kazanian; Russia); (D) oblique view of the surface of the trepostome *Stenophragmidium buckhornensis* showing stout styles (Carboniferous, Moscovian, Boggy Fm.; Sulphur, Oklahoma, USA); (E) branch of the cryptostome *Rhabdomeson* with styles located at wall triple junctions (Carboniferous; UK); (F) corroded style of the cryptostome *Worthenopora spinosa* showing concentric laminae surrounding the core (Lower Carboniferous, Warsaw Beds; Warsaw, Illinois, USA). Scale bars: A, B = 200 µm; C = *c.* 200 µm; D = 100 µm; E = 200 µm; F = 10 µm.

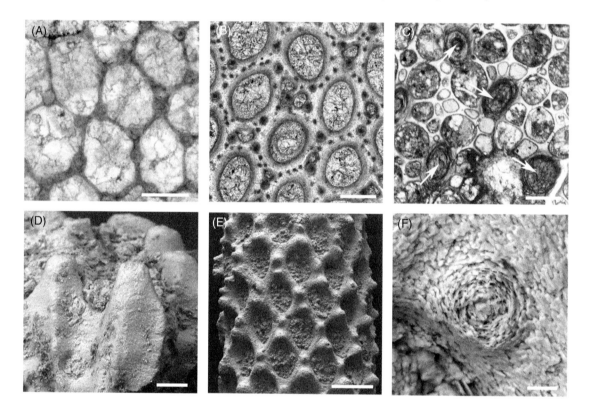

'styles' consisting of fibrous, iron-rich skeleton (Ernst, Königshof, and Schäfer 2009). Termed **cryptostyles**, these structures were interpreted as functioning in weight reduction, as their fibrous skeletons were of low relative density, as well as in skeletal strengthening.

Styles that projected as spines on colony surfaces (Figure 2.8D–F) could have performed a defensive function and/or have regulated the height of the hypostegal body cavity (e.g. Tavener-Smith 1975). It is conceivable that styles may have had a role in strengthening the skeleton by binding the skeletal laminae (e.g. Blake 1973), functioning almost like rivets to disrupt the lateral continuity of wall laminations. This hypothesis deserves to be tested using a modelling technique such as finite element analysis.

2.6 Mineralogy

Bryozoans as a phylum are comparable to molluscs in having the capability of making their skeletons from more than one of the mineral polymorphs of calcium carbonate – calcite and aragonite – either alone or together (Rucker and Carver 1969; Smith, Key, and Gordon 2006; Taylor et al. 2009). Unlike molluscs, however, calcite rather than aragonite is the primitive biomineral in the two independently evolved biomineralizing clades of bryozoans, and only cheilostomes seem to have evolved the ability to biomineralize aragonite; the few claims of aragonite in the skeletons of recent and fossil stenolaemates must be viewed with scepticism and may be due to contamination by aragonitic epibionts or substrates. The taxonomic distribution of cheilostome species having aragonite in their skeletons leaves no doubt that aragonite biomineralization

has evolved polyphyletically in this order. For instance, aragonite occurs in the malacostegine *Membranipora*, the convergent free-living 'anascan' genera *Cupuladria* and *Seleneria*, and in several genera such as *Hippoporidra* belonging to numerous families of ascophoran-grade cheilostomes that also have calcitic representatives (Taylor et al. 2009). Determining the exact number of times that aragonite has evolved in cheilostomes awaits the construction of a more comprehensive phylogeny for the order.

Several regional surveys of 'bryomineralogy' have been undertaken (Poluzzi and Sartori 1975; Smith, Nelson, and Spencer 1998; Kuklinski and Taylor 2009; Smith and Clark 2010; Krzeminska et al. 2016; Taylor et al. 2016; Loxton et al. 2018). A strong latitudinal pattern exists today in the proportion of aragonitic species within cheilostome faunas (Figure 2.9). High latitude faunas

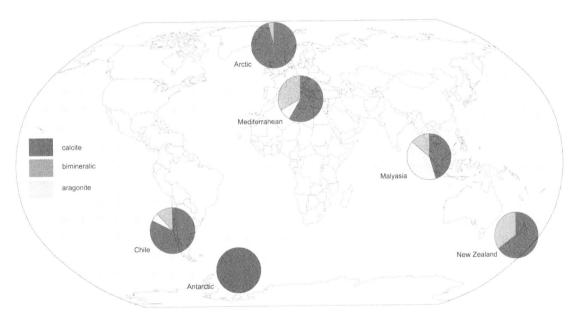

Figure 2.9 Mineralogical compositions of cheilostome bryozoans in biotas from different latitudes. Note the increasing proportion from the poles to the tropics of species biomineralizing aragonite, either as the only biomineral or bimineralically with calcite. Data sources are: Arctic, 76 species (Kuklinski and Taylor 2009); Mediterranean, 94 species (Poluzzi and Sartori 1975); Malaysia, 22 species (Taylor et al. 2016); Chile, 17 species (Smith and Clark 2010); New Zealand, 49 species (Smith, Nelson, and Spencer 1998); Antarctic, 71 species (Krzeminska et al. 2016).

(>60°) contain no or very few species with aragonitic or bimineralic (aragonite + calcite) skeletons; mid-latitude faunas (30–60°) contain significant proportions of bimineralic and a smattering of aragonitic species; low latitude faunas (<30°) are dominated by aragonitic and bimineralic species. For example, a study of 71 Antarctic bryozoan species found all to have monomineralic calcite skeletons (Krzeminska et al. 2016). In contrast, among 22 species from tropical Malaysia, nine were entirely aragonitic and three bimineralic (Taylor et al. 2016). Molluscs exhibit much the same pattern, in which it is usually attributed to the enhanced adaptive advantage of having a calcite skeleton in the cold waters of high latitudes because the solubility of calcium carbonate increases with decreasing water temperature. In some bimineralic cheilostome species with broad geographical ranges, the proportion of aragonite in the skeleton increases with ambient temperature increase (Rucker and Carver 1969, text-fig. 4). Furthermore, bimineralic species typically secrete a greater proportion of aragonite relative to calcite during warm summer months, resulting in mineralogical growth banding that may match morphological growth bands visible externally in the skeleton (Lombardi et al. 2008).

A palaeontological implication of the higher proportion of aragonitic species in low latitudes is that the preservation potential of tropical cheilostome faunas in the fossil record may be less than those from higher palaeolatitudes (Taylor and Di Martino 2014). This is because of the tendency for aragonite to be leached during diagenesis, resulting in partial or complete loss of the skeleton, although preservation of the diagenetically calcitized skeletons of originally aragonitic cheilostomes has been found in some Miocene cheilostomes from tropical East Kalimantan, Indonesia (Di Martino et al. 2016).

2.6.1 Seawater chemistry and bryozoan mineralogy

The mineralogical evolution of bryozoans warrants discussion because of its possible relationship with secular changes in seawater chemistry. Sandberg (1983) identified long-term oscillations in the carbonate mineralogy of marine ooids and inorganic cements: times when ooids and cements were calcitic (roughly Tommotian–mid-Carboniferous; Jurassic–Eocene) alternated with times when they were aragonitic (earliest Cambrian; mid-Carboniferous–end Triassic; Oligocene–Recent). He believed these oscillations to have been driven by changes in global CO_2, with low CO_2 levels corresponding with so-called aragonite seas and high CO_2 levels with calcite seas. While subsequent research has corroborated Sandberg's observation of mineralogical oscillations, it is now thought that the driver is the Mg/Ca ratio of seawater which itself may be related to rates of seafloor spreading and the intensity of metasomatic processes at mid-oceanic ridges (Hardie 1996). High rates of seafloor spreading cause Mg/Ca in seawater to be low, favouring precipitation of the more stable mineral calcite, whereas low spreading rates elevate Mg/Ca, which favours precipitation of aragonite because Mg 'poisons' calcite growth by substituting for Ca and interfering with lattice structure (see Berner 1975).

But what impact did changes in seawater Mg/Ca have on organisms producing calcareous skeletons? Stanley and others (see Stanley 2006) have hypothesized that organisms responded to these secular changes in seawater chemistry by either acquiring skeletons *de-novo* with a mineralogy matching the ambient environment, or switching mineralogy when there was a change in seawater chemistry. The plausibility of the latter is demonstrated by mineralogical modifications that occur in marine animals with calcareous shells when cultured in artificial seawater with different Mg/Ca levels; for example, Checa et al. (2007) stimulated secretion of aragonite in calcitic bivalves grown in high Mg/Ca conditions. Turning to the fossil record, there is a generally good correspondence between the mineralogy of newly acquired skeletons and times of calcite or aragonite seas, particularly during the Cambrian radiation (Porter 2010). Indeed, the earliest fossils of both biomineralizing clades of bryozoans occur during times of calcite seas – stenolaemates in the Early Ordovician; cheilostomes in the Late Jurassic – and both have calcitic skeletons.

Of greater interest is the response of cheilostomes to the switchover from calcite to aragonite seas that occurred in the late Palaeogene. As mentioned above, aragonite biomineralization evolved independently in numerous clades of cheilostomes. If seawater chemistry played a significant role, a high proportion of these origins should occur at, or shortly after, the late Palaeogene transition from a calcite sea to an aragonite sea. Analysis of the skeletal mineralogies of bryozoan assemblages from the Eocene and Oligocene of the southeastern United States (Taylor, James, and Phillips 2014) failed to find the expected increase in aragonitic taxa from the Eocene into the Oligocene (Figure 2.10). While it is impossible to discount the influence of local palaeoenvironmental factors or inexactitude over the timing of the transition on this finding, some cheilostomes dating back to the Early Eocene have been shown by XRD analysis to biomineralize aragonite (Taylor et al. 2009), thus pre-dating nearly all estimates for the advent of an aragonite sea. Furthermore, the presence of voids possibly formed through aragonite dissolution in skeletons of bryozoan species as old as Campanian in age suggests that cheilostomes may have begun biomineralizing aragonite in the indisputably calcite sea of the Late Cretaceous. The conclusion to be drawn from this data is that while aragonite seas may have given a selective advantage to cheilostomes having aragonitic skeletons, changes in seawater chemistry did not act as a simple trigger for the evolution of aragonite biomineralization in cheilostomes.

Stanley (2006) argued that calcifiers with calcite skeletons became hypercalcified during times of calcite seas, and those with aragonite skeletons hypercalcified during times of aragonite seas. The correspondence between the mineralogies of dominant reef-builders and calcite vs. aragonite seas supports this proposition but does it apply to bryozoans? This was tested by Taylor and Kuklinski (2011) who compiled data on branch diameter and exozonal wall thickness in ramose trepostomes for the calcite seas of the Ordovician and Devonian and aragonite sea of the Permian. As all trepostomes have calcitic skeletons, the prediction was that branches would be wider and walls thicker in the Ordovician and Devonian than the Permian. However, this was found not to be the case. Indeed, wall thicknesses are on average greater in

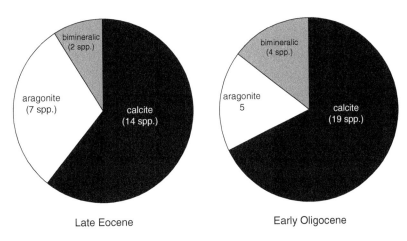

Figure 2.10 Mineralogical composition of cheilostome bryozoan skeletons in species from the latest Eocene and earliest Oligocene of Mississippi and Alabama, USA (based on Taylor, James, and Phillips 2014. © The Trustees of the Natural History Museum, London). The expected increase in aragonite biomineralization across this putative transition from a calcite sea to an aragonite sea is not evident.

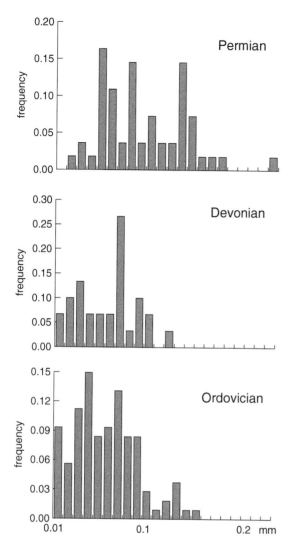

Figure 2.11 Frequency distributions of exozonal wall thickness in ramose trepostome bryozoans of Ordovician, Devonian, and Permian age (after Taylor and Kuklinski 2011. © The Trustees of the Natural History Museum, London). Mean values are: Ordovician, 0.068 mm; Devonian, 0.068 mm; Permian, 0.102 mm. The hypercalcification hypothesis predicts thinner walls in the aragonite sea of the Permian compared to the calcite seas of the Ordovician and Devonian, the opposite to the pattern seen in this data.

trepostomes that lived in the aragonite sea of the Permian (Figure 2.11).

2.6.2 Bimineralic bryozoans

Skeletons combining layers of calcite with those of aragonite are present in numerous cheilostomes. The frontal walls of many cheilostomes have an inner layer of calcite and an outer layer ('frosting') of aragonite (Taylor, Kudryavtsev, and Schopf 2008; Smith and Girvan 2010; Benedix, Jacob, and Taylor 2014), and in some species the autozooids have a framework of calcite while the adventitious avicularia are entirely aragonitic. Aragonite always succeeds calcite in any given skeletal wall. Once aragonite biomineralization has commenced, the wall concerned seemingly never reverts to biomineralizing calcite. Interlayering of calcite and aragonite has not been observed in bryozoans. However, calcite and aragonite deposition can occur simultaneously within the same colony, the same zooid, and in different parts of the same skeletal wall. A seemingly identical fibrous skeletal microstructure may persist as the wall transitions from a calcitic to an aragonitic mineralogy. The contact between the calcite and aragonite layer is either planar or undulose, and the Mg content in the calcite may increase as the transition to the aragonite layer is approached. Nothing is known about processes controlling the change from calcite to aragonite biomineralization in bryozoans, although equivalent changes in molluscs correlate with changes in the proteins associated with biomineralization (e.g. Belcher et al. 1996). A recent study of the bimineralic skeleton of *Anoteropora latirostris* by Jacob et al. (2019) used NanoSIMS to identify an organic layer between the calcite layer and succeeding aragonite layer, interpreted as a template for biomineralization of the aragonite.

Dissolution of aragonite during the diagenesis of bimineralic bryozoans can result in fossils preserved as partial moulds, with the originally aragonitic components (e.g. avicularia and outer

Figure 2.12 Dissolution of the outer aragonitic skeleton that formed the frontal shields of the zooids in the bimineralic ascophoran bryozoan *Laminopora dubia* results in an anascan-like appearance (Pliocene, Coralline Crag Fm.; Sudbourne, Suffolk, UK). Scale bar: 500 μm.

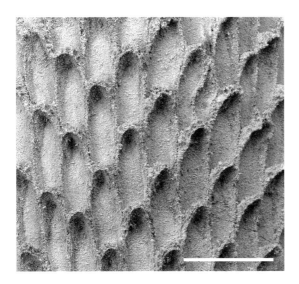

layers of frontal walls) represented by voids. Loss of the frontal shields in some ascophoran cheilostomes may leave a fossil resembling a simple anascan with only the calcitic basal and vertical walls of the zooids preserved, as in the cheilostome *Laminopora dubia* (Figure 2.12) which was originally assigned to the anascan genus *Flustra* (Zágoršek, Vávra, and Holcová 2007). Palaeogene specimens of so-called 'basally uncalcified' free-living cheilostomes were shown by Greeley (1969) to result from the dissolution during diagenesis of the aragonitic undersides of colonies but persistence of the calcitic frontal surfaces.

2.7 Geochemistry

Because of the technical difficulty of making precise analyses of skeletal walls that are typically less than 10 μm thick, relatively little has been published on the geochemistry of fossil bryozoan skeletons; most available information concerns the magnesium content of bryozoan calcite obtained through powder XRD analyses of bulk skeletal calcite. However, Hickey (1987) succeeded in analyzing the distributions of elements in the skeleton of the Ordovician trepostome *Peronopora*. Calcium was found to decline from the endozone into the exozone, which Hickey took to indicate a deceleration in growth rate, while the elevated levels of copper in the styles and median budding lamina he interpreted to indicate higher growth rates. These interpretations are consistent with growth models in stenolaemates with endozones and exozones (Chapter 4.2). Recrystallized calcite in some palaeostomates, particularly species from the Early Palaeozoic, may reflect skeletons originally with high-Mg calcite that suffered neomorphism during diagenesis. In some such cases, rhombic microdolomite inclusions are present in the neomorphosed skeletal walls (Taylor and Wilson 1999a; Ma, Taylor, and Feng-Sheng 2014). Neomorphic textures in the endozones of some palaeostomates which preserve pristine lamellar calcite in exozones (e.g. Ma, Taylor, and Feng-Sheng 2014) are consistent with the notion that Mg levels were higher in the fast-growing endozone, as has been demonstrated for molluscs and brachiopods in quickly growing parts of their shells (e.g. Rosenberg and Hughes 1991).

Numerous analyses of Mg in the calcite of modern bryozoans, both cheilostomes and cyclostomes, have been published (Figure 2.13). Low Mg-calcite (0–4 mol% Mg) to intermediate Mg-calcite (4–12 mol% Mg) characterizes most of the species that have been analyzed (Smith, Key, and Gordon 2006; Taylor et al. 2009; Loxton et al. 2012; Figuerola, Kuklinski, and Taylor 2015; Krzeminska et al. 2016). However, some cheilostomes have been found to secrete high-Mg calcite in excess of 12 mol% Mg, and in one species (*Membraniporopsis tubigera*) the skeleton comprises 14 mol% Mg (Gordon, Ramalho, and Taylor 2006). With regard to magnesium, analyses of 28 fossil cheilostome species showed a range of Mg of 0.7–7.2 mol%, with 82% of species

Figure 2.13 Skeletal mineralogy of 1051 specimens of marine bryozoans plotted according to their Mg content and carbonate mineralogy (after Smith, Key, and Gordon 2006). In species with less than 100% calcite, the remainder of the skeleton consists of aragonite. The 'mineralogical space' occupied by the phylum Bryozoa is extensive.

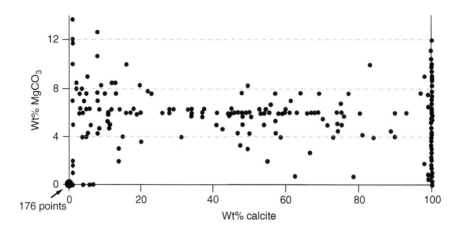

having low-Mg calcite and 18% intermediate-Mg calcite (Taylor et al. 2009).

In the cheilostome *Cellaria sinuosa*, the primary granular layer of the skeleton consists of low Mg-calcite, whereas the secondary lamellar layer is intermediate Mg-calcite, both layers showing seasonal cyclicity in the exact Mg content along the branches, with higher levels of Mg in parts of the skeleton formed during the summer than the winter (Schäfer and Bader 2008). The occurrence of two types of calcite distinguished by differing amounts of Mg has also been found in other cheilostomes (Smith and Lawton 2010), demonstrating control by the bryozoan of skeletal chemistry, whereas seasonal variations reveal the presence of an environmental overprint.

Across taxa, the level of Mg in bryozoan skeletons increases from the poles towards the equator (Taylor et al. 2009), presumably driven by higher water temperatures (Chave 1954). This is evident, for example, when comparing species belonging to the same genera from the Arctic, temperate latitudes and the tropics (Kuklinski and Taylor 2008). Nevertheless, neither Smith and Girvan (2010) nor Loxton et al. (2014b) were able to detect the

expected variations in Mg (or aragonite) within individual species across narrower latitudinal ranges. Superimposed over global trends in Mg content are local variations. For example, Loxton et al. (2014a) found significant differences within species between eight sites around Adelaide Island on the Antarctic Peninsula which did not correlate with temperature differences. And Krzeminska et al. (2016) found extreme variations in Mg levels between individuals of the same species at King George Island, Antarctica, suggestive of physiological control by the bryozoans. The predicted decrease in Mg with depth was not found by Figuerola, Kuklinski, and Taylor (2015) in a study of four common Antarctic bryozoans (three cheilostomes and one cyclostome) over a range of 185–660 metres in East Antarctica.

2.7.1 Stable light isotopes
Precise analyses of stable light isotopes in the thin-walled skeletons of bryozoans have been difficult to obtain, a problem exasperated by contamination from epitaxial diagenetic cements or adherent sediment particles in fossil material. Consequently, studies on the stable isotopes of

oxygen and carbon, which are commonly used as palaeoenvironmental proxies, have been relatively few in number (Pätzold, Ristedt, and Wefer 1987; Bone 1991; Rao 1993; Bone and James 1997; Brey et al. 1998, 1999; Bader 2000; Machiyama et al. 2003; Smith and Key 2004; Smith et al. 2004; Knowles et al. 2010; Key, Zagorsek, and Patterson 2013; Key et al. 2013, 2018). Nevertheless, this field holds considerable potential for the future as the precision of analytical techniques increases.

An early study (Forester, Sandberg, and Anderson 1973) of $\partial^{13}C$ and $\partial^{18}O$ values in 20 species of living cheilostomes and two cyclostomes concluded that cheilostomes secrete their skeletons in isotopic equilibrium with sea water. However, a subsequent study of 10 bryozoan species from New Zealand (Crowley and Taylor 2000) showed fractionation of $\partial^{18}O$ in two cheilostome species (*Celleporina grandis* and *Hippomonavella flexuosa*), whereas $\partial^{13}C$ was depleted in all but one (*Hippomenella vellicata*) of the species analyzed. While estimation of absolute temperatures using isotopic compositions of bryozoan skeletons may be problematical, seasonal changes in relative temperature consistent with growth banding can be obtained (e.g. Key et al. 2018).

Plate 1 Recent bryozoans in their natural habitats. (A)–(B) Rottnest Island, Western Australia; (C) Aleutian Islands, North Pacific. Photographs courtesy of Piotr Kuklinski.

Bryozoan Paleobiology, First Edition. Paul D. Taylor.
© 2020 Natural History Museum. Published 2020 by John Wiley & Sons Ltd.

Plate 2 Fossil bryozoans in the field. (A) Palaeozoic sequences of interbedded limestones and shales are often bryozoan-rich, as in this example from the Ordovician Bromide Formation of the Arbuckle Mountains, Oklahoma, USA. (B) Interpreted as a tidal channel deposit, the Pleistocene Setana Formation at Kuromatsunai in northern Japan contains bryozoan-encrusted clasts in a matrix rich in fragments of erect bryozoans. (C) *Steginoporella* bryolith (right of centre) along with bivalves, barnacles, and echinoid fragments in the Pleistocene Nukumaru Limestone, near Whanganui, New Zealand. (D) Silicified compound colony of *Nekhorosheviella*, the oldest known reef-associated bryozoan, from the Early Ordovician of central China. (E) Flint bands pick out the geometry of Early Danian bryozoan mounds at Stevns Klint in Denmark. (F) Weathered surfaces provide good opportunities for collecting fossil bryozoans, as in this Owl Creek Formation (Cretaceous, Maastrichtian) locality in Mississippi littered with branches of the cheilostome *Dysnoetopora*. (G) Late Miocene bryozoan/algal bioherms outcropping on the Crimean coast (courtesy of Tomas Koci).

Plate 3 Bryozoan-rich carbonates. (A) Bedding plane of Late Ordovician limestone from the Cincinnati region, USA, covered with branches of trepostome bryozoans. (B) Silurian Rochester Shale from New York State with abundant fenestrate bryozoans. (C) Hash of fenestellid and coarsely meshed polyporid fenestrates from the Carboniferous Warsaw Formation of Columbia, Illinois, USA. (D) Branches of the cheilostome *Heteroconopeum janieresiense* forming a monospecific bryozoan limestone in the Cenomanian Pindiga Formation of Nigeria. (E) Miocene bryozoan limestone containing abundant cheilostomes from the York Peninsula of South Australia. (F) Composed almost entirely of branches of delicate ramose bryozoans, the late Oligocene–early Miocene Gambier Limestone from the South Australia/New South Wales border. Scale bars: A = 5 cm; B, C, D, E = 2 cm; F = 1 cm.

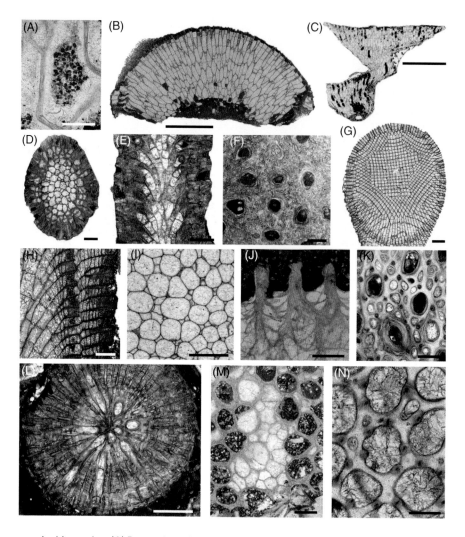

Plate 4 Bryozoans in thin section. (A) Brown deposits in the zooidal chamber of an unidentified trepostome are probably remnants of brown bodies formed during polypide degeneration (Ordovician, Katian; Vormsi Island, Estonia). (B) Vertical section through the dome-shaped trepostome *Mesotrypa* (Ordovician, Sandbian, Bromide Fm.; Arbuckle Mountains, Oklahoma, USA). (C) Vertical section of the esthonioporine *Dianulites fastigiatus* showing a new conical subcolony originating from the edge of an older subcolony (Ordovician, Darriwilian, Kunda stage; Putilovo Quarry, Leningrad Oblast, Russia). (D)–(F) Trepostome *Orbiramus* sp. (Ordovician, Tremadocian, Fenhsiang Fm.; Yichang, South China); (D) transverse section; (E) longitudinal section; (F) shallow tangential section. (G) Transverse section of the trepostome *Rhombotrypa quadrata* showing cyclic budding of quadrate zooids (Ordovician, Katian, Whitewater Fm.; Richmond, Indiana, USA). (H) Longitudinal section of the trepostome *Anomalotoechus lui* showing closely spaced basal diaphragms in the exozone and an overgrowth (Upper Devonian; China). (I) Tangential section through the trepostome *Mesotrypa* showing autozooids and smaller mesozooids (Ordovician, Sandbian, Bromide Fm.; Arbuckle Mountains, Oklahoma, USA). (J) Detail of transverse section through the trepostome *Kanoshopora droserae* showing the very shallow, sediment-filled living chambers of zooids at the edge of the branch (Middle Ordovician, Kanosh Fm.; Ibex, Utah, USA). (K)–(L) *Tibetopora orientalis*, a Triassic palaeostomate that may be the youngest cryptostome (Triassic, Ladinian; Tulung, Tibet): (K) tangential section showing autozooids, smaller exilazooids, and styles; (L) transverse section revealing the very narrow endozone forming a locus for spiral budding, and the broad thick-walled exozone containing styles. (M) Macula of a trepostome (?*Diplotrypa*) containing abundant mesozooids (Middle Ordovician, Kanosh Fm.; Ibex, Utah, USA). (N) Tangential section of the trepostome *Pseudobatostomella xizangensis* with styles indenting the autozooidal apertures (Triassic, Ladinian; Tulung, Tibet). Figure (A) courtesy of Andrej Ernst, (G) courtesy of Caroline Buttler. Scale bars: A, F, J, K, M = 200 μm; B = 5 mm; C, D, E, H, I, L = 500 μm; G = 1 mm.

Plate 5 Lophophores. The tentaculate feeding structures of some living bryozoans (A–G) and the phoronid *Phoronis ovalis* (H) which may be the closest living relative of Bryozoa. (A) Plan view of cheilostome *Membranipora membranacea* showing rectangular zooids with closely spaced lophophores, those in the bottom centre of the image leaning outwards around an excurrent chimney. (B) Single lophophore of the freshwater ctenostome *Paludicella* showing the long tentacle sheath with anus (arrowed); the shimmering along the edges of the tentacles is caused by the beating cilia. (C) Cyclostome *Crisulipora occidentalis* with small lophophores and tentacle sheaths hidden within the peristomes. (D) Two lophophores in a colony of the cheilostome *Thalamoporella californica*, the open opercula indicated by arrows. (E) Horseshoe-shaped lophophore of the phylactolaemate *Stephanella hina*. (F) Lophophores in two zooids of a small, unidentified cheilostome. (G) Bell-shaped lophophores of the marine ctenostome *Flustrellidra hispida*. (H) Emerging from a boring, a lophophore of *Phoronis ovalis*. Images (B), (E)–(H) are courtesy of Thomas Schwaha.

Plate 6 Bryozoan colony-forms. (A)–(C) Encrusting: (A) sheet-like cheilostomes *Aplousina*, *Escharella*, and *Puellina* on the interior surface of a bivalve (Pliocene, Waccamaw Fm.; Shallotte, North Carolina, USA); (B) runner-like cyclostome *Zigzagopora wigleyensis* on the interior surface of a brachiopod (Ordovician, Sandbian, Bromide Fm.; Fittstown, Oklahoma, USA); (C) ribbon-like cyclostome *Idmonea triquetra* on the exterior of the dorsal valve of a brachiopod (Jurassic, Bathonian, Bradford Clay; Wiltshire, UK). (D) Dome-shaped colony of the cheilostome *Celleporaria emancipata* in profile and plan view (Recent; Otago Shelf, New Zealand). (E) Articulated colony of the cheilostome *Cellaria* sp. (Recent; Otago Shelf, New Zealand). (F)–(G) Delicate ramose: (F) cyclostome *Diaperoecia purpurascens* (Recent; Otago Shelf, New Zealand); (G) cyclostome *Entalophora cellarioides* (Jurassic, Bathonian; Ranville, Calvados, France). (H)–(I) Robust ramose: (H) cyclostome *Ceriocava corymbosa* (Jurassic, Bathonian; Ranville, Calvados, France); (I) cheilostome *Celleporina grandis* (Recent; Otago Shelf, New Zealand). (J)–(K) Fenestrate: (J) cheilostome *Iodictyum phoeniceum* (Recent; Adelaide, Australia); (K) fenestellid '*Fenestella*' sp. (Lower Carboniferous; Halkyn, Flint, UK). (L)–(M) Free-living cheilostome *Lunulites tenax* (Cretaceous, Campanian Chalk; East Harnham, Wiltshire, UK): (L) upper, convex surface; (M) lower concave surface. (N)–(O) Palmate: (N) cyclostome *Siphoniotyphlus plumatus* (Cretaceous, Campanian Chalk; West Harnham, Wiltshire, UK); (O) cheilostome *Pentapora fascialis* (Recent; Mediterranean near Marseille). (P) Foliose cheilostome *Pentapora foliacea* (Recent; British waters). Scale bars: A, D, E, G, H, I, K = 1 cm; B, C, J, L, M, N = 5 mm; F, O = 2 cm; P = 4 cm.

Plate 7 Bryozoan colony disparity 1. (A) Flabellate colony of the cheilostome *Cellarinella roydsi* with conspicuous annual growth bands and a tuft of anchoring rootlets (Recent; Borge Bay, Signy Island, South Georgia). (B) Helicospiral colony of the iconic fenestrate bryozoan *Archimedes* (Carboniferous, Warsaw Fm.; Adams County, Illinois, USA). (C) Branch of the cheilostome *Celleporaria palmata* studded with corallites of the symbiotic coral *Culicia parasitica* (Pliocene, Coralline Crag Fm.; Ramsholt Cliff, Suffolk, UK). (D) Cribrate colony of the cheilostome *Adeona ? cellulosa* with basal rootlet system (Recent; Port Phillip Heads, Victoria, Australia). (E) *Demafinga pennsylvanica*, the fossil of a putative soft-bodied ctenostome (Carboniferous, Moscovian, Vanport Limestone; Elk County, Pennsylvania, USA). (F) Scimitar-shaped colony of the cryptostome *Ptilodictya lanceolata* with conical articulation process at the lower end (Silurian, Wenlockian, Much Wenlock Limestone Fm.; Dudley, West Midlands, UK). (G) Lyre-shaped colony of the fenestrate *Lyropora* with the proximal thickening at the bottom of the arched meshwork (Carboniferous, Chesterian; Chester, Illinois, USA). (H) The peculiar cystoporate *Evactinopora quinqueradiata* (Lower Carboniferous, Upper Burlington; Burlington, Iowa, USA). Scale bars: A, B, C, D, E, G, H = 1 cm; F = 2 cm.

Plate 8 Bryozoan colony disparity 2. (A) Fenestrate *Reteporina reticulata* showing elongate fenestrules through which multizooidal currents would have flowed (Silurian, Wenlockian, Much Wenlock Limestone Fm.; Dudley, West Midlands, UK). (B) Pinnate colony of the fenestrate *Arcanopora disticha* (Silurian, Wenlockian, Much Wenlock Limestone Fm.; Dudley, West Midlands, UK). (C)–(D) Cribrate colonies: (C) cryptostome *Stictoporellina* (Ordovician, Katian; Pechurki Quarry, Leningrad Oblast, Russia); (D) cheilostome *Adeona albida* (Recent; Australia). (E) Cyclostome *Kololophos* with radial fascicles of autozooids (Jurassic, Aalenian or Bajocian; southern England, UK). (F)–(G) Bryolith colonies of the cheilostome *Calpensia nobilis* (Pliocene; Palermo, Sicily, Italy): (F) exterior of spheroidal colony with partial exfoliation of the zooidal layers; (G) colony sectioned slightly off-centre of the nucleus. (H) Broken lichenoporid cyclostome colony consisting of multiple stacked and laterally joined subcolonies (Recent; Sea of Okhotsk, Russia). (I) Small cobble with an encrusting community dominated by sheet-like cheilostomes (Pleistocene, Setana Fm.; Kuromatsunai, Hokkaido, Japan). (J) Oncoid encrusted by sheet-like cyclostome bryozoans (Jurassic, Bathonian, Caillasse de la Basse-Ecarde; Luc-sur-Mer, Calvados, France). (K) Part of a colony of the cheilostome *Petraliella magna* with tubular branches (Recent; south of Broome, Western Australia). (L) Reticulate colony of the cyclostome Reticrisina that grew radially outwards from a cylindrical hollow (arrowed) representing the location of a perished stem-like substrate (Cretaceous, Maastrichtian; Fresville, Manche, France). (M) Broken colony of the cyclostome *Meandropora aurantium* with conspicuous growth banding (Pliocene, Coralline Crag Fm.; Suffolk, UK). (N) Branch of the cheilostome *Chiplonkarina dimorphopora*, a stenolaemate homeomorph (Cretaceous, Turonian, or Coniacian, Bagh Group; Madhya Pradesh, India). Scale bars: A, E, N = 5 mm; B, C, D, F, G, H, I, J, K, L = 1 cm; M = 2 cm.

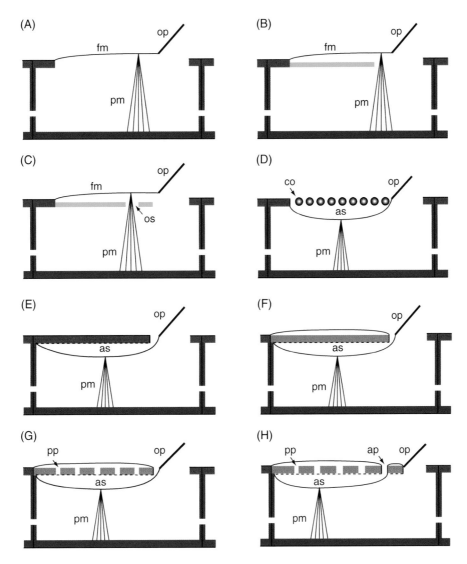

Plate 9 Cheilostome frontal wall types. (A) Anascan with a wide opesial opening ('membraniporimorph'). (B) Anascan with cryptocyst. (C) Anascan with cryptocyst perforated by opesiules. (D) Cribrimorph ascophoran with spinocystal frontal shield. (E) Gymnocystidean ascophoran. (F) Umbonuloid ascophoran. (G) Lepralioid ascophoran with entrance to the ascus via the orifice. (H) Lepralioid ascophoran with entrance to the ascus via an ascopore. Basal walls are shown in blue, vertical walls in purple, exterior frontal walls in red, anascan interior frontal walls (cryptocysts) in green, and umbonuloid and lepralioid frontal shields in pale red. Abbreviations: ap, ascopore; as, ascus; co, costa; fm, frontal membrane; op, operculum; pm, parietal muscles; pp, pseudopore.

Plate 10 Indicators of multizooidal current systems, substrates, and bryozoans from reefs. (A)–(E) Structures associated with multizooidal feeding currents: (A) elongate monticules in the trepostome *Spatiopora* growing on an orthoconic nautiloid (Ordovician, Katian, Cincinnati Group; Cincinnati, Ohio, USA); (B) stellate monticules in the cystoporate *Constellaria* (Ordovician, Sandbian, Bromide Fm.; Arbuckle Mountains, Oklahoma, USA; (C) maculae appearing as dark spots devoid of apertures in the cystoporate *Fistulipora asteria* (Carboniferous, Chesterian; Chester, Illinois, USA); (D) in the cystoporate *Hexagonella dendroides*, irregularly polygonal subcolonies centred on maculae are bounded by slightly raised walls (Permian; Western Australia); (E) polygonal windows in a colony of the cyclostome *Reticrescis* through which multizooidal currents are inferred to have flowed (Cretaceous, Campanian, Aubeterre Fm.; Meschers, Charente Maritime, France). (F)–(H) Colonies growing around the stems of seagrasses (Recent; Blanche Point, Adelaide, Australia): (F) cheilostome *Thornleya*; (G) cheilostome *Adeonellopsis*; (H) cyclostome *Densipora*. (I) Trepostome bryozoan that grew on an aragonitic bivalve shell lost through dissolution, the mould of the shell preserved through 'bryoimmuration' (Ordovician, Katian, Cincinnati Group; Cincinnati, Ohio, USA). (J)–(L) Reefal bryozoans: (J) tangle of silicified acanthocladiid fenestrate bryozoans revealed by acid dissolution of reef rock (Permian; Glass Mountains, Texas); (K) section through a reef consisting almost entirely of multilayered cyclostome bryozoans into which some oyster-like bivalves have been incorporated during bryozoan growth (Jurassic, Tithonian, Portland Limestone Fm.; Isle of Portland, Dorset, UK); (L) cheilostome and cyclostome bryozoan fragments from well-cuttings through a subsurface bryozoan reef mound (Middle Eocene; Great Australian Bight). Scale bars: A, C, I = 1 cm; B, D, E, F, G, H, L = 5 mm; J, K = 2 cm.

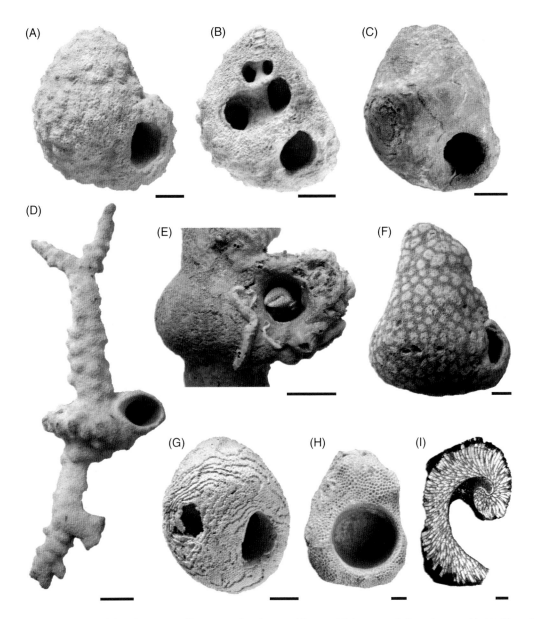

Plate 11 Conchicole symbiont bryozoans. Bryozoans forming symbioses with known or inferred paguroids (A–G), and with an unknown conchicole (H–I). (A)–(B) Pliocene fossils of the cheilostome *Hippoporidra edax* once believed to be parasites of gastropods: (A) external view with paguroid aperture on the lower right; (B) sectioned specimen showing a small, high-spired gastropod shell at the apex of the bryozoan colony which has constructed 2.5 whorls of living chamber for the paguroid symbiont (Pliocene, Coralline Crag Fm.; Suffolk, UK). (C) Cyclostome *Reptomultisparsa incrustans*, the oldest known inferred paguroid symbiont, with a worn patch ('pagurid facet') on the left shoulder of the colony (Jurassic, Bathonian, Caillasse de la Basse-Ecarde Fm.; Calvados, France). (D) 'Texas Longhorn' colony of *Hippoporidra edax* (Recent; Gulf of Mexico, USA). (E) Chela of the paguroid occupant visible in the aperture of another 'Texas Longhorn' (Recent; Gulf of Mexico, USA). (F) Pigmented surface of a worn colony of *Hippoporidra edax* (Recent; Margarita Island, USA). (G) Abraded and broken colony of the cheilostome *Odontoporella* showing layers of zooids (Pliocene, San Diego Fm.; San Diego Co., California, USA). (H)–(I) Trepostome *Leptotrypa calceola* which constructed a curved tube, seen most clearly in the thin section (I), presumably around a conchicole (Ordovician, Katian, Cincinnati Group; Cincinnati, USA). Scale bars: A, B, C, F, G = 5 mm; D, E = 1 cm; H, I = 1 mm.

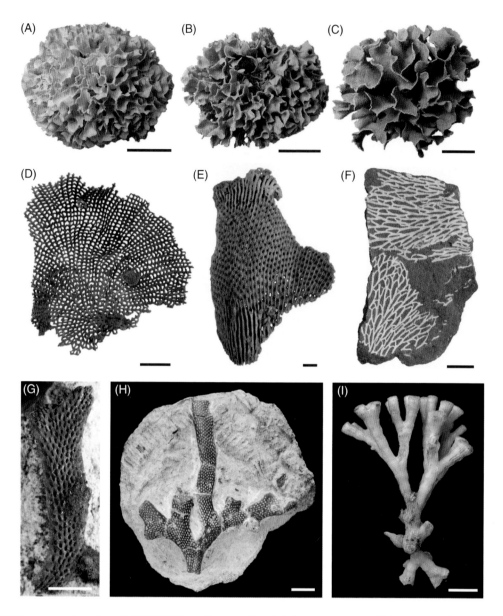

Plate 12 Bryozoan colony disparity 3. (A)–(C) Three distantly related cheilostome genera which convergently evolved foliose colonies: (A) malacostegine anascan *Acanthodesia* sp. (Recent; Zanzibar); (B) umbonuloid ascophoran *Porella* sp. (Recent; Point Barrow, Alaska, USA); (C) lepraliomorph ascophoran *Microporella bifoliata* (Pliocene, Yorktown Fm.; York County, Virginia, USA). (D) Silicified fenestrate (?*Septopora*) with irregular fenestrules formed by branch division and anastomosis (Permian; Glass Mountains, Texas). (E) Silicified fenestrate *Semicoscinium obliquatum* with superstructure consisting of thickened carinae visible at the top and bottom of the fragment (Devonian, Hamilton Group; Falls of the Ohio, Indiana, USA). (F) Two colonies of *Chasmatopora furcata* embedded in matrix (Ordovician, Darriwilian, Kuckers Shale; Estonia). (G) The oldest known unequivocal bryozoan, the phylloporinid cryptostome *Prophyllodictya simplex*, coarsely silicified (Ordovician, lower Tremadoc, Nantzinkuan Fm.; Liujiachang, central China). (H) A younger phylloporinid cryptostome, *Pachydictya holmi* (Silurian, Llandovery, Visby Beds; Gotland, Sweden). (I) Fragment of a ramose colony of the cyclostome *Fasciculipora ramosa* in which the zooids open only at the distal tips of the branches, branch sides being formed entirely of calcified exterior walls (Recent; Antarctica). Scale bars: A, B = 5 cm; C = 2 cm; D = 1 cm; E, G = 2 mm; F, H, I = 5 mm.

3

Zooid Morphology and Function

Zooids are the fundamental modules constituting bryozoan colonies. They are homologous with the individuals of unitary organisms (e.g. Mackie 1986, p. 177), such as the closely related brachiopods and molluscs. In the most primitive bryozoans, the zooids within a colony are clearly distinct, identical or nearly so, and capable of sustaining themselves in all life functions (feeding, respiration, reproduction, defence etc.). At the other end of the spectrum are species in which the physical boundaries between the zooids are vague, different zooids in the same colony differ in appearance and function, and some of the zooids are incapable of surviving independently. Functional differences between zooids are reflected in differences in their morphologies, resulting in polymorphism. However, not all variations in zooid morphology within colonies reflect polymorphism: some are due to differences in zooid ages (ontogeny), the age of the colony at the time the zooid was budded (astogeny), or small-scale, microenvironmental variations (Boardman and Cheetham 1973).

Primitively, each zooid in a bryozoan colony is capable of undertaking all of the life functions required of a suspension-feeding animal: in addition to acquiring food and disposing of solid waste (faeces), these include respiration, excretion, defence, and reproduction, both as a male and as a female. These functions, and the organs associated with them, were inherited from the solitary (non-colonial) ancestor of the colonial bryozoans. Through evolutionary time, however, bryozoans have partitioned life functions between the constituent zooids in the colony, resulting in colonies that contain zooids of differing morphologies, i.e. polymorphism. Zooid polymorphs retaining the capability to feed are called *autozooids*, while

Bryozoan Paleobiology, First Edition. Paul D. Taylor.
© 2020 Natural History Museum. Published 2020 by John Wiley & Sons Ltd.

non-feeding polymorphs are termed *heterozooids*. There may also be polymorphism within autozooids according to their ability to perform other functions; for instance, some autozooids in cheilostome bryozoans not only feed but also perform a female reproductive function, producing eggs and brooding the embryos that result from fertilization, whereas other autozooids in the same colony feed but do not reproduce as females. A wide range of heterozooids can be distinguished in recent bryozoans. Their functions are not always well established and even in some living bryozoans they are unknown as, for example, the heterozooids of the cyclostome *Jullienipora* with striated exterior walls (Reverter-Gil and Fernández-Pulpeiro 2005). In fossil bryozoans, the functions of polymorphic zooids can often be inferred by comparison with living species, but for some extinct groups their functions remain uncertain in the absence of soft part preservation.

3.1 Autozooids

All bryozoan colonies must contain autozooids in order to feed. Autozooids typically outnumber other polymorphs in the colony, or at least occupy the greatest area on the surface of the colony. They have two principal components: the protective **cystid** and the **polypide** with a crown of tentacles (lophophore) and digestive tract. In all bryozoans with the exception of phylactolaemates and ctenostomes, the cystid has a calcified layer that imparts a rigid outer shape to the zooid. At the distal end of the cystid is the aperture or orifice through which the tentacle crown is protruded. When the tentacle crown is withdrawn, the opening of the cystid is closed by a terminal membrane in most stenolaemates but by a hinged, lid-like **operculum** in most cheilostomes (Figure 3.1A). As is so often the case among bryozoans, there are exceptions: Eleidae, a Cretaceous–Paleocene family of cyclostome stenolaemates, evolved calcareous opercula (Figure 3.1F) to seal the zooidal apertures (Chapter 9.11), while cheilostomes belonging to the extant family Bugulidae have secondarily lost their opercula. Cheilostome opercula are usually thickened cuticular structures, sometimes double layered (Gordon, Voje, and Taylor 2017), and in most cases lack calcification apart from a few taxa scattered across the order that have calcified opercula (Koromyslova 2014a; Figure 3.1B–E). Prominent among these is the anascan cheilostome *Macropora* in which the opercula are thickly calcified in all known species (Gordon and Taylor 2008; Figure 3.1B).

3.1.1 Zooid size

The size of autozooids varies in bryozoan species. Among cheilostomes there is more than an order of magnitude variation in zooid length, equivalent to an enormous disparity in the biovolume of cheilostome zooids. The smallest zooids are only 0.2 mm in length (e.g. *Puellina gattyae* and *Mollia multijuncta*: Zabala and Maluquer 1988), whereas the largest can be 3 mm long and even longer in species that have narrow, tail-like proximal parts (**caudae**) such as *Herpetopora* where autozooids may be 8 mm long (Taylor 1988a). Many of the Recent cheilostome species with zooids longer than 1 mm live in the Antarctic, e.g. *Antarcticaetos bubeccata* and *Cellarinoides crassus* (Hayward 1995). Zooid area in a sample of 100 anascan-grade cheilostome species with sheet-like encrusting colonies was found to be strongly right-skewed: zooid area is small in the majority of species, with a steadily declining number of species having zooids of large surface area (Figure 3.2). A recent study of zooid size through time in cheilostomes has shown that, taking the order as a whole, there has been no long-term temporal change in autozooid surface area; however, within genera, younger species tend to have larger zooids than older ones (Liow and Taylor 2019).

No comprehensive survey exists of zooid size in stenolaemates but Anstey and Perry (1972) compiled data on the diameter of autozooidal chambers in 122 species of trepostomes, finding the range to be 0.10–0.46 mm (median 0.21 mm). In living cyclostomes,

Figure 3.1 Opercula closing the orifices of cheilostomes (A–E) and the aperture of an eleid cyclostome (F); (A) cuticular, non-calcified operculum of *Watersipora nigra* (Recent; Galapagos); (B) mineralized operculum of *Macropora retusa* with reticulate ornamentation indicative of cryptocystal calcification (i.e. interior wall) (Oligocene, Kakanui Limestone; North Otago, New Zealand); (C) calcified operculum of *Einhornia venturaensis* surrounded by dark-coloured frontal membrane (Recent; San Diego, California, USA); (D) calcified operculum of *Einhornia* sp. fossilized in situ despite loss of the supporting frontal membrane (Pleistocene, Norwich Crag Fm.; Covehithe Warren, Suffolk, UK); (E) another in situ calcified operculum in *Einhornia* sp. (Pleistocene, Red Crag Fm.; Ramsholt, Suffolk, UK); (F) calcified operculum filling the skeletal aperture of an autozooid of *Meliceritites royana* (Cretaceous, Campanian, Biron Fm.; Talmont, Charente Maritime, France). Scale bars: A = 200 μm; B, D, E, F = 100 μm; C = 50 μm.

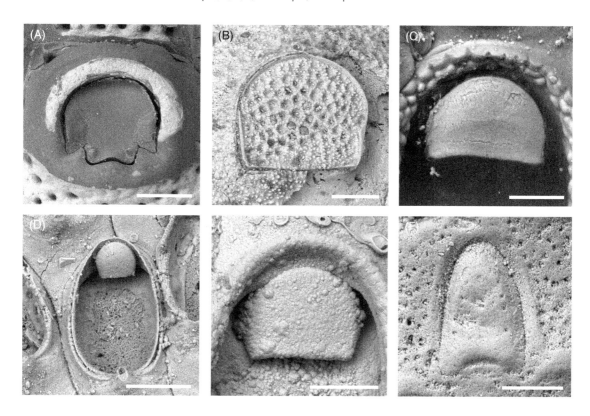

the diameter of the skeletal aperture ranges from about 0.06 mm (in some articulates: Hayward and Ryland 1985) to more than 0.5 mm (in cinctiporids: Boardman, McKinney, and Taylor 1992), almost equalling the range seen in cheilostomes.

3.1.2 Tentacle crown protrusion and withdrawal

The tentacle crown can be withdrawn into the safety of the cystid very rapidly by the contraction of retractor muscles connected to its base and anchored on the walls of the cystid. The retractor muscle in the cheilostome *Membranipora membranacea* was shown by Thorpe, Shelton, and Laverack (1975) to have a very fast rate of contraction of about 20 or more muscle lengths per second. It can be triggered to contract by touching the lophophore with the point of a needle, while lophophores of nearby zooids will also retract if

Figure 3.2 Frequency distribution of zooid size, expressed as a proxy of frontal area, in a sample of 100 species of encrusting anascan cheilostomes (redrawn from Grischenko, Taylor, and Mawatari 2002, fig. 25).

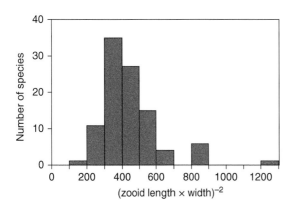

the frontal surface of the zooid is stimulated, the signal being transmitted through the nervous plexus that connects the zooids of the colony.

Whereas straightforward muscular contraction brings about withdrawal of the lophophore, the mechanism for protruding the lophophore is more complicated and varies between bryozoan groups. Muscular contraction provides a pulling force and therefore cannot be employed alone to push the tentacle crown out through the orifice. Instead, bryozoans employ a hydrostatic system in which muscular contraction pulls on membranes, increasing local hydrostatic pressure and forcing coelomic fluid to migrate into an organ called the **tentacle sheath** (Plate 5B). This is a tubular, sock-like membrane extending from the base of the lophophore to the cystid walls and surrounding the lophophore when it is retracted. Migration of coelomic fluid

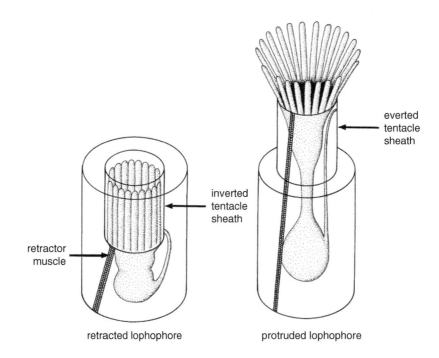

Figure 3.3 Basic mechanism of lophophore protrusion and retraction in bryozoans. Lophophore protrusion occurs through hydrostatic eversion – turning inside out – of the tentacle sheath using muscles (not shown) that squeeze or depress the body wall of the zooid. Retraction of the lophophore is brought about through inversion of the tentacle sheath by contraction of the retractor muscle (modified from Taylor 1981. © The Trustees of the Natural History Museum, London).

Figure 3.4 The four main modes of tentacle sheath eversion, and therefore lophophore protrusion, in marine bryozoans (modified from Taylor 1981. © The Trustees of the Natural History Museum, London). The tentacle sheath is stippled but the lophophore is omitted. Arrowheads show membranes that are squeezed or depressed to raise hydrostatic pressure and cause tentacle sheath eversion.

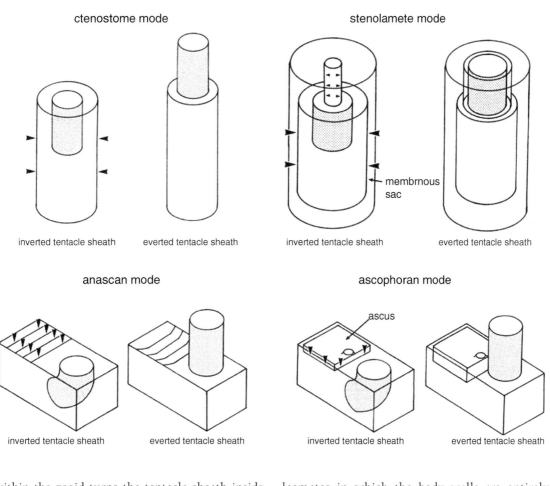

ctenostome mode

inverted tentacle sheath   everted tentacle sheath

stenolamete mode

membrnous sac

inverted tentacle sheath   everted tentacle sheath

anascan mode

inverted tentacle sheath   everted tentacle sheath

ascophoran mode

ascus

inverted tentacle sheath   everted tentacle sheath

within the zooid turns the tentacle sheath inside out, i.e. causes it to be everted (Figure 3.3). Eversion forces the lophophore attached to the end of the tentacle sheath through the zooidal orifice.

Four main modes of lophophore protrusion can be distinguished in bryozoans (Taylor 1981; Figure 3.4). The simplest, and probably the most primitive in evolutionary terms, entails muscles pulling the flexible outer body walls of the cystid inwards, like squeezing a tube of toothpaste. This mode is found in some ctenostomes and phylacto-laemates in which the body walls are entirely uncalcified and the zooids tend to be cylindrical in overall shape.

The same cylindrical shape is retained in steno-laemates but here the body walls are rigidly calcified and the polypide is enclosed within a structure called the **membranous sac**. The membranous sac functions like the cystid wall of a ctenostome and is equipped with circular muscles whose contraction exerts the necessary hydrostatic pressure to evert the tentacle sheath (Nielsen and Pedersen

1979). Unlike other bryozoans, stenolaemates have two fluid-filled body cavities – one within the membranous sac and the other outside it – termed the entosaccal and exosaccal cavities, respectively. While the entosaccal cavity is universally regarded as a true coelom, the exosaccal cavity is believed by many to be a pseudocoel. The exosaccal cavity extends distally beyond the point at which the tentacle sheath is attached to the zooid body walls and encloses a vestibule. When the muscles of the membranous sac contract, fluid in the distal part of the exosaccal cavity migrates proximally to occupy the space vacated by the tentacle crown as it is forced out through the skeletal aperture.

An alternative mode is employed by anascan-grade cheilostomes, in which the essentially box-shaped autozooids have rigidly calcified side walls. A flexible frontal membrane is pulled downwards by parietal muscles anchored to the basal and side walls of the zooid to increase hydrostatic pressure and force the tentacle sheath to evert. Fossil anascans sometimes have small pits (Figure 3.5A) in the basal walls of the autozooids representing the anchorage points of these muscles or of the more distally placed occlusor muscles that close the operculum. In some groups of anascans, a shelf-like interior frontal wall (**cryptocyst**) develops beneath the frontal membrane and the parietal muscles either share the distal opening (**opesia**; pl. **opesiae**) through which the lophophore is protruded, or pass through separate holes (**opesiules**) in the cryptocyst (Figure 3.5B). Typically, opesiules are paired on either side of the frontal wall and are few in number, although in *Macropora* each zooid has numerous opesiules distributed over the entire frontal wall (Banta, Gray, and Gordon 1997; Figure 3.5C). Both gymnocystal (i.e. exterior walls) and cryptocystal (i.e. interior walls) frontal walls may fulfil protective functions in anascans but the latter confer the benefit of not diminishing the area of the frontal membrane which is depressed to bring about protrusion of the lophophore (Ristedt 1991).

Ascophoran-grade cheilostomes evolved an internal frontal membrane, protected beneath the calcified **frontal shield**. The space above the frontal membrane constitutes a compensation sac called the **ascus**, hence the name ascophoran. When the parietal muscles pull on the frontal membrane to evert the tentacle sheath, they depress the floor of the ascus which fills with seawater, compensating for the volume once occupied by the lophophore. When the lophophore is withdrawn, the ascus empties of seawater. The entrance to the ascus varies among different ascophoran groups. In cribrimorphs, gaps between the costae forming the frontal shield (Figure 3.5D) allow ingress and egress of seawater. In other ascophorans with more complete frontal shields the entrance to the ascus is more restricted. Often it is marked by a sinus on the proximal edge of the orifice (Figure 3.5E), in which case the operculum acts like a double valve, the distal part opening outwards to allow the tentacle crown to protrude while the proximal part opens inwards to permit water to flow into the ascus. Alternatively, the opening into the ascus may take the form of a separate pore (or pores) in the frontal shield, termed an **ascopore** (Figure 3.5F) or **spiramen** (Figure 3.10D), depending on its exact mode of development and anatomy (see Chapter 9.14).

These four basic modes of tentacle sheath eversion have important consequences for zooidal and colonial morphology. The mode employed by many ctenostomes and phylactolaemates requires that the outer body wall is deformed by muscles. This precludes rigid biomineralization and requires that the wall must be at least partly free (i.e. not completely contiguous with the walls of neighbouring zooids), which means that fully multiserial colonies are less easy to construct. Accordingly, ctenostomes and phylactolaemates both lack calcified skeletons and show a more limited variety of colony-forms than stenolaemates and cheilostomes, with a greater proportion being either uniserial or biserial.

Internalization of the deformable membrane in stenolaemates is hypothesized to have evolved through fission of the body wall into two components: an outer body wall and the deformable membranous sac, with the exosaccal cavity (or pseudocoel) between (Taylor 1981). It then

Figure 3.5 Morphological features of the zooidal skeletons in some anascan (A–C) and ascophoran (D–F) cheilostomes: (A) pair of scars (arrowed) left by the opercular occlusor muscles on the basal wall in *Copidozoum planum* (Plio-Pleistocene, Burica Fm.; Burica Peninsula, Panama); (B) paired opesiules (arrowed) for the passage of muscles attached to the frontal membrane in *Calpensia nobilis* (Recent; Ischia, Italy); (C) zooid of *Macropora levenseni* with small pores in the frontal wall through which muscles pass during life (Recent; Cook Strait, New Zealand); (D) zooid of the cribrimorph *Pelmatopora suffulta* showing the frontal shield formed of overarched spines between which are spaces allowing seawater to enter or exit the ascus below (Cretaceous, Coniacian or Santonian, Chalk Gp.; Northfleet, Kent, UK); (E) sinus (arrowed) along the proximal edge of the orifice providing a different route into the ascus in *Myriapora subgracilis* (Pleistocene, Setana Fm.; Kuromatsunai, Hokkaido, Japan); (F) in *Microporella stenoporta* the entrance to the ascus is via a separate opening, the kidney-shaped ascopore (arrowed), located proximally of the orifice (Recent; Cape Hallett Canyon, Antarctica). Scale bars: A, F = 100 μm; B, D = 200 μm; C = 500 μm; E = 50 μm.

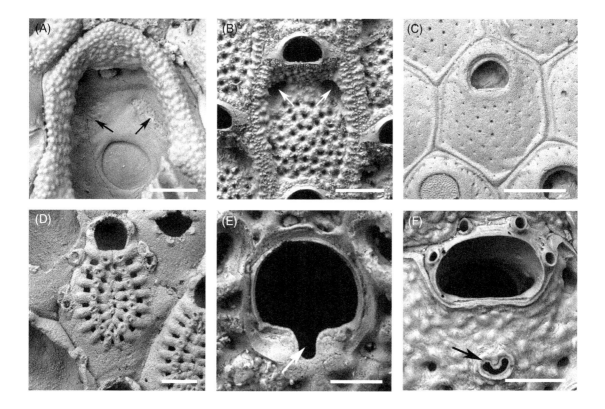

became possible to calcify the outer body wall beneath the cuticular layer, and to construct fully multiserial colonies with complete contiguity of the zooids, opening the way for a much greater range of colony-forms. In order to maintain a large surface area of deformable membranous sac, stenolaemate zooids have essentially tube-shaped skeletons that grow in length through accretion around the terminal aperture. Often the length of the tubular zooidal skeleton is more than 10 times the width, and the soft parts are able to extend distally as growth progresses (Figure 3.6). Bending of the tube through an angle of approximately 90° during growth is typical in stenolaemates (Plate 4E, H). This has particular relevance for species with erect ramose colonies, as in many trepostomes. Here the first-formed part of the tubular zooidal skeleton is oriented parallel to branch growth direction and its growth contributes to branch lengthening in the central **endozone** of the branch. As the zooid passes into the peripheral **exozone**, its orientation changes to become roughly perpendicular to the branch axis. Continued growth contributes to branch thickening – and therefore strengthening – rather than branch lengthening. A decrease in skeletal growth rate away from the branch axis is responsible for the cylindrical shape of the branches, not only in stenolaemate bryozoans but also in other animals and plants with similar ramose morphologies (Key 1990). This can be seen in longitudinal sections where growth increments, delimited by bands of diaphragms or wall thickenings, are deepest at the centre of the branch and taper in height outwards (Plate 4H).

Another important correlate of the tentacle sheath eversion mode characteristic of stenolaemates is that the point of polypide attachment – at least in the extant cyclostomes – is situated well beneath the skeletal aperture in order to leave space for the distal exosaccal cavity which forms a reservoir of fluid that can be moved proximally to fill the everting tentacle sheath. Therefore, the tentacle crown does not protrude as far beyond the skeletal aperture as in other groups of bryozoans, with major consequences for feeding behaviour (Chapter 4.5).

Like stenolaemates, the modes of tentacle sheath eversion used by other groups of marine bryozoans are optimized by having zooids with rigid vertical body walls that are not deformed as the lophophore is extruded. Rigidity is achieved by biomineralization in cheilostomes but in some ctenostomes, such as *Alcyonidium*, a thick gelatinous body wall performs this function. However, a significant part of the frontal surface of the zooid – the frontal membrane – must remain flexible enough to be deformed by muscles to bring about the increase in hydrostatic pressure necessary for tentacle sheath eversion. This frontal membrane is either exposed on the outer surface of the zooid, as in anascan-grade cheilostomes, or overgrown by a calcified frontal shield, as in

ascophoran-grade cheilostomes. The optimal zooid shape to maximize the area of deformable frontal membrane is closer to a box, contrasting with the tubes seen in stenolaemates (Figure 1.4). The rigid lateral walls of these box-like zooids can be juxtaposed with those of neighbouring zooids, allowing for the development of fully contiguous zooids and a wide range of colony-forms. However, unlike the

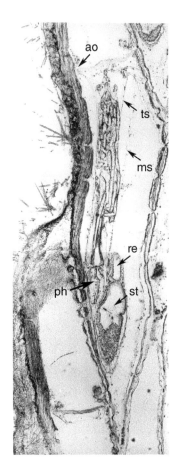

Figure 3.6 Longitudinal section of a polypide situated within the elongate living chamber defined by the tubular zooidal skeleton of the cyclostome bryozoan *Attinopora zealandica*. Abbreviations: ao, attachment organ; ms, membranous sac; ph, pharynx; re, rectum; st, stomach; ts, tentacle sheath. See Boardman, McKinney, and Taylor (1992, fig. 10) for a fully annotated diagram of a closely related bryozoan.

continuously growing tubular zooids typical of stenolaemates, the size and box-like shape of the zooids is generally fixed in early development. Therefore, thickening of colonies or widening of the branches of ramose colonies cannot be achieved by simply lengthening the zooids as in stenolaemates. Instead, overgrowth by new layers of zooids is normally required. Hence, dome-shaped and thick-branched cheilostomes typically have multilamellar colonies with layers of zooids piled one on top of the other (see Hara 2001 for good examples from the Eocene of Antarctica). It should be noted, however, that a few groups of cheilostomes did succeed in evolving long tubular zooids that mimic those of stenolaemates, including the Late Cretaceous genus *Chiplonkarina* (Chapter 9.11).

3.1.3 Living chamber shape in stenolaemate autozooids

The optimal shape of zooidal living chambers for the mode of tentacle eversion employed by stenolaemates is an elongated tube oriented perpendicular to the colony surface. This shape allows for accommodation of a distal vestibule from which fluid can be displaced proximally to compensate for the space vacated by the extruded lophophore, as well as a long, cylindrical membranous sac with a large surface area for muscular squeezing (Figure 3.4). In most stenolaemates, zooidal skeletons are long tubes. This reaches an extreme in some late Palaeozoic stenoporid trepostomes with zooids that are many tens of times longer than they are wide (e.g. Madsen 1987). However, some ramose trepostome bryozoans of early Palaeozoic age appear to have had anomalously shallow living chambers (Anstey 1990) in which the distance between the colony surface and the outermost basal diaphragm defining the floor of the living chamber is often scarcely greater than the diameter of the zooidal aperture (Plate 4J). Accommodating a long polypide similar to those found in modern stenolaemates (cyclostomes) into such a short chamber is problematical, leading to reconstructions of these trepostomes with tiny polypides oriented tangentially to the colony surface, as in the box-shaped zooids of cheilostomes. An alternative interpretation is that such shallow chambers only developed after the final cycle of polypide degeneration and never housed a functional polypide, with the basal diaphragms close to the colony surface functioning in structural strengthening of the colony instead (Boardman 2001). This interpretation is supported by the presence in some trepostomes of diaphragms that become so closely spaced that they form a thick covering on the colony surface (Plate 4E, F). If this interpretation is correct, autozooids with lophophores would have been confined to the distal tips of the branches, with branch sides being devoid of active autozooids, recalling the living cyclostome *Fasciculipora* where autozooids open only at distal branch growth tips and are absent on the sides of the branches (Plate 12I).

3.1.4 Suspension feeding

Bryozoans are active suspension feeders, i.e. filter feeders. Individual autozooids have the ability to create a feeding current that drives particles towards their mouth (see McKinney 1990). Coordinated movements – beating – of the lateral cilia on either side of the tentacles provide the force to drive this current of water. A metachronal wave passes along the lateral cilia up one side of the tentacle and back down the other. This propels water from inside the lophophore to the outside, establishing an overall flow of water that enters the open top end of the lophophore and departs at the sides between the tentacles (Figure 3.7). Water travels fastest at the very centre of the lophophore, propelling entrained particles of food here directly towards the mouth. And because the flow velocity at the centre of the lophophore can be up to six times greater than at the edges, a high proportion of particles travel within this stream. Particles not in the mainstream of flow towards the mouth can be captured and transported by the frontal cilia or stopped by the laterofrontal cilia and batted back into the central flow by flicking of the tentacles in marine bryozoans (Riisgård and Manríquez 1997; Nielsen and Riisgård 1998; Larsen and Riisgård 2002). The fundamental processes involved in particle capture are very similar in all bryozoans and closely resemble those seen in the related brachiopods and phoronids (Riisgård, Okamura, and Funch 2009).

Figure 3.7 Schematic vertical and horizontal sections through a bryozoan zooidal lophophore to show the direction of the feeding currents created by the cilia (shown only on the horizontal section), which drive water through the open end of the lophophore, towards the mouth (M) and outwards between gaps in the tentacles.

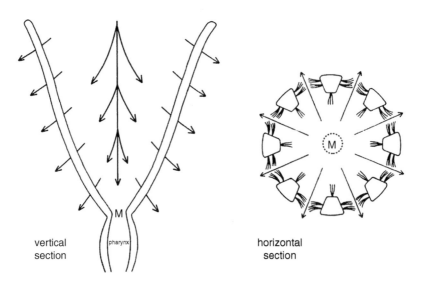

vertical section

pharynx

horizontal section

The feeding behaviour of individual zooids varies considerably (Winston 1978; Shunatova and Ostrovsky 2001; Tamberg and Shunatova 2016). For example, individual tentacles may flick, sometimes to direct food particles into the fast flow regime at the centre of the lophophore but at other times to expel unwanted particles. Reversal of the direction of ciliary beating is another way of removing unwanted particles from within the lophophore. Formation of a cage by bending of the tentacles has been seen in many different bryozoans but its function is unclear, although protists have been observed trapped in these cages.

Interaction between the feeding currents of individual autozooids leads to various patterns of multizooidal currents that are subcolony- or colony-wide in extent. These have great importance in shaping morphology at the level of the colony and will be described in the next chapter on colony structure (Chapter 4.5). However, it is worth noting here that the typical close packing of lophophores that are circular in plan view can act as a constraint on zooid geometry in multiserial bryozoans (Ryland and Warner 1986). A small number of either hexagonal or rectangular zooidal outline shapes are compatible with the quincuncially arranged lophophores – the two most equidimensional of these have proportions of 1:0.87 and 1.72:0.5 relative to lophophore diameter (Figure 3.8). Also using the premise of close packing of lophophores, Starcher and McGhee (2000) undertook a geometric analysis of lophophore shape and arrangement in extinct fenestellid bryozoans by creating a theoretical morphospace of aperture positioning in the colonial meshwork. This allowed them to infer the likely presence of obliquely truncate lophophores in zooids adjacent to the fenestrules in some taxa.

While all known marine bryozoans feed mostly on phytoplankton, there are differences between species in the patterns of behaviour of the lophophores (e.g. flicking), velocity of the feeding current generated (which is directly proportional to lophophore size: see McKinney 1990, fig. 6), size of the mouth, and the presence or absence of a gizzard. These variables may correlate with dietary

Figure 3.8 Hexagonally close-packed arrays of lophophores with basic zooid outline shapes fitted to show how the length/width proportions of the zooids are determined geometrically when the zooids are rectangular (left) or hexagonal (right) in shape. Arrows indicate proximal–distal growth directions of the colony (based on Ryland and Warner 1986, fig. 4).

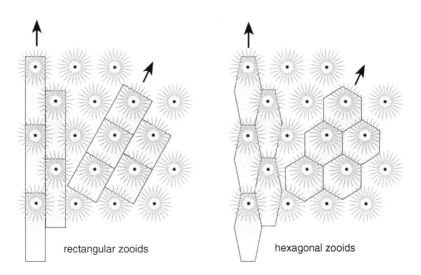

rectangular zooids

hexagonal zooids

selectivity. For example, a gizzard has evolved convergently in a few species (Markham and Ryland 1987) and this apparently allows the frustules of some diatoms to be broken open and the contents ingested. The effectiveness of gizzards in crushing diatom frustules suggests that species possessing gizzards subsist on a diet that includes diatoms as well as the naked phytoplankton (e.g. dinoflagellates) that are utilized more generally by bryozoans. Observations have also been made of bryozoans with large lophophores capturing protists and small arthropods.

In addition to acquiring planktonic food, bryozoans must dispose of faecal pellets efficiently in order to circumvent fouling of the colony surface with their own waste products. It has been shown that different species use contrasting behaviours to accomplish this task (McKinney 1997). Pellets in some species are drawn into the lophophore and expelled from the open top, while in others they are carried away along the colony surface beneath the level of the lophophores until they reach the edge of the colony.

3.2 Ontogenetic and Astogenetic Variations

Despite their genetic homogeneity, the morphology of the zooids within a single bryozoan colony is never uniform. There are four main sources of variability: ontogeny, astogeny, polymorphism, and microenvironment (Boardman and Cheetham 1973). Polymorphism has already been mentioned above and is described in more detail below. Ontogeny and astogeny relate, respectively, to the developmental stage of the individual zooid and that of the colony. Microenvironmental variation essentially encompasses any remaining 'ecophenotypic' variability in zooid form caused, for example, by obstructions on the substrate or the ambient temperature at the time when a zooid was budded (see Chapter 6.7).

3.2.1 Ontogeny

Like unitary organisms, the constituent zooids in bryozoan colonies undergo changes during their development or ontogeny. These ontogenetic changes in living colonies can be tracked by observing individual zooids through time. However, it is also possible to reconstruct ontogeny in preserved colonies, including fossils, by comparing zooids at ever greater distances from the colony growing edge as these were budded progressively earlier in time and therefore chronicle successively more advanced stages of development. Thus, a **gradient of ontogenetic change** extends in a proximal direction from the growing edge towards the earliest formed parts of the colony.

Ontogenetic changes in bryozoan zooidal skeletons include increases in the lengths of zooids in taxa with tubular zooids, thickening of skeletal walls, and occasionally more profound changes that can continue over the entire lifespans of the zooids.

Lengthening of the tubular zooidal skeletons during ontogeny is continuous and extreme in some Palaeozoic trepostomes, with the oldest zooids being several centimetres long and thus far longer than those of newly budded zooids at colony growing edges. In such bryozoans, the polypides must migrate distally as the tubular skeleton elongates, vacating the older parts of the tube, which may become sealed off by a **basal diaphragm** (Figure 3.9A). In some fixed-walled cyclostomes (see Figure 8.7), zooids have tubular **peristomes** (Figure 3.9D) that grow upwards around the apertures. Peristome height increases during zooidal ontogeny as long as the zooid has an active polypide. Following final polypide degeneration, the peristome breaks off and the aperture of the zooid becomes sealed by a **terminal diaphragm** (Figure 3.9B, C). Such colonies may exhibit three ontogenetic zones in their skeletons (Silén and Harmelin 1974). From distal to proximal these are: (i) a zone formed by the growing edge with skeletally immature zooids without frontal walls or peristomes; (ii) a zone of zooids possessing peristomes that increase in length proximally; and (iii) a zone of zooids with broken-off peristomes and apertures sealed by terminal diaphragms (Figure 3.9E). Ontogenetic changes of this sort must obviously be borne in mind when attempting to identify fragmentary fossils of stenolaemate bryozoans. Fragments of young parts of colonies

near distal growing edges may look very different from those of old, proximal parts of colonies.

Cheilostomes may likewise exhibit substantial ontogenetic changes in the morphology of their zooidal skeletons (Figure 3.10). These can be particularly marked in ascophoran-grade species where thickening of the frontal shield with age is often accompanied by changes in its surface texture and porosity, as well as the development of associated polymorphs, particularly adventitious avicularia (e.g. Lidgard 1996). In addition, areolar pores that are confined to the circumference of young zooids may multiply through branching and 'migrate' into the more central parts of the frontal shield in older zooids. Thus, the frontal shield can change from being porous only at the margins to porous over most of the surface. Furthermore, progressive thickening of the frontal shield in ascophoran cheilostomes can obscure the primary orifice, a key character in taxonomy, making it difficult to identify fragments of old colonies.

Like cyclostomes, ontogenetically aged, moribund zooids of cheilostomes may become sealed by a calcified wall, here termed a **closure plate**. In anascan-grade cheilostomes the closure plate often bears an impression of the operculum (Figure 3.11), making it possible in fossil species to determine the size of the orifice, which in turn allows inference of tentacle crown dimensions as there is a good correlation between orifice diameter and tentacle crown size in cheilostomes (Winston 1981). Zooids of the anascan *Conopeum seurati* with closure plates were misinterpreted as polymorphs – 'kleistozooids' – by Poluzzi and Sabelli (1985) who, however, showed these zooids to be packed with protein-rich globules suggesting that they functioned in nutrient storage after demise of the final polypide.

3.2.2 Astogeny

Colonial animals uniquely show a second kind of age-related change in zooidal morphology. This is due to the developmental stage of the colony at the time the zooids were budded and is manifested as a **gradient of astogenetic change** that extends distally from the oldest towards the youngest part of the colony (i.e. the reverse of the direction of the gradient of ontogenetic change). All bryozoan colonies

Figure 3.9 Diaphragms and peristomes in some cyclostome bryozoans: (A) longitudinally fractured zooid (distal is towards the top right) of *'Heteropora' neozelanica* with the floor of the living chamber formed by a basal diaphragm deflected distally where it meets the walls of the zooid (Recent; New Zealand); (B) pseudoporous terminal diaphragm occluding an autozooidal aperture in *'Hastingsia' whitteni* (Recent; Stewart Island, New Zealand); (C) pseudoporous terminal diaphragms occluding several autozooidal and a smaller kenozooidal aperture in *Ceriocava hakapaensis* (Cretaceous, Campanian, or Maastrichtian, Kahuitara Tuff; Pitt Island, Chatham Islands, New Zealand); (D) long peristomes of an unidentified cyclostome preserved in a recess on the substrate (Pleistocene, Setana Fm.; Kuromatsunai, Hokkaido, Japan); (E) oblique view of a colony of *Plagioecia* sp. showing ontogenetic zonation, the young zooids close to the edge of the colony having long peristomes whereas old zooids near the centre have lost their peristomes and are closed by terminal diaphragms, usually with a hole indicating the presence of a secondary nanozooid (cf. Figure 3.36C) (Recent; New Zealand). Scale bars: A = 50 μm; B = 100 μm; C, D = 500 μm; E = 1 mm.

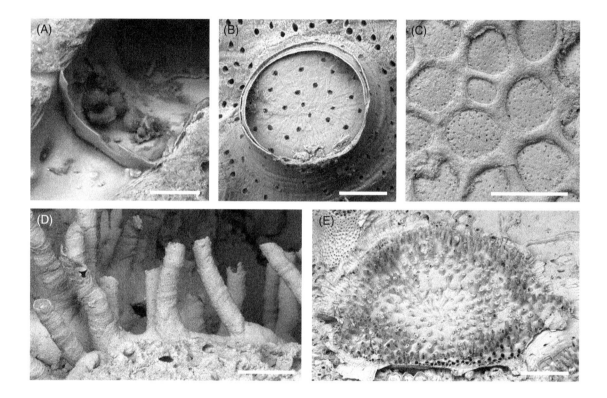

have a **primary zone of astogenetic change** that begins from the ancestrula and normally encompasses the early generations of budded zooids (Figure 3.12A, C–D). Zooid size typically increases progressively through the primary zone of astogenetic change (e.g. Taylor and Furness 1978). In the Plio-Pleistocene cheilostome *Floridina regularis*, for instance, Knowles (2008) found the primary zone of astogenetic change to comprise the first 8–10 generations of zooids. Qualitative changes in zooid morphology also occur through the primary zone of astogenetic change in some taxa. For instance, the early zooids of anascan-grade cheilostomes may have oral spines that decrease in number and eventually disappear altogether in later budded zooids. Ascophoran-grade cheilostome zooids commonly lack adventitious avicularia in the primary zone of astogenetic change, these defensive

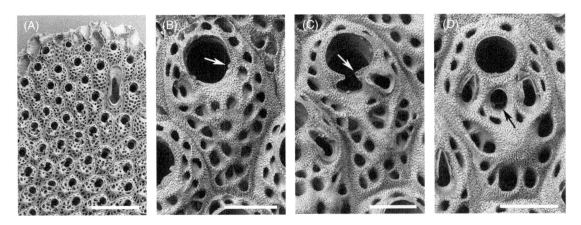

Figure 3.10 Ontogenetic changes in autozooids of the ascophoran cheilostome *Adeonella lichenoides* (Recent; Langkawi, Malaysia): (A) distal part of a bifoliate colony with growing edge where new zooids are budded at the top; (B) young zooid from close to growing edge showing primary orifice with small condyles (arrow) at the proximolateral corners; (C) slightly older zooid with tongues of calcification (arrow) beginning to construct a bridge over the now sunken primary orifice (note also the complete adventitious avicularium developed on the righthand tongue); (D) skeletally mature zooid, more distant from the growing edge, in which the bridge is complete and separates the circular secondary orifice from an elliptical spiramen (arrow) that opens into the ascus between a pair of adventitious avicularia. Scale bars: A = 1 mm; B, C, D = 100 μm.

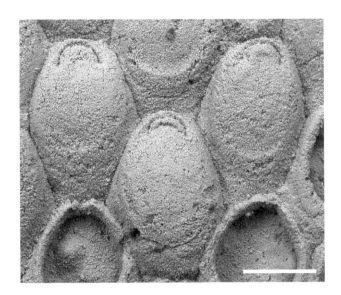

Figure 3.11 Anascan cheilostome *Wilbertopora* sp. with the opesiae of three autozooids sealed by closure plates bearing crescent-shaped scars of the opercula, thus providing evidence of the size and exact location of the unmineralized opercula in this fossil bryozoan (Cretaceous, Coniacian Chalk; Chatham, Kent, UK). Scale bar = 200 μm.

Figure 3.12 Zones of astogenetic change: (A) early astogeny in the uniserial cyclostome *Stomatopora* sp. showing increase in zooid size through the zone of primary astogenetic change beginning with the small ancestrula (arrowed); note also the decrease in branch bifurcation angle during astogeny (Jurassic, Bathonian, White Limestone Fm.; Foss Cross, Gloucestershire, UK); (B) secondary astogenetic change in the uniserial cheilostome *Rhammatopora gaultina* manifested by the increase in zooid length in the distal direction (arrowed) in two new branches developing from the sides of older branches (Cretaceous, Albian, Gault Clay; Burwell, Cambridgeshire, UK); (C) zone of primary astogenetic change in the multiserial cheilostome *Aechmellina anglica* (ancestrula labelled 'a') (Cretaceous, Maastrichtian; Hanover, Germany); (D) zone of primary astogenetic change in a multiserial onychocellid cheilostome (ancestrula labelled 'a') (Cretaceous, Coniacian or Santonian Chalk; Northfleet, Kent, UK); (E) secondary astogenetic change in a colony of the multiserial cheilostome *Rhagasostoma* in a lobate subcolony forming at the growing edge of the main colony (Cretaceous, Santonian Chalk.; Croydon, Surry, UK). Scale bars: A, B, E = 1 mm; C, D = 500 μm.

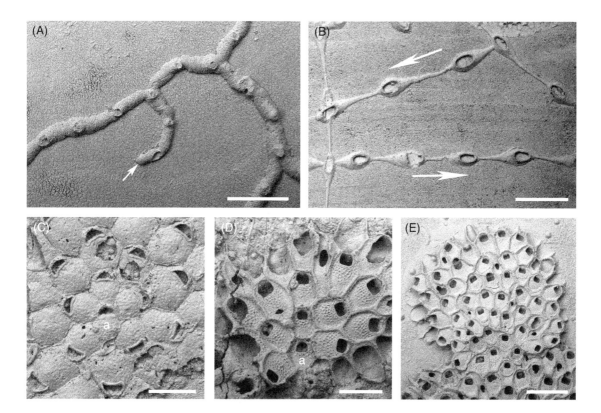

polymorphs appearing only in zooids budded later in colony development. The extent of the zone of change may vary according to the zooidal polymorph in question. Among cyclostome bryozoans, the zone can be more extensive for gonozooids than it is for autozooids, the former continuing to enlarge and change shape well after autozooid size and morphology have stabilized.

A stable zooid morphology and size is eventually attained in most bryozoan species during colony growth, defining a **primary zone of astogenetic repetition**. There are, however, some taxa in which the primary zone of astogenetic repetition is followed by secondary zones of astogenetic change. For example, some bryozoans have 'subcolonies' in which the initial zooids are small, forming a secondary zone of astogenetic change (Figure 3.12B, E). An extreme example of continued astogenetic change is found in the cheilostome *Herpetopora* from the Late Cretaceous (Taylor 1988a). Colonies of *Herpetopora* are encrusting, with zooids arranged in uniserial branches. New branches arise on one or both sides of existing branches, resulting in a cruciate branching pattern. The first zooids in the new branches are relatively short and broad, but zooids budded subsequently become progressively longer and narrower, forming a secondary zone of astogenetic change (cf. Figure 3.12B). In fact, while the first few generations of zooids in the new branch are 'normal' autozooids, those from the most distal generations lack an opening and constitute kenozooids that, nevertheless, may give rise to new lateral branches consisting initially of autozooids. Astogenetic change pervades the entire *Herpetopora* colony and it appears that a zone of astogenetic repetition is never developed. Another bryozoan with runner-like colonies, the primitive Ordovician cyclostome *Corynotrypa delicatula*, develops similar zones of secondary astogenetic repetition manifested as a progressive increase in the length of the autozooids distally from the points of origin of lateral branches (Taylor and Wilson 1994). As with *Herpetopora*, the elongated zooids of the lateral branches facilitate 'exploration' of the substrate by the runner-like colonies of *Corynotrypa* (see Chapter 4.3).

Extreme changes in zooidal morphology occur during astogeny in a few bryozoans. In the Australian cheilostome *Corbulipora tubulifera* (Bock and Cook 1994), autozooids forming the encrusting colony base of this species have costate frontal shields, the so-called *Acanthocella*-phase. Erect branches growing from the colony base initially contain weakly calcified autozooids without frontal shields – the *Watersia*-phase – before switching to autozooids – the *Corbulipora*-phase – with costate frontal shields but that are slightly different from those of the *Acanthocella*-phase. The fact that separate parts of colonies if found in

isolation would be assigned to different genera is reflected by the names of the three growth phases.

Another cheilostome, the cribrimorph *Cribrilina mutabilis*, has autozooids with three different morphologies, named 'R-type', 'I-type', and 'S-type' by Ito, Onishi, and Dick (2015). R-type zooids possess 8–12 well-separated costae, each with an oval lumen where they fuse along the midline of the zooid. S-type zooids have 12–15 costae, closely spaced and with two or three lateral fusions as well as the mid-line fusion. As the name implies, I-type zooids are intermediate in morphology. The proportion of these autozooidal polymorphs varies between colonies, even when colonies are growing next to one another, and also at different times of the year. Although the function of the three types is unclear, increased budding of S-type zooids in late summer suggests that these more heavily calcified zooids may have a role in overwintering.

Nothing is known about the processes determining astogenetic variations in zooid morphology in bryozoan colonies. It is tempting to suggest that they are governed by morphogenetic gradients, of the kind originally proposed for graptolites (Urbanek 2004). However, a simple gradient of morphogens within the colony cannot explain the complex patterns of astogenetic variation seen in bryozoans such as *Herpetopora* with new zones of astogenetic change initiated at the origins of each new branch.

3.3 Ancestrulae

As the only zooid formed through metamorphosis of a larva and not by budding, the ancestrula – or founder zooid – is unique. Not surprisingly, therefore, the ancestrula usually differs in morphology from the other zooids in the colony (Figure 3.13). Ancestrulae are typically smaller than the asexually budded zooids and often have a simpler morphology.

Stenolaemate ancestrulae begin with a low dome-shaped structure called the **protoecium** or

Figure 3.13 Ancestrulae in stenolaemate (A–F) and cheilostome (G–J) bryozoans: (A) small colony of an unidentified cystoporate with ancestrula lower left (Silurian, Llandovery, Visby Beds; Axelsro, Gotland, Sweden); (B) small colony of the cyclostome *Disporella hispida* with protoecium of ancestrula lower left (Recent; Stoke Point, Devon, UK); (C) cyclostome

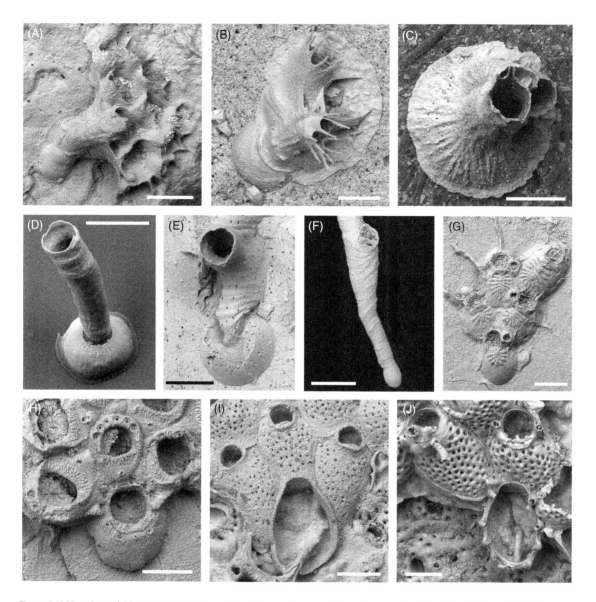

Figure 3.13 (Continued) *Hornera* sp. with ancestrular tube growing erect from a large protoecium (Recent; Otago Shelf, New Zealand); (D) cyclostome *Filicrisia geniculata* with erect ancestrular tube articulated with the protoecium, the elastic joint visible as a narrow ring (Recent; Wembury, Devon, UK); (E) cyclostome *Entalophoroecia deflexa* in which the ancestrular tube is largely prostrate on the substrate surface (Recent; English Channel); (F) long ancestrula of an unidentified cyclostome genus with an ovoidal protoecium (lower right) suggesting larval settlement and metamorphosis in the absence of a substrate (Miocene, Serravallian, Lower Sarmatian; Polupanivka, Ukraine); (G) ancestrula of the cheilostome *Tricephalopora larwoodi* with a spinocystal frontal shield similar to those of the budded zooids (Cretaceous, Maastrichtian, Prairie Bluff Fm.; Montpelier, Mississippi, USA); (H) ancestrula of the cheilostome *Chaperiopsis* sp. unusual in being larger than the first budded zooids (Miocene, Aquitainian, White Rock Limestone; Karetu River, North Canterbury, New Zealand); (I) featureless ancestrula of the cheilostome *Powellitheca waipukurensis* lacking the frontal shield seen in the budded zooids (Pleistocene, Nukumaru Limestone; Waiinu Beach, Whanganui, New Zealand); (J) tatiform ancestrula of *Mucropetraliella thenardii* with a ring of spines but no frontal shield (Recent; Mediterranean Sea, Israel). Scale bars: A = 500 μm; B, D = 100 μm; C, E, G, H, I, J = 200 μm; F = 250 μm.

primary disc (e.g. Nielsen 1970; Figure 3.13A–F). The flat base of the dome is attached to the substrate. Small colonies of an unidentified cyclostome from the Miocene (Sarmatian) of Ukraine are unique in having spheroidal or ellipsoidal protoecia (Taylor, Hara, and Jasionowski 2006; Figure 3.13F). This has been interpreted as resulting from metamorphosis and calcification of the floating larva not attached to a substrate. Normally, a single tube forming the chamber of the ancestrular zooid emerges from the protoecium, either on one side and adpressed to the substrate (Figure 3.13A, B, E) or from the top and oriented perpendicularly to the substrate (Figure 3.13C, D). However, one to three additional zooids emerge from the protoecium in the cyclostome *Filicrisia* (Jenkins and Taylor 2014). These are kenozooids (Chapter 3.7), probably originating later in development following resorption of small patches of the protoecial skeleton to form new sites for zooidal budding.

The ancestrulae of cheilostomes (Figure 3.13G–J) are more diverse in morphology than those of stenolaemates. Many are provisioned with numerous spines which can vary in number and distribution even within populations of a single species (Rodgers and Woollacott 2006). Ancestrulae of ascophoran-grade cheilostomes often resemble an anascan zooid, with an exposed frontal membrane ringed by spines but no ascus or frontal shield, and are termed **tatiform** (Figure 3.13J). In a few cheilostomes multiple zooids all derived directly from metamorphosis of the larva constitute an **ancestrular complex**. For example, in the common epiphytic anascan *Membranipora membranacea* a pair of mirror-image zooids forms a twinned ancestrula (Atkins 1955), as in other Membraniporidae. A similar twinned ancestrula is found in the unrelated ascophoran *Myriapora truncata* (Ferretti, Magnino, and Balduzzi 2007), whereas the ancestrular complexes of the ascophorans *Stylopoma* and *Pseudocelleporina* comprise clusters of five and three zooids (Cook 1973; Mawatari 1986), respectively, and that of *Reptadeonella* has six zooids (Cheetham, Sanner, and Jackson 2007). In anascan cheilostomes belonging to the family Bugulidae the ancestrula is elongated, grows erect

and has long rhizoids (Cook 2001). Very occasionally the ancestrula in cheilostomes is a kenozooid, as in *Hesychoxenia* (Gordon and Parker 1991a).

In order to trace patterns of growth and astogeny in bryozoan colonies, it is very useful to be able to locate the ancestrula. Unless the ancestrula has been overgrown by later zooids (as in most palaeostomates) or lost, this is relatively straightforward. Not only is the ancestrula typically the smallest zooid in the colony, but tracking the succession of zooids backwards towards the ancestrula is possible when the polarity of the zooids can be discerned: the end of the zooid opposite to the aperture or orifice is always closest to the ancestrula. Additionally, in most cheilostomes a distinctive triad of zooids is budded from the ancestrula, one at its distal end plus a pair of distolateral buds (e.g. Gordon 1971; Kahle, Liebezeit, and Gerdes 2003, fig. 8; Figure 3.13H, I).

3.4 Polymorphism

The great majority of bryozoan colonies contain two or more kinds of zooids. These polymorphs have, by definition, different morphologies, which are known, or inferred, to correlate with different functions. A review of polymorphism in bryozoans by Silén (1977) has recently been updated by the more comprehensive study of Schack, Gordon, and Ryan (2018b) who considered polymorphism to have been very important for the evolutionary success of the phylum. These authors echoed the view of Ryland (1979) that the evolution of polymorphism in bryozoans correlates with the extent of zooid dissociation or compartmentalization: phylactolaemates have poorly compartmentalized zooids and lack polymorphism, whereas stenolaemates and especially gymnolaemates have more compartmentalized zooids endowing the potential to develop a variety of polymorphs. The reasoning is that compartmentalization allows selective pressures to act on individual zooids. On the other hand, according to Mackie (1986), zooidal compartmentalization has prevented bryozoans from achieving a 'true colonial individuality' resembling that seen, for instance, in the highly polymorphic siphonophore cnidarians. Differences in the levels of polymorphism among bryozoan groups may also be correlated with the presence of particular morphological traits that can form the basis for various kinds of polymorphs. Notably, autozooidal opercula are a pre-requisite for the evolution of polymorphs possessing mandibles and setae as these structures represent modified opercula.

3.5 Reproductive Polymorphs

Polymorphs with reproductive roles are widespread among marine bryozoans. Most reproductive polymorphs are females functioning in the incubation or brooding of embryos before they are released into the plankton as larvae. The nature of female zooids is very different in living stenolaemates and cheilostomes. Male polymorphs are less common but are known in a few genera of cheilostomes.

Based on multilevel selection theory, Simpson (2011) hypothesized that polymorphs associated with reproductive division of labour should evolve before other kinds of polymorphism in colonial and social organisms. This should be testable using bryozoans with better knowledge of their phylogeny.

3.5.1 Brooding in stenolaemates
Almost all modern stenolaemates (cyclostomes) are known to possess **gonozooids**, enlarged zooids that contain numerous embryos simultaneously. Only a few gonozooids may be present in each colony. Indeed, finding any gonozooids at all can be difficult, hampering identification as these polymorphs have a high taxonomic value. Cyclostome gonozooids are heterozooids that do not feed, at least once mature. The aperture – known as an **ooeciopore** – functions as a point of exit for the larvae. Cyclostome gonozooids vary in shape (Figure 3.14), some being elongate (Figure 3.14A), others broad (Figure 3.14E), and yet others having

complex shapes with chambers penetrated by autozooidal tubes (Figure 3.14F). The frontal walls of cyclostome gonozooids characteristically have a greater density of pseudopores that those of the autozooids (Chapter 2.2).

Recent genetic research (Hughes et al. 2005; Jenkins et al. 2017) has confirmed the finding based on anatomical studies made more than a hundred years ago (Harmer 1893) that all of the larvae within a single cyclostome gonozooid are clones. After fertilization of the egg, the embryo undergoes multiple divisions, a process termed **polyembryony**. Few species in the animal kingdom show such monozygotic polyembryony – armadillos are another example – and there has been debate about why such a method of reproduction should be employed as it partly negates the advantage of the genotypic variation associated with sexual reproduction (Craig et al. 1997). One suggestion has been that fertilization events are rare in cyclostomes – the rapid dilution of sperm in seawater with distance is a known problem for sessile animals employing external fertilization (e.g. Bishop 1998) – and polyembryony could be a way of taking advantage of the rare fertilization events that do occur. However, Pemberton, Hansson, and Bishop (2011) failed to find the predicted greater number of gonozooids in densely packed vs. sparsely distributed colonies of *Crisia denticulata*, calling into question this idea.

3.5.2 Brooding in gymnolaemates
The great majority of species of living and extinct cheilostome brood their embryos. Various modes of brooding are employed in cheilostomes – for example, external membranous sacs in *Aetea* (Cook 1977b) and clusters of brood sacs attached to the tentacle sheath in *Carbasea* (Stach 1938) – but the commonest is brooding in a calcified capsule-like structure called an **ovicell** located at the distal end of the zooid just beyond the orifice (Figure 3.15). Ovicells are double-layered structures (Figure 3.15H), with an inner entooecium and an outer ectooecium separated by a narrow ooecial coelom. Primitively, both of these layers are fully calcified (Figure 3.15B, D) but in

Figure 3.14 Cyclostome gonozooids for larval brooding showing variation in shape, location of the ooeciopore (arrowed) from which the larvae are released, and surface texture: (A) large, longitudinally elongate gonozooid of *Reptomultisparsa incrustans* (Jurassic, Bathonian, Caillasse de la Basse-Ecarde; Ranville, Calvados, France); (B) heart-shaped gonozooid of *Hyporosopora* sp. (Jurassic, Bathonian, Frome Clay; Watton Cliff, Dorset, UK); (C) gonozooid with surface formed of alveoli (?kenozooids) in *Patinella radiata* (Recent; Ischia, Italy); (D) dendritic gonozooid of *Crescis dumetosa* (ooeciopore not apparent) (Jurassic, Bathonian, Caillasse de la Basse-Ecarde; St Aubin-sur-Mer, Calvados, France); (E) narrow, transversely elongate gonozooid of *Actinopora* sp. (Cretaceous, Coniacian or Santonian Chalk; Northfleet, Kent, UK); (F) transversely elongate gonozooid of *Plagioecia patina* with the roof penetrated by autozooids and a non-terminal ooeciopore (Recent; Skye, UK). Scale bars: A, D, F = 1 mm; B, C, E = 500 μm.

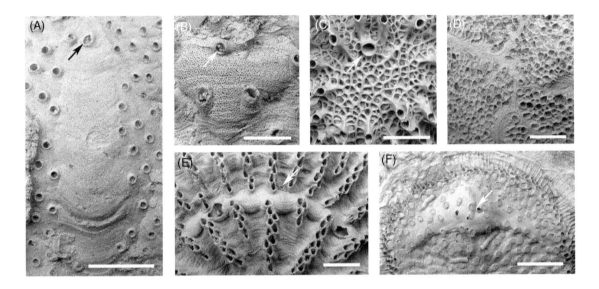

more advanced cheilostomes pores may be present in the ectooecium (Figure 3.15C), or this layer may be partly or entirely uncalcified, exposing the underlying entooecium at the skeletal surface (Figure 3.15G). Nielsen (1981) found embryos to be brooded for 11–12 days before release in two cheilostome species from California. However, a much longer brooding period of 10 months was reported by Barnes and Clarke (1998) for some Antarctic cheilostomes.

Cheilostome zooids possessing ovicells normally retain the capacity to feed and are therefore autozooids, in contrast to the gonozooids of cyclostomes that are heterozooids incapable of feeding. Often the zooids possessing ovicells are otherwise identical to those without ovicells, but they may

have a slightly enlarged orifice in some genera (e.g. *Fovoporella*: Gordon 2014), or very occasionally a smaller orifice (López Gappa and Liuzzi 2008). An additional difference with cyclostomes is that polyembryony does not occur in cheilostomes and each ovicell broods only one embryo at a time in most species. However, a succession of embryos can be housed in the same ovicell, with a new embryo being brooded after the first has been released as a larva. Whereas cyclostome gonozooids are notoriously rare in many species, ovicells are typically present in a moderate to high proportion of zooids within a cheilostome colony. Sometimes the ovicells are scattered seemingly at random through the colony but they may be concentrated in bands (Figure 3.15A) when reproduction is seasonal.

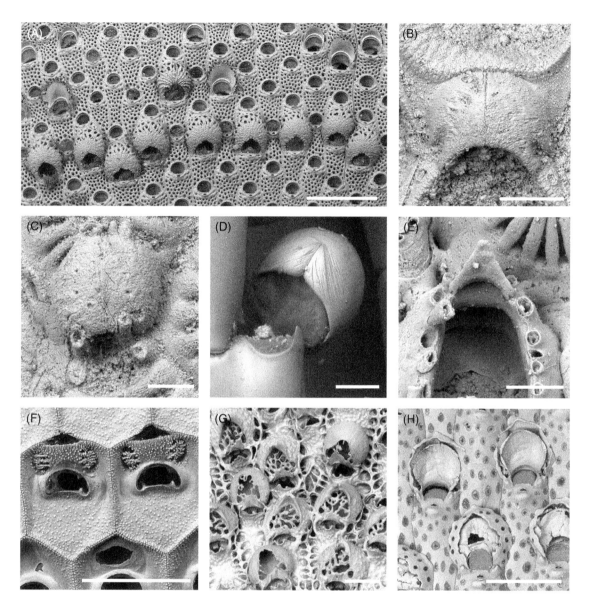

Figure 3.15 Cheilostome ovicells used for larval brooding: (A) band of ovicells in *Primavelans insculpta* indicating a pulse of reproduction (Recent; San Diego, California, USA); (B) simple ovicell of *Wilbertopora hayesi* with fully calcified ectooecium and a median suture (Cretaceous, Maastrichtian, Prairie Bluff Chalk; Jefferson, Alabama, USA); (C) ovicell of *Aeolopora catillus* with a porous ectooecium (Cretaceous, Maastrichtian, Prairie Bluff Chalk; Livingston, Alabama, USA); (D) ovicell of *Bugula neritina* showing ooecial vesicle closing the opening (Recent; Hadera, Israel; (E) ovicell of *Cauloramphus disjunctus* with calcification reduced to a small triangular structure at the distal end of the zooid (Pleistocene, Setana Fm.; Kuromatsunai, Hokkaido, Japan); (F) two fertile zooids of *Melicerita blancoae* with immersed ovicells (Recent; Antarctica); (G) ovicells of *Parasmittina protecta* in which a window in the incompletely calcified ectooecium reveals the reticulate calcification of the entooecium (Recent; Achviz Canyon, Israel); (H) partly formed ovicells in *Haplopoma* sp. showing the two layers (entooecium and ectooecium) that form the calcified part of the ovicell (Recent; Sagres, Portugal). Scale bars: A = 1 mm; B, C, D, E = 100 µm; F = 500 µm; G, H = 200 µm.

Research described in detail by Ostrovsky (2013) has illuminated the variability of brooding patterns among cheilostomes. Ovicells evolved from hollow, basally articulated spines borne on the zooid distal of the egg-producing maternal zooid (e.g. Santagata and Banta 1996; Ostrovsky and Taylor 2004). These spines initially formed a loose cage over the embryo that rested on the proximal gymnocyst of the distal zooid (Figure 3.16). Subsequent evolutionary changes entailed loss of the basal articulations of the spines, flattening and lateral fusion of the spines to close the gaps in the cage, and reduction in spine number to two, one on each side. Spinose ovicells can be found in a taxonomically diverse range of cheilostomes, mostly fossil but some extant (Ostrovsky and Taylor 2005a; Figure 3.17). Primitive non-spinose ovicells of early cheilostomes (e.g. Ostrovsky and Taylor 2005b) have a medial suture (Figure 3.16B–D), betraying their origin as two flattened spines that fused during the formation of the **ooecium** (the calcified roof of the brood chamber). In a separate evolutionary trend, several cheilostome clades have independently evolved placentas allowing continued nutrition of the developing embryo after its emplacement in the ovicell (Ostrovsky 2013).

Another evolutionary trend in cheilostomes has been towards immersion of the ovicells (e.g. Ostrovsky et al. 2006, 2009a). Ovicells in early and primitive cheilostomes are conspicuous raised structures (hyperstomial ovicells) on the colony surface, seemingly sitting targets for predators seeking to devour the nutritious, yolk-laden embryos within, but in others they are more cryptic (Figs 3.15E, 4.18C). In more advanced cheilostomes still the embryo is housed in an internal sac deep within the zooid and there is little or no indication of its presence on the surface of the colony. However, there can be a trade-off involved in such internalization: during internal brooding in some species insufficient space remains in the zooidal chamber for a functional polypide and the zooid is therefore unable to feed (e.g. Mawatari 1952), thus reducing the food-gathering capacity of the colony as a whole.

In a few cheilostomes with internal brooding, such as *Adeona* (Bock and Cook 2004a), the brooding zooids are dimorphic (Figure 3.18A), differing conspicuously in skeletal morphology from the non-brooding zooids in the colony, and are often referred to as gonozooids. Seemingly unique among cheilostomes is the Arctic genus *Harmeria* in which the discoidal encrusting colonies switch

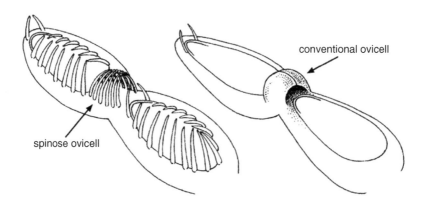

conventional ovicell

spinose ovicell

Figure 3.16 Primitive ovicell comprising a cage of several spines formed by the zooid distal to the maternal zooid (as in *Distelopora*), compared with a more conventional ovicell which is a solid structure consisting of two flattened spines united along a median suture (as in *Wilbertopora*) (based on Ostrovsky and Taylor 2004. © The Trustees of the Natural History Museum, London).

Figure 3.17 Spinose cheilostome ovicells: (A) crescent of circular bases on which spines would have articulated to form a cage-like ovicell (cf. Figure 3.16) in *Distelopora langi* (Cretaceous, Cenomanian; Barrington, Cambridgeshire, UK); (B) ovicell formed of laterally fused spines in *Leptocheilopora* sp. (Cretaceous, Campanian, Weybourne Chalk; Norwich, Norfolk, UK); (C) ovicell of *Macropora leeae* comprising a ring of fused spines bordering a solidly calcified area (Oligocene; North Otago, New Zealand); (D) four flattened, incompletely fused spines forming the ovicell of *Monoporella* sp. (Recent; Alaska). Scale bars: A, B = 100 µm; C, D = 500 µm.

during growth from budding autozooids to small brooding polymorphs that form a band around the perimeter of the colony (Kuklinski and Taylor 2006). The dwarf female zooids (Figure 3.18B) are more thickly calcified than the autozooids and form at the end of the summer growing season. It is believed that some survive the strong wave action and ice scour of the winter and release their larvae in the spring.

Although no ctenostomes have specialized brooding polymorphs, many do brood their embryos.

For example, Reed (1988) described the reproductive biology of the ctenostome *Amathia* [as *Bowerbankia*] *gracilis* in which the embryo is brooded within the tentacle sheath of the autozooid following degeneration of the polypide.

3.5.3 Putative brood chambers in palaeostomates

A variety of putative brooding polymorphs, often referred to as 'ovicells', have been described from palaeostomates, especially species of Fenestrata (Bancroft 1986; Gautier, Wyse Jackson, and McKinney

2013) but also the cystoporate *Botryllopora* (Ernst, Bohaty, and Taylor 2014). Many have the form of bowl-like depressions on the colony surface opening proximally into a modified zooidal chamber (Figure 3.19A). Interestingly, structures of broadly similar morphology occur in a few genera of mid-Cretaceous cyclostomes where they are interpreted as symbiont galls as the colonies in which they occur also possess conventional gonozooids for brooding (Chapter 5.3.1). Dendritic tunnels (Figure 3.19B) described in the Upper Ordovician and Silurian cystoporates *Lichenalia* and *Rhinopora* more strongly resemble the gonozooids found in a few post-Palaeozoic cyclostomes (e.g. '*Entalophora*' *vassiacensis*: Walter 1977, fig. 2; Figure 3.14D).

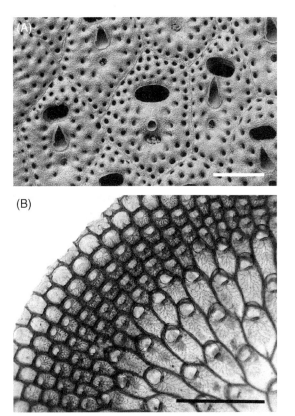

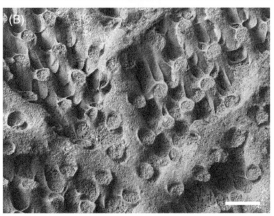

Figure 3.18 Female polymorphs in cheilostomes: (A) gonozooid (centre) of the cheilostome *Reptadeonella violacea* in which the orifice is larger and a raised pore replaces the avicularium present in the autozooids on either side (Recent; Herceliya, Israel); (B) transmitted light image of *Harmeria scutulata* showing a zone of dwarf brooding zooids at the edge of a colony with the larger autozooids visible at lower right (Recent; Spitsbergen). Scale bars: A = 200 μm; B = 1 mm.

Figure 3.19 Possible larval brood chambers in palaeostomates: (A) shallow depressions (arrowed) located proximally of the apertures in a fenestellid that have been interpreted as 'ovicells' (Lower Carboniferous; UK); (B) dendritic structure in the cystoporate *Lichenalia* cf. *concentrica* resembling the gonozooids of some cyclostomes (e.g. Figure 3.14D) (Silurian, Wenlock, Much Wenlock Limestone Fm.; Dudley, West Midlands, UK). Scale bars: A = 200 μm; B = 500 μm.

The most reasonable interpretation of these large structures is that they functioned as chambers for brooding clonal embryos (Buttler 1991).

3.5.4 Male polymorphs

Male polymorphs – **androzooids** – are much less common in bryozoans than are female polymorphs. Production of sperm is normally shared among all or most of the autozooids in the colony, with the sperm being released into the water through tiny pores in the tips of the tentacles (Silén 1972). However, a few species of cheilostomes possess dwarf male zooids that are incapable of feeding. These dwarf males have small orifices through which is protruded a tentacle crown with a reduced number of tentacles. Male zooids in the hermit crab symbiont *Odontoporella bishopi* have four long and four short tentacles, compared to the autozooids which have 15–16 tentacles (Carter and Gordon 2007). The gut is reduced in male zooids of *Odontoporella* but the skeletal orifice is indistinguishable from that of the autozooids, which would render their distinction impossible in fossils as other features of the skeleton are also the same.

Interestingly, some of the genera with male zooids have colonies that are mobile, either by virtue of possessing appendages (setae) for walking, as in *Selenaria* (Chapter 3.11), or by colonizing the shells occupied by living gastropods (Ryland 2001) or hermit crabs (Chapter 5.3), as in *Odontoporella* and *Hippoporidra*. It is tempting to speculate that the long tentacles of the male zooids in *Selenaria* and the hermit crab symbionts are used to transfer sperm to other colonies during times when crab movements bring them together. Whether or not this is true, male zooids may be located in areas of excurrent water flow which should enhance sperm discharge (Ryland 2001).

In another cheilostome – *Celleporella hyalina* – which has male zooids with orifices of reduced size (Figure 3.20), as well as dwarf female zooids (Ostrovsky 1998) that are also incapable of feeding – it is possible to estimate reproductive allocation by simply counting the number of polymorphic zooids of the three kinds, a procedure

Figure 3.20 Sexual polymorphism in the living cheilostome *Celleporella* sp., showing autozooids with large orifices (az), female zooids with spindle-shaped orifices (fz) and porous ovicells, and male zooids with tiny orifices (mz). Scale bar = 100 μm.

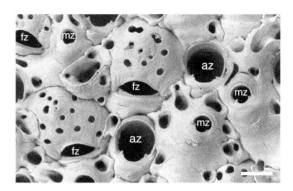

that is potentially equally applicable to fossil specimens of this genus. Different genotypes show variation in the allocation between feeding, male reproductive and female reproductive functions (Hughes and Hughes 1986): female reproduction (egg growth) is triggered by the presence in the water of sperm from conspecific colonies (Bishop, Manríquez, and Hughes 1989), and various forms of stress (e.g. incipient competition for substrate space) can result in an increase in the proportion of male zooids budded (Hughes et al. 2003). A related southern hemisphere species – *Antarctothoa bougainvillei* – shows a greater number of female zooids in colonies that are obstructed by conspecific neighbours (Liuzzi and López Gappa 2008).

3.6 Active Defensive Polymorphs

One of the unsolved enigmas of bryozoology concerns the exact functional role – or more likely, roles – of **avicularia**. Widespread among cheilostome, these polymorphs are, with a few exceptions, generally considered to perform broadly defensive functions. The key feature of an avicularium is its moveable appendage. This represents a hypertrophied

Figure 3.21 Comparison of an autozooid and an avicularium in the anascan cheilostome *Flustra foliacea*, showing the absence of a polypide in the avicularium but well-developed muscles (redrawn after Silén 1977).

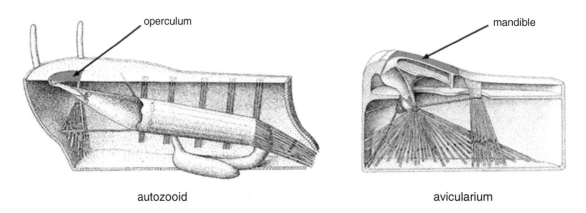

operculum

mandible

autozooid

avicularium

autozooidal operculum developed into either a powerful **mandible** or a hair-like **seta**. Strong muscles are associated with both mandibles and setae (Figure 3.21). However, as avicularia lack polypides and cannot feed, the energy to operate these muscles must be supplied by other zooids in the colony. Despite non-preservation of mandibles and setae, avicularia are generally easy to distinguish in fossil cheilostomes because of the contrast in the skeletal morphology of these zooids, particularly the orifice, with respect to the autozooids. They have an opening that is large relative to the size of the zooid as a whole and comprises a semi-circular opesia proximally, separated by a pair of lateral **condyles** or a calcified **hinge bar** from a long distal rostrum on which the closed mandible rested.

Among present-day bryozoans, avicularia vary considerably in morphology, ranging from those with a similar size and morphology to the autozooids in the same colony, to others that are extremely different from the autozooids in both size and morphology, as well as in their patterning across the colony surface (Figures 3.22, 3.23). Autozooid-like avicularia that retain the ability to feed were described by Cook (1968b) in the calloporid anascan *Crassimarginatella*. Relatively

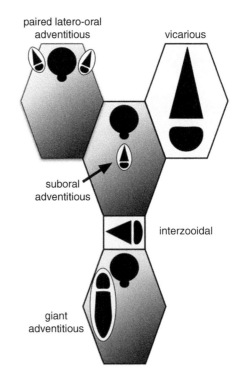

paired latero-oral adventitious

vicarious

suboral adventitious

interzooidal

giant adventitious

Figure 3.22 Schematic diagram of some of the main types of avicularia found among cheilostome bryozoans. Autozooids are shaded.

Figure 3.23 Cheilostome avicularia: (A) 'bird's head' avicularium with intact mandible in '*Bugula*' sp. (Recent; Hadera, Israel); (B) vicarious avicularium (left of centre) in *Wilbertopora* sp. (Cretaceous, Coniacian Chalk; Hope Gap, Sussex, UK); (C) curved vicarious avicularium in *Onychocella angulosa* (Recent; Madeira); (D) spatulate vicarious avicularium in *Adeonella lichenoides* (Recent; Langkawi, Malaysia); (E) interzooidal avicularia on either side of an autozooid in *Puellina* sp. (Recent; Puerto Rico); (F) adventitious avicularium positioned distolaterally of the orifice in *Hippopodina feegeensis* (Recent; Langkawi, Malaysia); (G) adventitious avicularium incorporated into the calcification of an ovicell in *Micropora brevissima* (Recent; Pennel Bank, Antarctica). Scale bars: A = 50 μm; B, C = 200 μm; D, E, F, G = 100 μm.

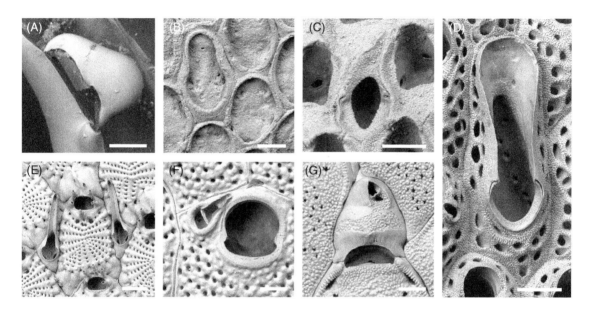

unmodified, autozooid-like avicularia with enlarged, 'duckbilled' opercula are also found in *Sinoflustra* (e.g. Liu 1992, figs 32, 33). These avicularia are reasonably interpreted as early stages in polymorph differentiation.

The name avicularium refers to the superficial similarity of these polymorphs in some cheilostomes to a bird's head situated atop a flexible stalk that vaguely resembles an avian neck (Figure 3.23A). However, stalked bird's head avicularia are far less common than unstalked, 'sessile' avicularia (Figure 3.23B–G). Sessile avicularia are often classified into three types: (i) **vicarious**, which substitute for an autozooid in the pattern of zooids (Figure 3.23B–D); (ii) **interzooidal**, which are significantly smaller than the autozooids between which they are budded (Figure 3.23E); and (iii)

adventitious, which are usually much smaller than the autozooids and are budded onto the surface of an autozooid (Figure 3.23F) (see Carter, Gordon, and Gardner 2010). Adventitious avicularia are often positioned very close to the orifice of an autozooid, seemingly as sentinels guarding the entrance. Avicularian mandibles and the skeletal opening into which they close range from sharply pointed to ligulate or spatulate in shape. Most are straight but in many species of the anascan family Onychocellidae they are curved (Figure 3.23C). Rarely, the avicularium is intimately associated with an ovicell and is incorporated into its calcification (Figure 3.23G). Vibracula are a subcategory of avicularia characterized by their very long, whip-like setae sometimes with serrated edges, discussed below.

Schack, Gordon, and Ryan (2018a) published a detailed classification of cheilostome polymorphs showing the considerable disparity found among avicularia and other polymorphs.

Even some highly modified avicularia retain vestiges of their origins from autozooids. Carter et al. (2011) identified a series of homologous characters between the autozooids and feeding zooids of *Bugulina flabellata*. Many of the characters in the avicularia are vestigial equivalents of those seen in autozooids, while others are augmented compared to the autozooids. For example, the bundle of retractor muscles that withdraw the lophophore in the autozooids are reduced to single fibres attached to each side of the vestigial polypide. On the other hand, the mandible of the avicularium is a heavily reinforced homologue of the autozooidal operculum.

Observations of avicularia have been made in a range of bryozoans, particularly by Winston (1984, 1986, 1991, 2009, 2010). Both mechanical and chemical stimuli may cause avicularian mandibles to close. Syllid polychaetes can be captured and held for long periods by the avicularia in species of *Celleporaria* and *Reptadeonella* (Winston 1986). Even the small avicularia of *Micropora* can capture and hold polychaete cirri and the legs of amphipods and pycnogonids (Winston 2010). The 'bird's head' avicularia of bugulids have been observed to capture and hold isopods, which build tubes on the colony surface that are detrimental to the bryozoan (Kaufmann 1971), as well as copepods (Cook 1985).

It is not uncommon for a single colony to have avicularia of more than one kind. Indeed, in some species of *Stylopoma*, as many as five types of avicularia can be found (Simpson, Jackson, and Herrera Cubilla 2017), distinguished by differences in size, shape, and sometimes position. Ascophoran-grade cheilostomes often have small **suboral adventitious avicularia** located immediately beneath (proximal of) the autozooidal orifice, which in some zooids of the colony are substituted by **giant avicularia** covering half or more of the frontal shield of the autozooid, as in *Parasmittina* (Soule 1973) and several species of *Pentapora* (Lombardi, Taylor, and Cocito 2010).

In the free-living cheilostome *Cupuladria*, autozooids can be progressively 'replaced' by avicularia in a sequence from the oldest parts of the colony outwards (Cook and Chimonides 1994a). A predictable succession of changes is observed in the host autozooid, including absorption of the operculum, leading up to the completion of the vicarious avicularium which develops a *de-novo* mandible. Therefore, these avicularia can be interpreted as representing **intrazooidal polymorphs**, i.e. zooids that change function – and morphology – during their development.

Normally, avicularia are smaller than the autozooids in the same colony. In *Steginoporella*, however, the mandibulate polymorphs (Figure 3.24A, B) are the same size, or even somewhat larger, than the other autozooids (Pouyet and David 1979). These polymorphs differ from typical avicularia in possessing a polypide and having the ability to feed, making them autozooidal polymorphs. They are referred to as **B-zooids,** whereas the other autozooids in the colony, which are invariably more numerous, are called **A-zooids** (Harmer 1900; Banta 1973). As the mandibles of the B-zooids are only slightly larger than the opercula of the A-zooids, their function is uncertain. Observations of living *Steginoporella* colonies (Winston 2010) showed that the opercula of B-zooids remained open for longer than the A-zooids when disturbed, perhaps allowing the B-zooids to perform a sensory role after retraction of the A-zooids. In Ghanaian species of *Steginoporella*, however, Cook (1985, p. 36) observed that B-zooids rarely protruded their lophophores to feed even though they had the capability to do so.

Polymorphs with enlarged mandibles are also characteristic of most species of the anascan genus *Macropora*. These 'avicularia' (Figure 3.24C) are typically very infrequent within colonies and have calcified mandibles only slightly larger than the calcified opercula of the autozooids but with a cuspate distal edge (Gordon and Taylor 2008). Despite the fact that *Macropora* is extant, nothing seems to be known about the function of the avicularia, although they do apparently possess polypides in at least one species, suggesting a capacity to feed.

Figure 3.24 Polymorphic cheilostome zooids with modified opercula/mandibles: (A) B-zooid of *Steginoporella vickburgica* surrounded by normal A-zooids with smaller opesia (Oligocene; Conecuh River, Alabama, USA); (B) unbleached specimen of *Steginoporella haddoni* showing two B-zooids (left and right of centre) with quadrate opercula larger than those of the normal autozooids (Recent; Darwin, Australia); (C) *Macropora leeae* 'avicularium' (left) and autozooid (right), both with calcified opercula, that in the avicularium having a scalloped distal edge (Oligocene; North Otago, New Zealand). Scale bars = 500 μm.

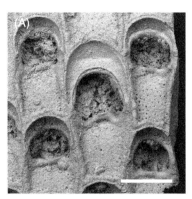 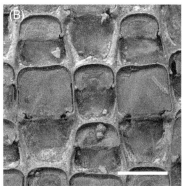 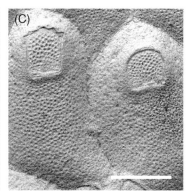

Figure 3.25 Variability in the avicularia of mid-Cretaceous species of the early neocheilostome *Wilbertopora*: (A) avicularium of *Wilbertopora mutabilis* differing from the surrounding autozooids only in having a constriction of the opesia probably corresponding to the hingeline of the mandible (Cretaceous, Albian, Fort Worth Limestone; Kaum, Texas, USA); (B) avicularium of *Wilbertopora listokinae*, this example having an ovicell which lacks the normal opening and would not have been functional (Cretaceous, Cenomanian, Grayson Fm.; Lake Waco Dam, Texas, USA); (C) avicularium of *Wilbertopora acuminata* with a narrow pointed rostrum indicating the presence of a gracile mandible (Cretaceous, Cenomanian, Del Rio Fm.; Austin, Texas, USA). Scale bars: A = 200 μm; B, C = 100 μm.

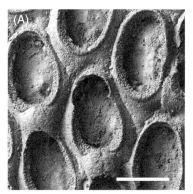 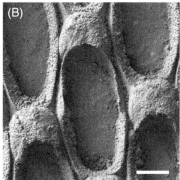 

The oldest known avicularia occur in the anascan cheilostome *Wilbertopora* from the Late Albian–Cenomanian Washita Group of the southern USA (Cheetham et al. 2006). Eight species of *Wilbertopora* have been recognized in the Washita Group, with avicularia of varying morphologies (Figure 3.25). In *W. listokinae* and *W. tappanae*, the avicularia are scarcely distinguishable from the autozooids (Figure 3.25A): avicularia and autozooids are of about the same size but the former have a slight constriction in the gymnocyst about half way along the zooid presumably marking the position of the hinge of the mandible. In some other species (e.g. *W. spatulifera*) the avicularia are large, vicarious, and have a well-defined rounded rostrum, making them more obviously different from the autozooids in the colony. Two species have interzooidal avicularia substantially smaller than the autozooids – *W. hoadleyae* in which the avicularia have rounded rostra, and *W. acuminata* with pointed rostra – indicating greater degrees of polymorphic differentiation (Figure 3.25C). Indeed, a cladistic analysis of the inter-relationships between the eight species of *Wilbertopora* found the more highly differentiated species to be placed further from the base of the phylogenetic tree (Figure 3.26). Interestingly, among the species of *Wilbertopora* having avicularia of similar size to the autozooids, ovicells were occasionally found in association with these avicularia (Figure 3.25B). Sometimes

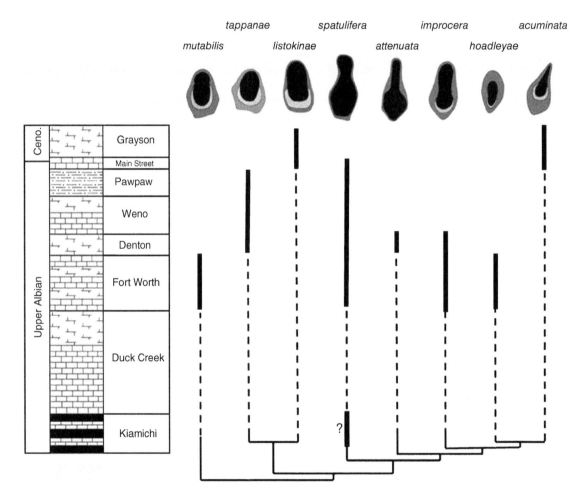

Figure 3.26 Cladistic phylogenetic tree of species of the early neocheilostome *Wilbertopora* from the Washita Group of Texas and Oklahoma showing the evolution of avicularian morphology (based on figures in Cheetham et al. 2006).

these avicularian ovicells are closed by calcification and non-functional but in other instances they have a conventional appearance, suggesting that they were able to brood embryos and, by extension, possessed a feeding polypide and ovary.

While mandibulate polymorphs at the present day are found exclusively in cheilostomes, the Cretaceous–Paleocene cyclostome family Eleidae evolved polymorphs – **eleozooids** – that are similar in morphology to cheilostome avicularia and presumably were also defensive in function (Taylor 1985; Figure 3.27). The evolution of these mandibulate polymorphs was made possible because eleid autozooids, like those of cheilostomes, had

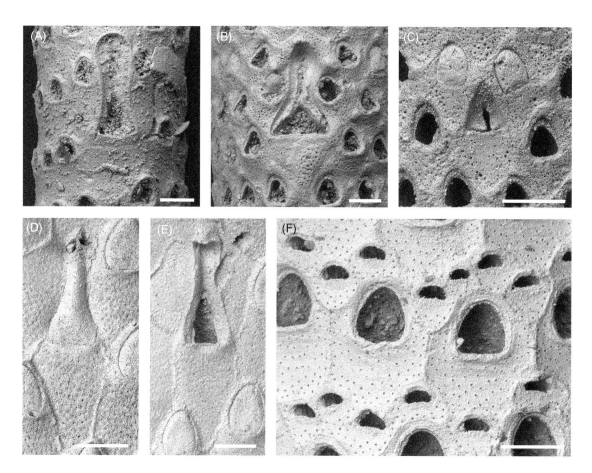

Figure 3.27 Eleozooids of eleid cyclostomes: (A) *Meliceritites* sp. with a rostrum resembling that of the cheilostome avicularium depicted in Figure 3.23D (Cretaceous, Campanian, Segonzac Fm.; Genté, Charente, France); (B) *Meliceritites ornata* showing the torqueing of the autozooidal apertures adjacent to the rostrum (Cretaceous, Campanian, Biron Fm.; Talmont, Charente Maritime, France); (C) *Meliceritites royana*, a species in which the eleozooids have inverted T-shaped openings (Cretaceous, Campanian, Aubeterre Fm.; Archiac, Charente Maritime, France); (D), (E) *Reptomultelea levinseni* showing an eleozooid with intact mandible (D) and another with the mandible missing (E) (Cretaceous, Coniacian or Santonian; Vattetot, Haute Normandie, France); (F) *Atagma* sp. with numerous small eleozooids in irregular transverse bands between the autozooids (Cretaceous, Santonian; Vesterival, Seine Inférieux, France). Scale bars = 200 µm.

hinged opercula to close the aperture of the feeding zooid when the tentacle crown was retracted, a preadaptation for the evolution of mandibulate polymorphs. Whereas cheilostome opercula are seldom calcified, those of eleids are calcified. Likewise, the mandibles of eleozooids, when known, are calcified and may be found preserved in situ in fossil eleids (Figure 3.27D). Several species of eleids have more than one kind of eleozooid – usually large and small – roughly corresponding with the large vicarious and small adventitious avicularia found in many species of cheilostomes.

Avicularia-like structures – **aviculomorphs** – have been described from the Devonian fenestrate genus *Fenestratopora* (McKinney 1998; Ernst 2016b). These openings in the skeleton are arch-shaped with a straight proximal edge and an acute tip that is raised above the surface of the colony. It has been hypothesized that thickened cuticle formed a mandible hinged along the proximal edge. Unlike cheilostome avicularia, however, they were regarded as colonial rather than zooidal structures.

3.7 Structural Polymorphs

A great variety of polymorphs exist across almost the entire phylum that serve structural functions in supporting and strengthening the colony, or occupying spaces between other zooids to regulate the distribution of lophophores over the colony surface. The nomenclature of these structural zooids is complex and varies between groups. In the extant cyclostomes and cheilostomes they are generally termed **kenozooids** (Figure 3.28). Lacking a polypide or a mandible, the kenozooids of cyclostomes and cheilostomes have no moving parts and are essentially coelomic sacs. They are usually smaller than the autozooids in the same colony (Figure 3.28A, F). A single opening is usually present more or less at the centre of the kenozooids in cheilostomes (Figure 3.28B) as well as some cyclostomes (Figure 3.28F). In other cyclostomes the kenozooids are totally sealed by an exterior wall

(Figure 3.28E), which in some taxa develops only during late ontogeny. Idmidroneid and similar cyclostomes (Hinds 1975) have kenozooids on the dorsal sides of their branches where autozooids do not open (Figure 3.28G), seemingly to strengthen these branches. Kenozooids are also often developed at the edges of branches, including the axils were branches bifurcate (Figure 3.28D, E).

Some cyclostomes have stilt-like kenozooids growing from the dorsal sides of the branches (Voigt 1992; Figure 3.28H). These not only provide additional points of attachment of the colony to the substrate, but can also support the colony slightly above the surface of the substrate, thereby ameliorating competition from other organisms closely encrusting the substrate.

In addition to their structural roles, for example in strengthening the dorsal sides of the erect branches of cyclostomes, it is possible that kenozooids are also capable of storing nutrients such as lipids (cf. the so-called kleistozooids described above), but this function has yet to be demonstrated.

A plethora of names has been introduced for the structural zooids of extinct palaeostomates. Of particular note are the **mesozooids** and **exilazooids** that characterize many trepostomes (Figure 3.29). Both are tubular zooids intercalated between the larger autozooids and often present in higher proportions in the maculae, which are loci of colonial exhalent currents (Chapter 4.5). Whereas mesozooids are packed with diaphragms (Figure 3.29A, B), exilazooids lack or contain only a few diaphragms (Figure 3.29C). Neither contain brown deposits (see p. 3), implying that they did not possess polypides. Aside from being typically smaller in size than the autozooids, mesozooids and exilazooids may have different cross-sectional shapes. A striking example can be seen in the trepostome *Mesotrypa orientalis* where the mesozooids are triangular and are positioned at the six corners of the hexagonal autozooids in tangential sections of colonies (Ernst and Nakrem 2011, fig. 7G–I).

The function, or functions, of mesozooids and exilazooids have been debated, as indeed has their status as zooidal polymorphs: they are sometimes

Figure 3.28 Kenozooids in some cheilostomes (A–D) and cyclostomes (E–H): (A) small interzooidal kenozooids surrounding an autozooid of *Akatopora* sp. (Eocene, Lutetian, Bracklesham Group; Selsey, Sussex, UK); (B) vicarious kenozooid with small, circular opesia in *Wilbertopora cheethami* (Cretaceous, Campanian, or Maastrichtian, Simsima Fm.; Jebel Buhays, Al Ain, UAE); (C) group of kenozooids with reduced costate frontal shields located at the confluence of two lobes in *Pelmatopora* sp. (Cretaceous, Campanian Chalk; Norwich, Norfolk, UK); (D) kenozooids with reduced opesiae at a branch bifurcation in *Melicerita blancoae* (Recent; Antarctica); (E) kenozooids (k) in the branch axils and at the edges of the branches of ?*Oncousoecia* sp. (Jurassic, Bathonian, Fullers Earth Rock; East Cranmore, Somerset, UK); (F) autozooids of *Heteropora* sp. surrounded by kenozooids with smaller apertures (Pleistocene, Setana Fm.; Kuromatsunai, Hokkaido, Japan); (G) growth tip of a branch of *Erksonea* sp. seen from the obverse (abfrontal) side which comprises kenozooids closed by pseudoporous frontal walls (Recent; New Zealand); (H) underside of a colony of *Apsendesia cristata* showing broken strut (centre) seemingly formed by a small group of kenozooids (Jurassic, Bathonian, Fullers Earth Rock; East Cranmore, Somerset, UK). Scale bars: A, B, F = 200 μm; C = 1 mm; D, E, G, H = 500 μm.

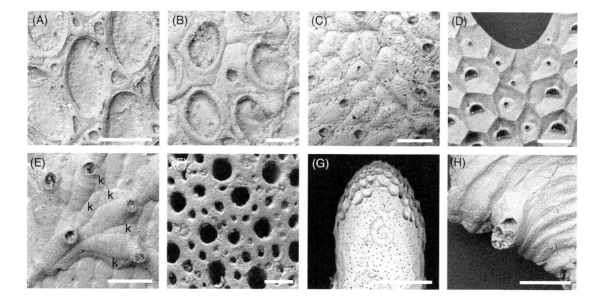

referred to in the early bryozoan literature as 'mesopores' and 'exilapores'. Boardman and Buttler (2005) suggested that they may be extra-zooidal structures (Chapter 3.14), and revived the two terms mesopore and exilapore. Comparable polymorphs present in many cryptostome bryozoans have been called **metapores** (Plate 4L) for much the same reason. An argument against the idea that these structures are pores is provided by the occurrence of mesozooids that transformed into autozooids through ontogeny, as in *Hallopora*. Similar ontogenetic polymorph transitions are known in some extant cyclostomes in which autozooids are replaced in late development by nanozooids (see below) with tiny apertures (Silén and Harmelin 1974; Gordon and Taylor 2001).

It is more likely that the main functions of trepostome mesozooids and exilazooids, as well as cryptostome metapores, were primarily structural. Tavener-Smith (1973b, p. 357) described them as 'residual spaces left between diverging autozooecia'. A detailed study of branch growth in a colony of the Permian trepostome *Tabulipora* showed the exilazooids to have a predominantly

Figure 3.29 Thin sections of palaeostomate bryozoans showing internal morphological features. Courtesy of Andrej Ernst: (A) longitudinal section of the ramose trepostome *Hallopora* sp. showing zooids bending through 90° from the axial endozone into the surrounding exozone (Ordovician, Katian, Vormsi Stage; Vormsi Island, Estonia); (B) longitudinal section of the trepostome *Diplotrypa bicornis* with abundant basal diaphragms in the narrow mesozooids (Ordovician, Dariwillian, Lasnamäe-Stage; Väo, Estonia); (C) three narrow exilazooids separating autozooids in a longitudinal section of the exozone of the trepostome *Dyscritella* sp. (Permian, Sakmarian–Artinskian, Callytharra Fm.; Blair's Camp, Western Australia); (D) longitudinal section of the trepostome *Anisotrypa* sp. with hemisepta on the proximal walls of the zooids (Carboniferous, Namurian; Algeria); (E) two ring diaphragms in a longitudinal section of the trepostome *Tabulipora* sp. (Carboniferous, Namurian; Algeria); (F) tangential section of *Tabulipora* sp. intersecting a ring diaphragm showing the constriction of the zooidal chamber (Carboniferous, Namurian; Algeria). Scale bars: A = 1 mm; B, C, D = 500 µm; E, F = 100 µm.

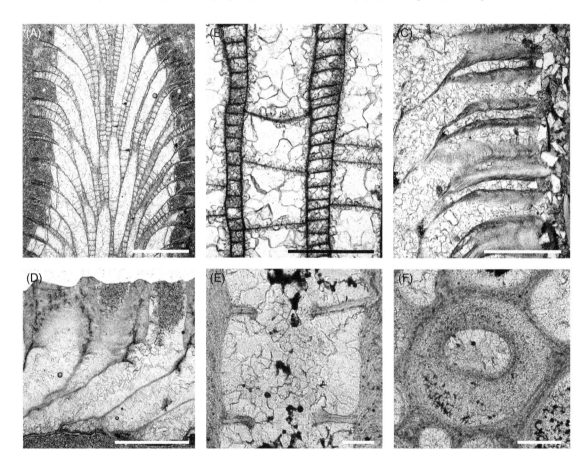

space-filling role, including regulation of the gaps between the autozooidal lophophores (Key, Thrane, and Collins 2001). Further evidence for such a role comes from the related cyclostome bryozoans. In two genera – *Diaperoecia* and *Heteropora* – which have been shown to be closely related using molecular sequence data (Taylor et al. 2015; see Chapter 8.5), the spacing between lophophores is achieved either using the calcified frontal walls of the autozooids (in *Diaperoecia*) or using kenozooids where these frontal walls are lacking (in *Heteropora*).

Axial zooids occur in some trepostomes (e.g. *Coelotubipora*: Gorjunova 2011), rhabdomesine cryptostomes (e.g. *Ascopora*: Gorjunova 2010) and cyclostomes (e.g. *Kukersella*: Buttler 1989). These may be either single or in bundles and can be extremely long. Apparently lacking brown deposits, axial zooids probably performed a structural function, forming a scaffolding over which the later zooids grew.

3.8 Spinozooids

Hollow, calcified spines are present in many cheilostomes. Most are basally articulated and intimately associated with the autozooids, often arranged around the opesia (in anascans) or the orifice (Figure 3.30A). They are probably defensive against spatial competitors or micropredators (Chapter 5). The presence of a pore plate at the base of these spines in some species (e.g. *Membraniporopsis tubigera*: Gordon, Ramalho,

and Taylor 2006) has led to their classification as polymorphic zooids called **spinozooids**. Not all hollow spines in cheilostomes are necessarily spinozooids: Bobin (1968) argued that the spines of *Electra* are simple outgrowths of the body walls of the autozooids.

Apparent interchangeability between spines and pedunculate avicularia, which are undoubted polymorphs, encircling the opesia of the anascan genus *Cauloramphus* has been used to argue for the zooidal nature of spines (e.g. Silén 1977). However, as it is clear that the avicularia originate from a slightly different location (e.g. Min et al. 2017, fig. 7), a positional homology is not fully justified.

The distal tip of the spinozooid may have a small opening or it may be closed by calcification. Rarely, the spinozooid is infundibuliform, expanding in diameter distally and with a wide opening, as in *Membraniporopsis tubigera* (Gordon, Ramalho, and Taylor 2006). Some spinozooids bifurcate, while others are flattened to form a structure called a **scutum** (Figure 3.30B) that overlies and protects the frontal membrane of the autozooids in anascan cheilostomes such as *Scrupocellaria*.

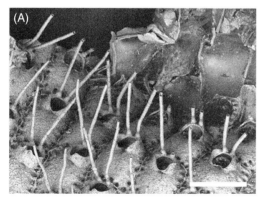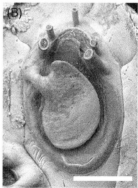

Figure 3.30 Cheilostome spinozooids: (A) oblique view of long, paired, basally articulated oral spines in *Parasmittina alba* (Recent; Ilha de Caloo Frio, Rio de Janeiro State, Brazil); (B) plate-liked scutum of *Scrupocellaria* sp. protecting the underlying frontal membrane (Recent; Sdot-Yam, Israel). Scale bars: A = 200 μm; B = 100 μm.

3.9 Rhizooids

Root-like structures functioning to anchor the colony are present in cheilostomes (Plate 7A, D) and a few cyclostomes. Many have been categorized as polymorphs termed rhizooids (Silén 1977). They take the form of long tubular structures which are uncalcified (cuticular) in most cheilostomes in which they are present but have a mineralized outer layer in a few cheilostomes and in cyclostomes. In cheilostomes, rhizooids can develop as prolongations of the cuticle on the frontal surface of the autozooids, generally after final polypide degeneration, leaving no trace of their presence in the skeleton. Alternatively, in other cheilostomes a distinct 'rootlet pore' is present in the skeleton which can be recognized in fossils (Figure 3.31).

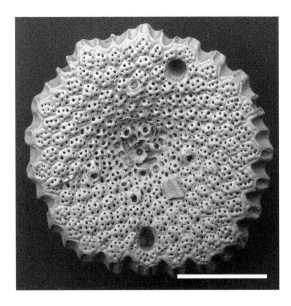

Figure 3.31 Underside of a fossil colony of *Schizorthosecos interstitia* with two large circular openings near the top and bottom edges interpreted as rootlet pores (Eocene, Bartonian, Moodys Branch Fm.; Jackson, Mississippi, USA). Scale bar = 1 mm.

Rhizooids provide the main mode of colony fixation in some cheilostomes: a tuft of rhizooids at the colony base can extend between sand grains and shell fragments on the seabed, anchoring the colony. In *Palmularia* a single stout rootlet, which may be extrazooidal, anchors the fan-shaped colony into the muddy or sandy seafloor. The larvae in this genus burrow into the sediment, attach to a sand grain, metamorphose below the surface, and produce the rootlet before budding any zooids (Cook and Chimonides 1985). Anchorage by rootlets can be secondary following initial fixation through cementation of the ancestrula and early zooids to a solid substratum. In the cheilostome *Synnotum*, the delicate erect part of the colony is easily lost but the rootlets remaining on the substrate are capable of regenerating erect growth (Marcus 1941), as they are in another cheilostome, *Bugula neritina*, after wintertime loss of the erect branches (Numakunai 1960). A ramifying mass of tubes made from calcified segments linked by thick cuticle develops at the base of colonies of *Adeona* and attaches the main part of the colony to the substrate often via a long connecting stem (Plate 7D). In some 'encrusting' cheilostomes, the underside of the colony is not closely adpressed to the substrate but instead is held a short distance above it by rootlets that emerge from chambers on the basal walls of the zooids (e.g. petraliellid ascophorans: Tilbrook and Cook 2005, fig. 10D). Fixation in this way can be particularly advantageous when the substrate is flexible. For example, *Fenestrulina commensalis* encrusts tubes of the anemone *Pachycerianthus* in Brazil, the rootlets imparting the flexibility necessary to grow around these tubes (Vieira and Stampar 2014). The presence of rootlet pores in the skeleton, as well as the typical slightly convex, non-planar shape of the basal walls, potentially allows fixation of this kind to be recognized in fossils in which the non-calcified rootlets themselves are not preserved.

Articulation above the colony base can be achieved in cheilostomes through rhizooids that link successive internodes, as in *Margaretta* (Keij 1972, pl. 3, fig. 8).

3.10 Cleaning Polymorphs

Although best known in the genus *Diplosolen*, several genera of cyclostome bryozoans develop dwarf zooids with small apertures (Figure 3.32A, B). Silén and Harmelin (1974) observed the behaviour of such **nanozooids**, which have a polypide with a single tentacle that sweeps the surface of the colony. It seems likely that they function in helping to remove sediment from the colony surface, as well as to discourage fouling by the larvae of other species. While the nanozooids in *Diplosolen* and a few other cyclostomes are budded at growing edges and are consequently interspersed among the autozooids, in some other cyclostomes (e.g. *Plagioecia*) – and apparently also a few fenestrate palaeostomates (McKinney 1994) – **secondary nanozooids** may develop within autozooids during late ontogenetic stages. They are recognizable as autozooidal apertures closed by a terminal diaphragm containing a small aperture (Figure 3.32C). Given the vulnerability of older parts of bryozoan colonies to fouling by sediment and epibionts, the presence of secondary nanozooids here may be particularly important.

A **vibraculum** (pl. **vibracula**) is another kind of cleaning polymorph found exclusively in cheilostomes (Figure 3.33). As noted above, vibracula are a kind of avicularium derived through modification of the operculum, in this case into a long, hair-like appendage – a **seta** (pl. **setae**) – that extends beyond the limits of the skeleton of the polymorphic zooid (Figure 3.33C). In cheilostomes such as *Scrupocellaria*, setae sweep the colony surface. Their motion may be in two dimensions, like that of avicularia, or in three dimensions. In the latter case, articulation with the basal skeleton is through a ball-and-socket joint. The vibracula of the Antarctic bryozoan *Nematoflustra* were observed by Winston (2010) to move in a wave along the branch from distal to proximal, dislodging any organisms and detritus.

3.11 Locomotory Polymorphs

Vibracula in free-living, lunulitiform bryozoans perform an additional function – supporting the colony above the substrate – and in some cases also acting as walking appendages. Lunulitiform

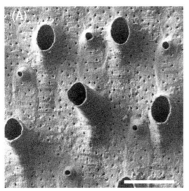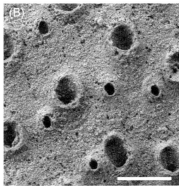

Figure 3.32 Nanozooids of cyclostomes: (A) primary nanozooids with tiny apertures dispersed among the autozooids of *Diplosolen obelium* (Recent; Port Gros Island, Marseille, France); (B) primary nanozooids in a fossil species of *Diplosolen* (Cretaceous, Campanian Chalk; Norwich, Norfolk, UK); (C) secondary nanozooid of *Plagioecia* sp. showing the minute peristome growing from the terminal diaphragm of an autozooid (Recent; New Zealand). Scale bars: A, B = 200 μm; C = 100 μm.

cheilostomes have disc-shaped colonies, typically less than one centimetre in diameter, with a concave underside and convex upper surface on which the autozooids and vibracula open. Colonies are free-living, or vagile, in the sense of not being anchored to a sizeable substrate. Instead, the larvae generally settle on tiny shell fragments or sand grains which are rapidly outgrown and may

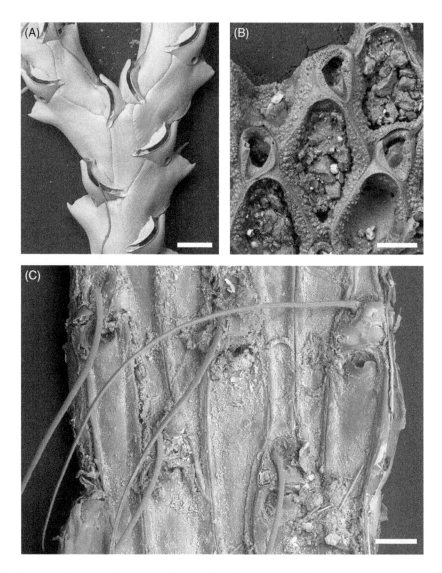

Figure 3.33 Cheilostome vibracula: (A) crescent-shaped vibracula on the dorsal side of a branch of *Cradoscrupocellaria* sp. (Recent; Azores); (B) three vibracula near the edge of a lunulitiform colony of *Cupuladria* (Eocene, Bartonian; Bende Ameki, Nigeria); (C) unbleached specimen of *Nematoflustra flagellata* showing the long vibracula setae (Recent; Antarctica). Scale bars: A, C = 200 μm; B = 100 μm.

be enveloped within the bryozoan skeleton. Alternatively, the colony (strictly the ramet) develops by regrowth from a fragment of a broken parent colony (Chapter 4.4). Vibracula with long, hair-like setae are regularly arranged among the autozooids. The setae of vibracula close to the centre of the colony are used to sweep objects from the colony surface, whereas vibracula around the margins have downward-oriented setae that function like legs to hold the colony above the seafloor. It is these marginal setae that are capable of providing the colonies of some lunulitiform species with automobility.

Detailed observations were made of living colonies of the lunulitiform bryozoan *Selenaria maculata* by Cook and Chimonides (1978). They described the slow, lurching movement of colonies maintained in an aquarium, with colonies being able to climb over one another. Setae of the avicularia moved sequentially in a clockwise direction. Maximum speed was one metre per hour. Colonies were able to right themselves using the vibracula if turned upside down. Movement towards the light was observed and later shown by Berry and Hayward (1984) to be correlated with increases in the frequency of bursts of neuroelectrical impulses that spread throughout the colony.

Another cheilostome family, Cupuladriidae, independently evolved mobile lunulitiform colonies (Chapter 9.11). Aquarium-maintained colonies moved much more slowly than *Seleneria*, attaining a maximum speed of only five millimetres per day. Differences in speed between different cupuladriid species were found to be directly proportion to the length of the vibracular setae (O'Dea 2009). Colonies of the cupuladriid genera *Cupuladria* and *Discoporella* when buried under two centimetres of sediment showed varying rates of emergence, species having the longer setae tending to be more adept at unburying themselves. It seems possible that the ability of the vibracula to allow buried colonies to regain the surface is their most important function, locomotion representing a secondary consequence of this capability.

3.12 Microenvironmental Variability

After subtracting ontogenetic, astogenetic, and polymorphic variations, there remains a residuum of within-colony ecophenotypic variability in zooid-level traits that is usually grouped as microenvironmental variation. Both spatially and temporally variable factors are potentially able to produce such microenvironmental variability in zooid morphology. In two species of runner-like Jurassic cyclostomes, *Stomatopora dichotomoides* and *S. bajocensis*, microenvironmental variation was found to be related more to the position of zooids on the substrate than to the time of their budding (Taylor and Furness 1978).

Steve Hageman and co-authors (e.g. Hageman 1995; Hageman, Bayer, and Todd 1999; Hageman, Needham, and Todd 2009; Hageman and Sawyer 2006; Hageman and Todd 2014) have made extensive and detailed studies of within-colony variability, comparing it to between-colony variability that can be either genetic or ecophenotypic in origin. Laboratory-reared colonies of the cheilostome *Electra pilosa* were used by Hageman, Bayer, and Todd (1999) to evaluate the relative influence of genotypic and environmental differences on six morphometric characters. Differences between the three genotypes in the study far outweighed environmental effects. The importance of genotype is not unique to this species. Genotypic variations between colonies of another cheilostome, *Celleporella hyalina*, have also been found to be important in determining colony growth rates (Hughes 1989), as well as responses to changes in water temperature and pH (Pistevos et al. 2011).

Hageman and Todd (2014) investigated variability in zooidal characters in natural populations of *Electra pilosa* at millimetre- to kilometre-spatial scales, which they categorized as: (i) microenvironmental (within and among immediately adjacent colonies); (ii) mesoenvironmental (small-scale positional effects among colonies within a common habitat); and (iii) macroenvironmental (among colonies from different environmental settings). Microenvironmental variation dominated, accounting for more than 60% of total variation, compared with about 30% for mesoenvironmental variation and a smaller amount for macroenvironmental variation.

In another malacostegine cheilostome, *Membranipora membranacea*, colonies growing in a vigorous tidal environment with relatively slow flow have elongate and more rectangular-shaped zooids than colonies living in faster flow regimes (Okamura and Partridge 1999). Lophophores in faster flow were also smaller in diameter and had fewer tentacles. As colonies from the two flow regimes have the same growth rates, these variations were interpreted as being local adaptations.

Hageman and Sawyer (2006) collected discrete piles of fragments ('microcommunities') of the ramose trepostome *Leioclema punctatum* from a Carboniferous locality in Missouri, USA. No significant morphometric differences were found between the microcommunities, and zooid variability was found to be larger within colonies. From this analysis they concluded that very small-scale environmental variations – both within and among colonies – play a greater role than larger scale environmental variability within this particular depositional setting. Relevant to this conclusion is the study of Keough (1989) on the cheilostome *Bugula neritina* who found colonies from the same habitats and parentage to show substantial differences in growth rate (and onset of sexual reproduction), apparently driven by plastic responses of colonies to very fine-scale environmental variations.

Comparisons of within- and between-colony variability in zooid-level morphology has been suggested as a proxy for environmental stability on the premise that colonies from more unstable environments should exhibit relatively lower levels of between-colony variability (Farmer and Rowell 1973). Key (1987) tested this by measuring the diameters of autozooids from both maculae (see Chapter 4.5) and intermacular areas in three trepostome species from the Late Ordovician of the Cincinnati region of the United States. The expected pattern of relatively low variability was found between colonies in environments believed to be increasingly unstable (shallower and with a greater frequency of storms).

3.13 Zooid-level Skeletal Structures

The often-complex skeletons of bryozoans contain a variety of external and internal structures. In some cases, the functions of these structures are known, but in many they are poorly understood.

3.13.1 Stenolaemates

The characteristically tubular zooids of stenolaemates contain a variety of mostly internal skeletal structures. Because of the ubiquity of thin sectioning as a study technique for Palaeozoic taxa (but see Hinds 1973), these are better known in palaeostomates where they may be of considerable taxonomic importance.

Diaphragms are planar walls oriented at right angles to the long axes of the zooidal tubes. Two main types are distinguished: basal and terminal. **Basal diaphragms** (Figures 3.9A, 3.29B) form the floors of zooidal living chambers (see Chapter 3.1). Because they are secreted from the distal (oral) side, skeletal laminae in basal diaphragm are deflected in a distal direction at the junction of the diaphragm and the zooidal walls. Basal diaphragms lack pores or pseudopores, and are mostly found in palaeostomates. In contrast, **terminal diaphragms** (Figure 3.9B, C), which are more common in cyclostomes than they are in palaeostomates, seal the zooid at or close to the aperture. They are secreted from the proximal (adoral) side and have skeletal laminae deflected proximally at their junctions with the zooidal walls. Most are exterior walls, possess pseudopores, and had an external covering of cuticle. Multiple basal and less commonly terminal diaphragms can be found within a single zooid. In the case of the basal diaphragms, they represent successive floors of the zooidal living chamber that formed as the zooid lengthened and the soft tissues migrated distally. In contrast, multiple terminal diaphragms represent a series of successively more proximally located seals formed as the soft tissues of the zooid retracted ever deeper into the zooidal chamber.

Unique to the trepostome *Hallopora*, the **cap-like apparatus** (Figure 3.34A, B) is a peculiar diaphragm-like structure above which the zooidal cavity is partitioned by radial septa into 5–7 chambers surrounding a reduced-diameter continuation of the original zooidal chamber. Conti and Serpagli (1987) studied phosphatic internal moulds of zooids with cap-like apparati impressed on their distal ends, believing them to

be phosphatized membranous attachment structures. Boardman (1999) thought that the reduced chamber at the centre may have housed a secondary polymorph. However, the cap-like apparatus can be more simply explained as a product of budding through **intrazooecial fission** (Figure 3.35). Described from some other stenolaemates (Hillmer, Gautier, and McKinney 1975; Taylor 1994c), intrazooecial fission entails the chamber of a parent zooid becoming subdivided into smaller chambers, each developing into a separate zooid. In the case of the cap-like

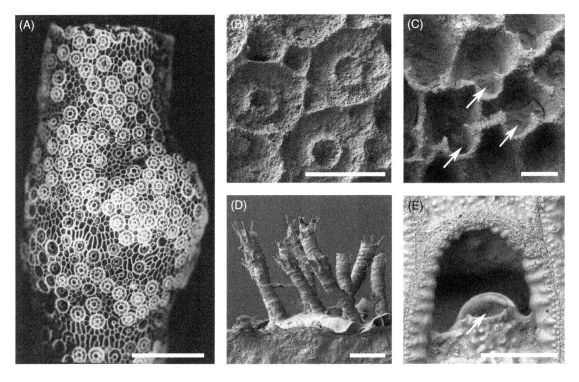

Figure 3.34 Some skeletal structures of bryozoan zooids: (A)–(B) cap-like apparatus in the trepostome *Hallopora*; (A) branch of *H. elegantula* in which a cap-like apparatus is present in all of the autozooids (Silurian, Rochester Shale; Rochester, New York State, USA); (B) cap-like apparati visible in three autozooids as slightly raised rings at the centres of the apertures in *Hallopora* sp. (Ordovician, Katian; Pechurki Quarry, Leningrad Oblast, Russia); (C) horseshoe-shaped lunaria (arrows) on the proximal edges of three autozooids of the cystoporate *Ceramopora* sp. (Ordovician, Sandbian, Bromide Fm.; Fittstown, Oklahoma, USA) SEM; (D) encrusting cyclostome with tall peristomes ending in spinose processes (Recent; Green Island, Taiwan); (E) polypide tube (arrow) opening on the proximal edge of an opesia in the cheilostome *Labioporella granulosa* (Recent; Gulf of Mexico, USA). Image in Figure 3.34A courtesy of Caroline Buttler. Scale bars: A = 2 mm; B–E = 200 μm.

Figure 3.35 Thin section of the trepostome *Stenoporella romingeri* (Carboniferous, Chaesterian; Cave Creek, Arkansas, USA) showing new buds developing through intrazooecial fission in two zooids (centre). Note also the styles at the corners of the zooids. Image courtesy of Caroline Buttler. Scale bar = 200 µm.

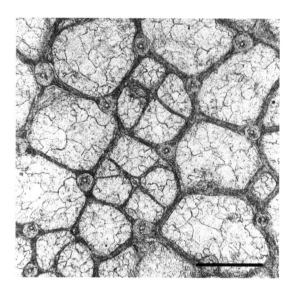

apparatus, the 5–7 chambers would be the bases of the new buds.

Cystiphragms are curved, cyst-like, orally convex skeletal walls attached to the insides of the tubular zooidal skeletons in many genera of palaeostomates. They are especially common in trepostome bryozoans such as *Monticulipora* and *Prasopora* where overlapping cystiphragms may pack the perimeters of the autozooids, severely restricting the central space that remains to accommodate the polypide. **Funnel cystiphragms** are structures that ring the skeletal chamber of the zooid making it flask shaped (Utgaard 1973). Whereas the roles of basal and terminal diaphragms respectively as floors and seals of zooidal chambers are clear, the function of cystiphragms is less obvious. Boardman (1971) regarded flask-shaped chambers in palaeostomates as a kind of intrazooidal polymorph possibly functioning as brooding structures but this idea is difficult to uphold.

Partial diaphragms occur in some stenolaemates. **Ring septa** are particularly characteristic of late Palaeozoic trepostomes belonging to the family Tabuliporidae. They resemble diaphragms but contain a central or slightly off-centred hole (Figure 3.29E, F). Often associated with ring septa are **monilae**, extreme annular thickenings in skeletal walls that can give the walls a beaded appearance in longitudinal sections (Boardman 1971; Bartley and Anstey 1987). **Hemisepta** (Figure 3.29D) and **hemiphragms** are partitions that extend only part of the way across the zooidal chambers. Widely distributed among palaeostomates, they are the reason for the ordinal name Cryptostomata – 'hidden mouth' – based on the notion that the true 'apertures' of these palaeostomates were coincident with the hemisepta and therefore positioned well inward of the openings visible on the colony surface. Cryptostomes, as well as fenestrates and occasional trepostomes (e.g. *Idioclema*), may possess a pair of offset hemisepta on opposite sides of the zooid, one on the proximal wall ('superior hemiseptum') and the other on the distal wall ('inferior hemiseptum'). Hemiphragms have been described from a few post-Palaeozoic cyclostomes in which the zig-zag zooidal chambers have hemiphragms at the points of inflection. Histological sections of living *Tubulipora* colonies with zig-zag zooids show the polypides snaking within these chambers (Harmelin 1976, pl. 28, figs 6 and 7, pl. 29, fig. 4; Schäfer 1985, fig. 3B).

In trepostomes, Ernst and Voigt (2002) distinguished Type A hemiphragms that are located alternately on opposite sides of the zooid, and Type B hemiphragms positioned along one side of the zooid, as in *Stenophragmidium* from the Lower Carboniferous (Cleary and Wyse Jackson 2007). Whereas ring septa would have restricted the retracted polypide to the centre of the zooidal chamber, Type B hemiphragms would have forced it to one side, and Type A hemiphragms have caused it to zig-zag from one side of the chamber to the other (Figure 3.36), as in the species of *Tubulipora* mentioned above. The function/s of these incomplete partitions remain unclear

Figure 3.36 Trepostome living chambers of different proportions and morphologies showing possible consequences for the size and morphology of the polypides within (after Ernst and Voigt 2002).

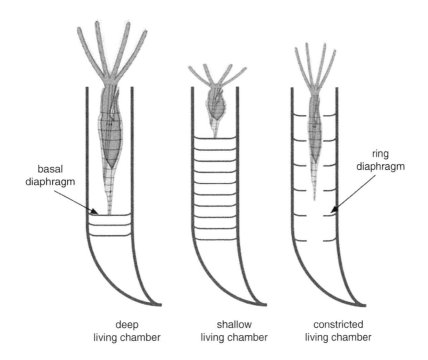

basal diaphragm

ring diaphragm

deep living chamber

shallow living chamber

constricted living chamber

although they may have provided extra support for membranes and in some cases hindered the access by particular kinds of predators (see Chapter 5.2) to the deep interiors of the zooids where the fully retracted polypides would be located.

Typical of Cystoporata, **lunaria** (Figures 3.34C, 3.37) are crescentic, hood-like projections over the apertures which in thin sections can be microstructurally distinct from the adjacent skeletal walls and commonly have a smaller radius than the rest of the zooid, with the two ends sometimes projecting slightly into the zooidal chamber. It seems likely that they deflected the expanded lophophore and assisted in the formation of multizooidal current systems (Chapter 4.5). Structures similar to lunaria, although not necessarily homologous, are also found in a few post-Palaeozoic cyclostomes and in the Devonian cryptostome *Lunostoma* (Ernst et al. 2012b).

Blades are internal ridge-like structures oriented parallel to the lengths of the zooid (Hinds 1973). Recorded only in a few Cenozoic fossil cyclostomes, their function is unknown, although it could be related to anchorage or positioning of ligaments or membranes. Apertures of the Carboniferous fenestrate *Thamniscus colei* contain a pair of small plates oriented parallel to growth direction (Wyse Jackson and Bancroft 1994). These enigmatic structures were suggested to be opercula, formed by calcification of the terminal membrane, that opened to allow lophophore extension.

Tiny dome-shaped structures (*c.* 10–50 µm in diameter) found on the inner surfaces of the basal walls of some living cyclostomes were termed **mural hoods** by Weedon (1997). They have openings directed proximally relative to the growth direction of the wall. The function of mural hoods is unknown, and it is even possible that they

Figure 3.37 Tangential thin section of dark, horseshoe-shaped lunaria in autozooids of the cystoporate *Fistulipora elegantula* (Permian, Dzulfian; Kuh-E-ali-Bashi, Iran). Vesicular skeleton occupies the area between the autozooidal apertures (see Figure 3.39). Courtesy of Andrej Ernst. Scale bar = 500 μm.

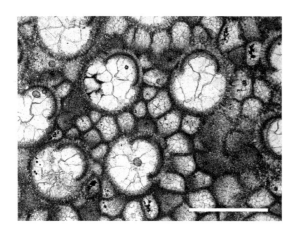

develop as a response to the presence of a microsymbiont. In particular, mural hoods resemble the 'igloo-shaped' structures constructed by bivalves around trematode parasites (e.g. Huntley and De Baets 2015), although they are at least an order of magnitude smaller in size.

Finally, tubular prolongations around apertures called **peristomes** are present in some free-walled taxa, including palaeostomates, and are ubiquitous among fixed-walled cyclostomes where they may rise a significant height above the surface of the colony surface (Figure 3.9D). As peristomes are delicate, they are invariably broken off in fossils, although Voigt (1987) figured peristomes almost one millimetre long in some Maastrichtian cyclostomes that encrusted the walls of hardened thalassinoid burrows. Functionally, peristomes elevate the tentacle crown into faster flow regimes where the flux of plankton is greater; in contrast to cheilostomes, this elevation cannot be achieved in cyclostomes simply by using the extroverted tentacle sheath (Chapter 3.1). In addition, the peristome may conceivably provide an insuperable impediment to some kinds of micropredators attempting to consume deeply retracted polypides, as well as a barrier to overgrowth by bryozoans and other organisms competing for substrate space. The peristomes of a living Taiwanese cyclostome have spinose tips (Taylor and Lewis 2003; Figure 3.34D), which might have enhanced any defensive function.

3.13.2 Cheilostomes

The elaborate skeletons of cheilostome zooids furnish the morphological characters on which the taxonomy of this order has been founded. Far less taxonomic importance has been accorded to colony-level morphology in cheilostomes than in stenolaemates. One consequence is that even tiny fragments of cheilostome colonies comprising just one or a few zooids may be adequate for genus or species identification.

The box-like zooids that are the norm for cheilostomes have an upper, frontal surface where most of the taxonomically and functionally important skeletal characters are expressed. In the simplest cheilostomes, the frontal surface is almost entirely membranous. In most species, however, calcification extends over a variable part of the frontal surface, producing a frontal wall that not only strengthens the zooid structurally but also provides protection for the retracted polypide. Frontal wall calcification in anascan-grade cheilostomes comprises exterior-walled gymnocyst, interior-walled cryptocyst, or a combination of both (Plate 9). So-called membraniporimorph anascans typically have a narrow gymnocyst around the margins of the zooid, frequently better developed proximally, enclosing an inner ring of cryptocyst that surrounds the opesia (Figure 1.12A). The boundary (**mural rim**) between gymnocyst and cryptocyst represents the point at which the flexible frontal membrane is attached to the calcified skeleton. Thus, the anascan cryptocystal frontal wall (Figure 3.12C–E) underlies the frontal membrane, and the muscles used to depress the frontal membrane, as well as operating the operculum, in most cases traverse the opesial opening. Alternatively, they may pass through holes (**opesiules**) in the cryptocyst, typically paired along the two lateral sides of the zooid (Figure 3.5B). The points of attachment of the muscles that depress the frontal membrane and operate the operculum can leave scars on the basal wall of the zooid which are visible in some fossil cheilostomes (e.g. Medd 1964; Figure 3.5A).

In thalamoporelline anascans the opesia leads into a **polypide tube** (Figure 3.34E) rather than opening into an unrestricted zooidal chamber. These tubes should make it more difficult for predators attempting to consume the fully retracted polypide, or may simply facilitate the correct positioning of the lophophore during protraction and retraction. A toothed structure called the **plectriform apparatus** lying beneath the frontal wall at the proximal edge of the opesia has been described in some primitive cheilostomes (Gordon and Parker 1991b). Its function is uncertain.

Frontal wall calcification is more complex and varied in ascophoran-grade cheilostomes (Plate 9). Here, the calcified wall, which lies above the flexible frontal membrane wall flooring the ascus, is known as a **frontal shield**. There are four types of frontal shield calcification – **spinocystal**, **gymnocystal**, **umbonuloid**, and **lepralioid** – which can occur alone or in combination with others (Figure 3.38). All four types serve both to strengthen the zooid (and hence the colony) and to protect the retracted polypide. Spinocystal frontal shields comprise overarched hollow spines that may be closely juxtaposed or separated by narrow gaps, each spine wrapped by cuticle. Gymnocystal frontal shields are continuous exterior walls covered by cuticle on their outer surfaces and regularly porous in a few genera. Umbonuloid frontal shields are also exterior walls but topologically reversed compared to gymnocystal walls (Figure 3.38B) in that the cuticle is on the underside facing the ascus. Lepralioid (or cryptocystidean) frontal shields are interior walls lacking a cuticular layer, at least when first formed (Tavener-Smith and Williams 1970, fig. 20). The evolution of these frontal shield types is discussed in Chapter 9.14.

Of great significance in ascophoran cheilostome taxonomy is the form of the **primary orifice** closed by the operculum. In many cheilostomes this is best seen in young zooids close to the growing edge as it can become hidden by the formation of a peristome defining a **secondary orifice** which is often significantly different in shape. Depending on taxon, structures associated with primary orifices include a sinus, condyles, and a lyrula. The proximal edge of the orifice often includes a typically U- or V-shaped **sinus** (Figure 3.5E). This represents the entrance to the ascus and is closed by a proximal extension of the operculum. When the main distal part of the operculum pivots outwards to allow protrusion of the lophophore, the proximal part over the sinus is rotated inwards to allow seawater to flow into the ascus and compensate for the volume vacated by the emerging lophophore. **Condyles** (Figure 3.39A) are teeth on which the operculum hinges. A **lyrula** (Figure 3.39A) is an anvil-shaped structure located just in front of the operculum, positioned medially on the proximal edge of the orifice. It is believed to function as a stop to limit the opening of the operculum to a position approximately vertical relative to the frontal surface of the zooid (Berning, Tilbrook, and Ostrovsky 2014). Phidoloporid ascophorans typically have a finely beaded distal edge to the orifice (Figure 3.39B), the function of which is unknown. While primary orifice morphology is of critical importance in ascophoran taxonomy, orifice shape was found to be unimportant in discriminating between two species of *Rhynchozoon* in a study using morphological characters combined with molecular sequence data (Dick and Mawatari 2005).

Several groups of ascophoran cheilostomes have evolved separate openings to the ascus that are proximal to the orifice and consequently are not closed by the operculum. Such openings in ascophorans with umbonuloid frontal shields are generally called **spiramina** and may consist of one or a cluster of openings, whereas in ascophorans with lepralioid frontal shield they are known as **ascopores** and are single openings (Figure 3.5F). Although ascopores are sometimes simple circular holes, they can also be slit-shaped, reniform, or crescentic. Some are covered by a perforated plate. Non-circular and perforated ascopores can be

Figure 3.38 Ascophoran cheilostome frontal shields: (A) spinocystal (cribrimorph) frontal shield in *Turnerellina periphereia* (Cretaceous, Maastrichtian, Prairie Bluff Fm.; Pontotoc Co., Mississippi, USA); (B) gymnocystal frontal shield in *Trypostega venusta*; note also the numerous small polymorphs ('zooeciules') and an ovicell (upper left) (Recent; Tel-Aviv-Gordom, Israel); (C) umbonuloid frontal shield in *Metrarabdotos thomseni* (Plio-Pleistocene, Kiritika Fm.; Esperia Faliraki, Rhodes); (D) underside of umbonuloid frontal shield in an unidentified bryozoan with arrows pointing to the ring scars (Eocene, Ypresian, Tumaio Limestone; Chatham Island, New Zealand); (E) leprialioid frontal shield developing at the growing edge of *Pentapora foliacea* (Recent; Cornwall, UK); (F) underside of leprialioid frontal shield in *Pentapora foliacea* with pseudopores (Recent; Plymouth, Devon, UK). Scale bars: A = 100 μm; B, D, E, F = 200 μm; C = 500 μm.

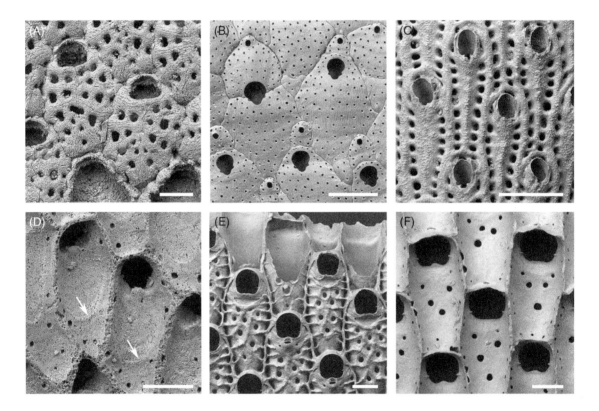

interpreted as adaptations to restrict the size of unwanted particles or organisms that can enter the ascus while maintaining a large cross-sectional area for the inflow and outflow of seawater. It would be interesting to model the flow dynamics through ascopores of different morphologies to compare their hydrodynamic properties. A recent molecular study (Orr et al. 2018) has shown that ascopores have evolved at least twice: two asco-

pore-bearing genera (*Microporella* and *Fenestrulina*) previously believed to be closely related belong within different clades dominated by genera lacking ascopores (Figure 8.8).

The frontal shields of umbonuloid and leprialioid ascophorans may contain pores of two kinds: **areolae** (Figure 3.38C) and **pseudopores** (Figure 3.38E). Areolar pores normally form a single row around the periphery of the zooid, are subcircular

Figure 3.39 Orificial characters in ascophoran cheilostomes: (A) paired condyles (short arrows) and a lyrula (long arrow) in *Parasmittina winstonae* (Recent; Langkawi, Malaysia); (B) beaded distolateral border in *Rhynchozoon* sp. (Recent; Susac Island, Croatia). Scale bars = 50 μm.

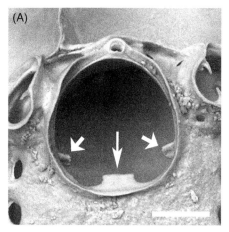 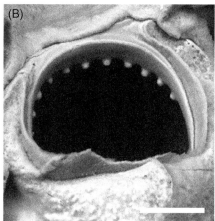

to elongate in shape, and sometimes act as budding sites for adventitious avicularia. They occasionally bifurcate and migrate towards the centre of the zooid as the frontal shield becomes thickened during ontogeny. Pseudopores tend to be scattered evenly over the surface of the frontal shield, forming a link between the hypostegal coelom above and roof of the ascus below (Tavener-Smith and Williams 1970). Analogues of both pore types can be found in the surface calcification of some ovicells. It seems probable that pseudopores in ovicells facilitate the respiration of the embryos being brooded.

3.14 Extrazooidal Structures

Some parts of the bryozoan skeleton cannot readily be attributed to a single zooid, often because they are located beyond the boundaries of recognizable zooids. In principle, these **extrazooidal** structures could have evolved from highly modified polymorphic zooids that lost their bounding walls, or *de-novo* as part of the colony as a whole.

Among stenolaemates, perhaps the best examples of extrazooidal skeleton are the superstructures of fenestrates, vesicular tissues of fistuliporoid cystoporates, and cancelli of cancellate cyclostomes. All of these taxa have free-walled skeletons (Chapter 9.12) and the extrazooidal structures are a product of the secretory epithelium associated with the hypostegal tissues enveloping the colony surface.

Fenestrate superstructures are prolongations of the skeleton on the obverse (aperture-bearing) sides of the branches. They cannot be attributed to individual zooids and instead are part of the secondary, laminated skeleton deposited beneath the surface investment of hypostegal tissues (Tavener-Smith 1969). Ranging from high keels that broaden laterally into flanges, to complete polygonal networks positioned above the main colonial meshwork (see Chapter 4.3), they have been recorded in about 16 genera (Ernst 2013). Superstructures first appeared in the Silurian, were abundant in the Devonian, but became rarer in the Carboniferous and Permian. They may have functioned to strengthen the colony and/or in defence (see below, p. 106).

Vesicular tissue (Figure 3.40) occupies the space between autozooids in fistuliporoid cystoporates such as the eponymous genus *Fistulipora*, which was globally widespread during the Silurian–Permian. It can also be found in basal parts of other bryozoan orders (e.g. Karklins 1983, fig. 225.4; Ernst, Taylor, and Wilson 2007; Suárez Andrés and McKinney 2010). The vesicles (sometimes referred to as cystopores) are interlocking, blister-like structures stacked vertically with their convex sides facing upwards. In the peculiar palaeostomate *Kanoshopora* from the Ordovician of Utah the endozone of the thick ramose branches is packed with vesicles that become aligned into longitudinal files at the boundary with the surrounding exozone (Ernst, Taylor, and Wilson 2007).

As with fenestrate superstructures, vesicles were evidently secreted from the epithelium flooring the hypostegal tissues that covered the surface of the colony (Utgaard 1973). In the absence of interzooidal pores, the lumens of the vesicles had no soft tissue connections with the zooids in the colony, or with one another. It seems clear that vesicles performed a structural function, providing skeletal continuity between the zooids and maintaining an optimal spacing between their lophophores, along with a strengthening role particularly when they became tightly packed to form stereom near the surface of the colony, as in the Permian genus *Ramipora* (Nakrem and Spjeldnaes 1995).

In older branches of some trepostomes (e.g. *Orbiramus*), outer layers of lamellar calcite, often containing styles, may develop, occluding the apertures and presumably strengthening the basal parts of the colony where feeding zooids are lacking. This skeletal covering is formed by the lateral expansion of the interzooidal walls shared by adjacent zooids. These walls normally have inflections in the laminations delineating the boundaries between the zooids. Obvious zooidal boundaries are lost in the covering laminae, producing a mass of skeleton that may be considered to be extrazooidal.

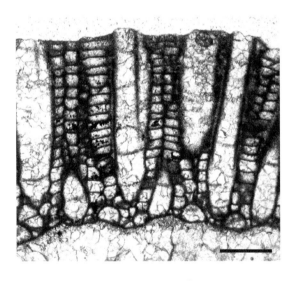

Figure 3.40 Longitudinal thin section of the cystoporate *Fistulipora elegantula* showing vesicular skeleton between the autozooids (Permian, Dzulfian; Kuh-E-ali-Bashi, Iran). Courtesy of Andrej Ernst. Scale bar = 500 μm.

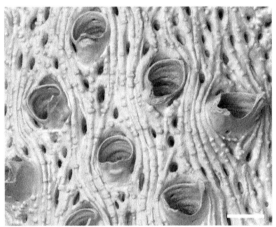

Figure 3.41 Surface of a branch of the cyclostome *Hornera erugata* showing autozooidal apertures surrounded by 'extrazooidal' calcification containing the small openings of the cancelli (Recent; Cape of Good Hope, South Africa). Scale bar = 100 μm.

More questionably interpreted as extrazooidal structures are the **cancelli** (Figure 3.41) of cancellate cyclostomes exemplified by the extant genus *Hornera*. Cancelli are long, narrow tubes appearing on the colony surface as small circular openings between the autozooidal apertures on the frontal sides of branches and covering the reverse sides of the branches. While it is possible that cancelli are highly modified versions of the kenozooids found in many free-walled cyclostomes, they do not clearly originate as zooidal buds, may be sinuous, and are perhaps better understood as highly elongated interzooidal pores than as polymorphic zooids.

Skeleton formed after the coalescence of the hypostegal coeloms of neighbouring zooids in cheilostomes is commonly regarded as extrazooidal (e.g. Cheetham and Cook in Boardman et al. 1983, p. 184). This is particularly well seen in ascophoran-grade cheilostomes where it forms the thickenings of basal branches in erect colonies overgrowing the orifices of old zooids that have ceased to feed (Figure 4.12).

3.15 The Cormidial 'Zooids' of Advanced Cheilostomes

The zenith of polymorphism among bryozoans occurs in ascophoran-grade cheilostomes (Lidgard et al. 2012). What at first sight appears to be the skeleton of a single ovicellate autozooid can in fact be cormidial structures derived in evolutionary terms from an association of diverse polymorphs. The frontal shield of the autozooid is considered to be a product of several fused kenozooids: the alveolar pores around the edges of the zooid are the openings of these kenozooids, while the shield itself consists of the splayed-out walls of the kenozooids (Figure 3.42). Adventitious avicularia

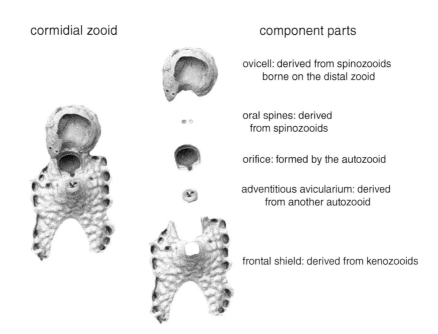

cormidial zooid   component parts

ovicell: derived from spinozooids
  borne on the distal zooid

oral spines: derived
  from spinozooids

orifice: formed by the autozooid

adventitious avicularium: derived
  from another autozooid

frontal shield: derived from kenozooids

Figure 3.42 Cormidial skeleton of an ascophoran cheilostome zooid formed of component parts derived evolutionarily from several different zooids.

originate as buds from some of these alveolar pores, which means that they are, effectively, frontal buds from kenozooids. They can be seamlessly integrated with the kenozooidal frontal shield skeleton. The ovicell in which embryonic brooding takes place is roofed by a double-walled structure (the ooecium) derived from flattened and laterally fused spine-like kenozooids which in the most primitive ovicellate cheilostomes can be seen to be borne not on the brooding zooid but on its distal neighbour (see Ostrovsky and Taylor 2004, 2005a; Figure 3.16). In anthropomorphic terms, this neighbour constructs the 'crib' in which the mother keeps its baby. Of the skeleton visible on the frontal surface of the zooid, perhaps only the rim around the orifice can be attributed entirely to the autozooid, and in some species even this is formed in part by the adjacent zooids. Consequently, unlike most bryozoans in which the skeleton is built wholly by the same zooid as the polypide it contains, in these ascophoran cheilostomes the polypide is housed within a complex, cormidial skeleton that is a consortium of highly derived polymorphic zooids.

4

Colony Morphology and Function

The large variety of colony-forms found among bryozoans reflects differences in colony function and results mainly from variations in the arrangement of the modular zooids making up the colonies. These in turn are due to different patterns of zooidal budding. Thus, growth, form, and function are intimately linked. This chapter begins with a summary of the myriad of colony-forms found in bryozoans, before reviewing the basic budding patterns responsible for these colony-forms and the functional morphology of different colony-forms. Bryozoan colonies are not simply collections of zooids that function with total independence; instead, they are inseparably linked in various ways, and to differing degrees depending on the species concerned, a concept known as colonial integration.

4.1 Colony-form Classifications

One of the most remarkable features of bryozoans is the immense variety of colony-forms found in the phylum (Plates 6–8, 10–12). Such within-phylum plasticity in body shape contrasts with nearly all phyla of unitary invertebrates found in the fossil record including brachiopods, the lophophorate 'cousins' of bryozoans which are far more morphologically conservative. Most of the main colony-forms found in bryozoans have evolved convergently more than once during the evolution of the phylum (Chapter 9.11). Their utility in taxonomy varies according to group. Typically, colony-form is given greater taxonomic weight in stenolaemates than it is in cheilostomes where zooidal morphology is considered to be more important.

Descriptive classifications of colony-forms in bryozoans vary from the simple to the complex, and the terminology employed from general and geometrical to specific and based on exemplar bryozoan genera. Simple classifications recognize relatively few colony-forms, complex classifications a bewilderingly large number: these two end members cut the 'cake'

Bryozoan Paleobiology, First Edition. Paul D. Taylor.
© 2020 Natural History Museum. Published 2020 by John Wiley & Sons Ltd.

into large and small slices, respectively. Which classification to employ depends very much on the questions being asked. For example, especially in Cenozoic cheilostomes, research on colony-forms has focused on their potential as palaeoenvironmental indicators. Much of this work, however, lacks a sound empirical basis: our understanding of the distributions of living bryozoans across environments, as well as the specific factors determining these distributions, is far from adequate. Furthermore, it is clear that many different colony-forms are able to co-exist in the same environment (see Taylor 2005). Only a few bryozoan colony-forms may be diagnostic of particular environments. Instead, different colony-forms more often represent alternative strategies for survival that are viable in several different environments.

A basic division of bryozoan colony-forms is into encrusting, erect, and free-living. Encrusters are adpressed to a hard or firm substrate; erect colonies grow perpendicularly from the substrate into the water column; and free-living colonies attach, at least initially, to tiny substrates that do not anchor them in a fixed location, in contrast to encrusting and erect colonies. Beyond this tripartite division, perhaps the most basic classification is that introduced by Jackson (1979a) for benthic colonies as a whole. Jackson was interested in different colony-forms as alternative survival strategies for life on hard substrates. He recognized six basic colony shapes: runners, sheets, mounds, plates, vines, and trees. **Runners** and **sheets** are encrusting forms, with runners having a linear, branching morphology (Figure 4.1A, B), and sheets a two-dimensional morphology (Figure 4.3; Plates 1C, 6A, 8I). **Mounds** (Plates 4B, 6D) are 'semi-erect' forms attached along most of their basal surfaces but, unlike sheets, mounds show significant vertical growth. The remaining three shapes are fully erect colonies. **Plates** have flattened, frondose branches (Plates 6O–P, 12A–H), **vines** (Plate 1B) are the erect equivalent of runners, growing vertically from the substrate and delicately constructed, while **trees** (Plate 6F–I) are thicker branching forms. **Ribbons** (Figure 4.2A) are an intermediate colony-form between runners and sheets, their branching colonies resembling runners but the branches being broader and exceeding one zooid in width. An additional encrusting category – **spots** – was introduced by Bishop (1989) for diminutive colonies exhibiting precocious sexual reproduction (Figure 4.2B).

In a study of changes through geological time in colony-forms with particular reference to the varying contributions of bryozoan skeletons to sediments, Taylor and James (2013) found it useful to distinguish nine major colony-forms: encrusting, dome-shaped, palmate, foliose, fenestrate, robust branching, delicate branching, articulated, and free-living. Encrusting, dome-shaped, and free-living are equivalent to the categories already mentioned above. **Palmate** (Plates 6N–O, 12G–H) and **foliose** (Plate 12A–C) colony-forms both have flattened erect branches but in the case of palmate colonies the branches are strap-like and usually bifurcate, whereas those of foliose colonies are broad sheets that are often folded or corrugated. **Fenestrate** colonies (Plates 6J–K, 8A, 12D–F) are planar erect colonies with perforations (**fenestrules**) created by regular bifurcations and anastomoses of branches (Plate 12D, F) or, alternatively, by cross partitions (**dissepiments**) linking

Figure 4.1 Growth patterns in some encrusting bryozoan colonies: (A) runner cyclostome *Stomatopora* sp. with bifurcating branches and ancestrula at far left (Jurassic, Bathonian, Fullers Earth Rock; East Cranmore, Somerset, UK); (B) runner cheilostome *Pyriporoides uniserialis* showing cruciate branching pattern with the lower zooid producing a distal and left and right distolateral buds (Recent; Antarctica); (C) composite multiserial cheilostome *Charixa sexspinata* showing median branch of long, narrow autozooids (with centrally perforated closure plates) and broader zooids formed as distolateral buds (Cretaceous, Albian, Glen Rose Fm.; Hays Co., Texas, USA); (D) pivot point (arrow) at the end of a growing edge of a spiral overgrowth in the cyclostome '*Mesenteripora*' *undulata* (Jurassic, Bathonian, Caillasse de la Basse-Ecarde; Luc-sur-Mer, Calvados, France); (E) eruptive buds with pseudoancestrulae (arrows)

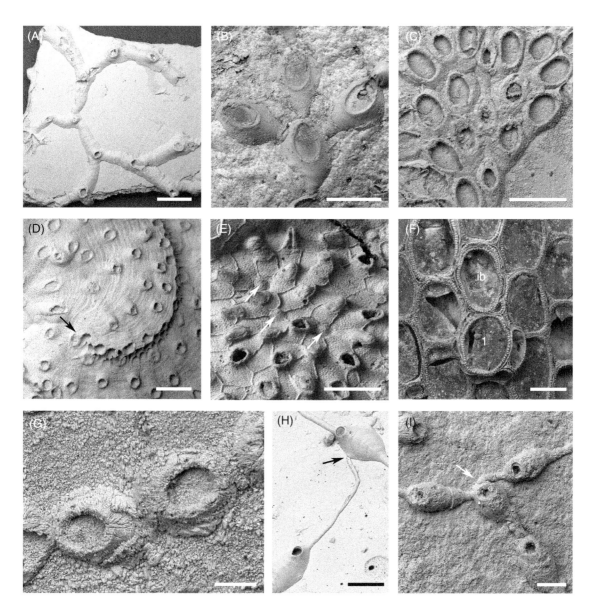

Figure 4.1 (Continued) forming coalesced overgrowths in the eleid cyclostome *Reptomultelea tuberculata* (Upper Cretaceous; Triquerville, Seine-Maritime, France); (F) intramural bud (ib) giving rise to a first generation zooid (1) of an overgrowth in an unbleached colony of the cheilostome *Conopeum seurati* (Recent; Bezirk Wadi, Tunisia); (G) bipolar pair of zooids oriented in opposite directions, indicating reparative growth from the severed ends of a broken branch in *Rhammatopora glenrosa* (Cretaceous, Albian, Glen Rose Fm.; Lakeway, Travis Co., Texas, USA); (H) branch reconnection in the cheilostome *Hippothoa flagellum*, the arrow pointing to a connection with an open pore (Recent; English Channel); (I) similar branch reconnection (arrow) in the primitive cyclostome *Corynotrypa* sp. (Ordovician, Sandbian, Bromide Fm.; Sulphur, Murray Co., Oklahoma, USA). Scale bars: A = 1 mm; B, C, D, E = 500 μm; F = 200 μm; G = 100 μm; H, I = 200 μm.

Figure 4.2 Small encrusting cheilostome colonies from the Pleistocene Red Crag Formation of Street Farm, Suffolk, UK: (A) oligoserial, ribbon-like colony of *Phylactella*; (B) spot colony of *Cribrilina puncturata* with precocious development of ovicells. Scale bars: A = 500 µm; B = 200 µm.

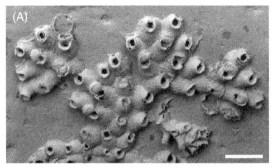

adjacent branches (Plate 12E). Importantly, and contrasting with cribrate colonies (see below), feeding zooids open only on the frontal faces of the branches and are lacking on the reverse sides. Erect colonies with bifurcating branches are split in this scheme into two categories: **robust branching** colonies have branches greater than 2 mm in diameter (Plate 6H–I), and **delicate branching** (Plate 6F–G) have branches less than 2 mm wide. The justification for this seemingly arbitrary division comes from the work of McKinney (1986b) who recognized a bimodality in branch diameter among bryozoans of various ages, correlating with the presence of maculae in robust branching colonies

but their absence in delicate branching colonies. Branch multiplication is typically dichotomous in branching colonies but some delicate branched colonies are **pinnate**, with secondary branches diverging from either side of a straight primary branch (Plate 8B). **Articulated** colonies (Plate 6E) have skeletal segments (**internodes**) linked by regular elastic joints (**nodes**). Taphonomic decay of the joints brings about disarticulation into internodes which can consist of anything from a single to numerous zooids depending on the frequency of the joints. Internodes may resemble branches broken from a delicate branching colony but lack the Y-shaped bifurcations found in the latter.

Some additional colony-forms were mentioned by Taylor and James (2013) that were insufficiently abundant for their analysis, or could be subsumed into one of the other colony-forms. **Fungiform** colonies (Figure 4.3A), as the name implies, resemble mushrooms in having a basal stalk supporting an expanded head (Taylor and Grischenko 1999). **Cribrate** colonies (Plate 8C–D) are planar erect forms with fenestrules but contrast with fenestrate colonies in having zooids opening on both sides of the branches (Taylor 2012), a difference that has important implications for the functioning of the colony with regard to feeding currents (see below). Tubular erect colonies (Plate 8K) are uncommon overall but can be locally abundant.

Free-living colonies (Plate 6L–M) are not secured by the substrate to which they are attached, which is typically small compared to the size of the mature colony, or else lack a substrate entirely. In both cases, colonies may potentially be mobile, some achieving through their own efforts (**automobility**), but mobility is precluded in others as rootlets anchor the colony into the substrate (see below).

Inverted cone-shaped bryozoan colonies with zooids opening only on the top of the cone are very occasionally found (Taylor and Wilson 1999a; Figure 4.3B), as are thin but extensive sheet-like colonies apparently free-lying on the substrate (McKinney and Jaklin 1993). Some enormous free-lying trepostome colonies have been described in Permian glaciomarine sediments of Tasmania (Reid 2012). Here, colonies of *Stenopora* up to 40 cm in diameter grew across the surface of soft sediments, extending

Figure 4.3 Fungiform and conical bryozoan colony-forms: (A) fungiform cyclostome of the cyclostome *Discantenna tumba* with ancestrula and encrusting base at the right (Recent; Chatham Rise, New Zealand); (B) lateral view of a conical colony of the esthonioporine *Dianulites fastigiatus* at the broad top of which the zooids open (Ordovician; near St Petersburg, Russia). Scale bars: A = 1 mm; B = 5 mm.

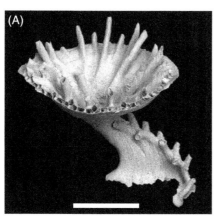

greatly from their primary substrates that comprised fragments of brachiopod shells or glacial dropstones.

The use of exemplar genera in naming bryozoan colony-forms has been widespread, particularly in palaeoecological studies of cheilostomes. It can be traced back to the work of Stach (1936) on the presumed relationships between colony-forms, functional morphology, and palaeoenvironments. Commonly employed colony-forms include **membraniporiform** (from *Membranipora*), **vinculariiform** (from *Vincularia*), **cellariiform** (from *Cellaria*), **celleporiform** (from *Cellepora*), **lunulitiform** (from *Lunulites*), and **reteporiform** (from *Retepora*) (Lagaaij and Gautier 1965; Schopf 1969; Nelson, Keane, and Head 1988; Moissette et al. 2007a). Only occasionally have these terms have been applied to stenolaemate bryozoans, and the parallel schemes devised for cyclostome stenolaemates have not been widely employed (Brood 1972; Nelson, Keane, and Head 1988). These classifications can be problematical for the nonspecialist unfamiliar with the eponymous genera. They are also flawed in that not all of the names closely match their respective genera: at least one of the genera does not have the colony-form to which it gave its name ('adeoniform' colonies are palmate, whereas colonies of most species of the genus *Adeona* are cribrate, e.g. Plate 7D), while other genera are obsolete (*Eschara*, *Retepora*). Table 4.1 lists the main colony-forms used in genus-based classifications, correlating these categories with more generalized colony shapes.

Hageman et al. (1998) introduced a detailed and comprehensive method for analyzing bryozoan colony-forms that focuses on colony-level morphological characteristics rather than overall colony-form. Eleven classes of characters were distinguished (Figure 4.4). The attributes of individual taxa can be scored for these characters and mapped in multidimensional morphospace. This opens the way for a more objective correlation between colony-forms and environments, and thus greater confidence in using the colony-forms of fossil bryozoans for inferring palaeoenvironments.

4.2 Growth and Colony-form

How can bryozoans produce so many different colony-forms employing zooids as building blocks that are either curved cylinders (most stenolaemates) or box-shaped (most cheilostomes)? The answer is mainly due to variations in budding patterns: the locations of new buds (**budding loci**),

Table 4.1 Bryozoan colony-form nomenclature derived from exemplar cheilostome genera, with references and summaries of the shapes they describe. Appendix 1 of Schopf (1969) lists additional genus-based colony-forms that have been used less frequently.

Colony-form	Source taxon	References	Colony shape
adeoniform	*Adeona*	Lagaaij and Gautier 1965; Schopf 1969; Nelson, Keane, and Head 1988; Moissette et al. 2007a	palmate erect
catenicelliform	*Catenicella*	Schopf 1969; Moissette et al. 2007a	articulated erect with short internodes of one or a few zooids
cellariiform	*Cellaria*	Lagaaij and Gautier 1965; Schopf 1969; Nelson, Keane, and Head 1988; Moissette et al. 2007a	articulated erect
celleporiform	*Cellepora*	Lagaaij and Gautier 1965; Schopf 1969; Nelson, Keane, and Head 1988; Moissette et al. 2007a	irregularly heaped zooids forming multilayered colony, often either robust branching or dome-shaped
conescharelliniform	*Conescharellina*	Schopf 1969; Cook and Lagaaij 1976; Moissette et al. 2007a	rooted conical
eschariform	*Eschara*	Schopf 1969; Nelson, Keane, and Head 1988; Moissette et al. 2007a	foliose erect
flustriform	*Flustra*	Schopf 1969	palmate erect but flexible due to weak calcification
lunulitiform	*Lunulites*	Nelson, Keane, and Head 1988; Moissette et al. 2007a	free-living, cap-shaped
membraniporiform	*Membranipora*	Nelson, Keane, and Head 1988; Moissette et al. 2007a	encrusting sheet
membraniporiform A	*Membranipora*	Lagaaij and Gautier 1965; Schopf 1969	encrusting sheet on hard substrate
membraniporiform B	*Membranipora*	Lagaaij and Gautier 1965; Schopf 1969	encrusting sheet on flexible substrate
petraliiform	*Petralia*	Schopf 1969; Lagaaij and Gautier 1965	foliose erect, unilamellar and rooted
pseudovinculariiform	*Vincularia*	Schopf 1969	encrusting sheet on 'algal' stem resulting in hollow cyclindrical colony
reteporiform	*Retepora*	Lagaaij and Gautier 1965; Schopf 1969; Nelson, Keane, and Head 1988; Moissette et al. 2007a	fenestrate erect
setoselliniform	*Setosellina*	Schopf 1969; Lagaaij and Gautier 1965	encrusting sheet on small substrate and with spiral patterning of zooids
vinculariiform	*Vincularia*	Lagaaij and Gautier 1965; Schopf 1969; Nelson, Keane, and Head 1988; Moissette et al. 2007a	robust or delicate branching erect

Figure 4.4 Character states for bryozoan colony-forms related to the six categories describing the orientation of the colony and its occupation of space recognized by Hageman et al. (1998; redrawn from their fig. 3).

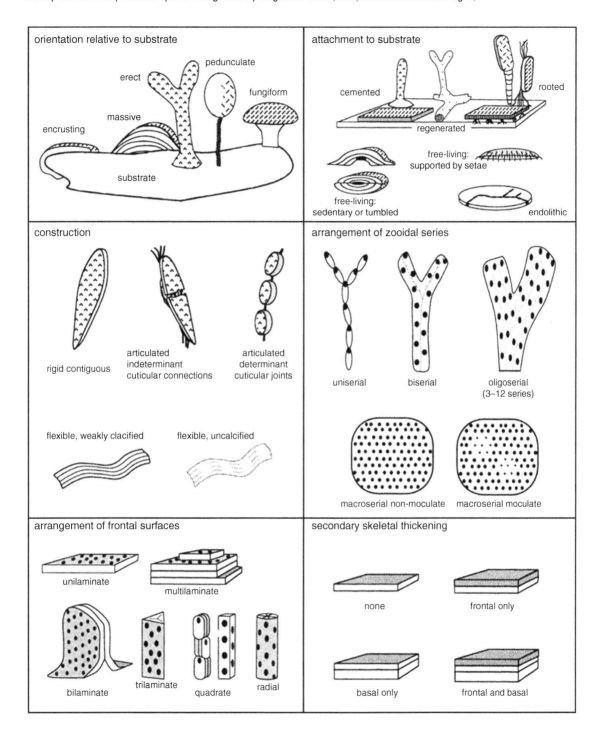

combined with differences in bud orientation and the rates of new bud formation, allows a wealth of colony-forms to be constructed.

The simplest illustration of budding pattern variability can be found among planar encrusting colonies with zooids arranged in single lines, i.e. uniserial colonies. In such colonies, the ancestry of individual buds is easier to observe than among the densely packed zooids of multiserial colonies. New buds may potentially arise anywhere around the free periphery of a parent zooid. However, budding loci tend to be restricted to particular sites. In the case of uniserial cyclostomes, there is typically only one budding locus, positioned at the distal end of the zooid. This normally produces a single daughter zooid in line with the parent zooid, or two diverging daughter zooids forming a branch bifurcation (Figure 4.1A).

Among uniserial cheilostomes, however, proximolateral, medial, and distolateral budding loci may be present as well distal loci. In addition, the ancestrula may also have a proximal budding locus (Silén 1987, fig. 19). Any of these budding loci is the potential source of a daughter zooid, although the common pattern is for either just the distal locus to form a new zooid, or for this locus and the two distolateral loci to produce buds, resulting in a cruciform branch trifurcation (Figure 4.1B). With a few exceptions (e.g. Cook 1985, figs 19, 39), proximal budding from the ancestrula is a trait characteristic of primitive cheilostomes, including malacostegines and *Aetea* (Cook 1977a).

During their growth, branches of uniserial and narrowly multiserial (**oligoserial**) colonies often collide with pre-existing branches. Collisions sometimes result in overgrowth of the older branch but more often the younger branch abuts the side of the older branch and its growth is halted. Such abutments are presumably to the detriment of the colony as they eliminate a site for zooidal budding. Some cyclostomes minimize the number of branch collisions through systematic decrease in the angle of branch bifurcation during colony growth. Computer modelling of growth in two Jurassic species of the runner-like cyclostome

Stomatopora (Gardiner and Taylor 1982) showed that this pattern was optimal because the high angles of bifurcation in early astogeny distributed branches around the full 360° arc around the colony origin, while the lower angles in later astogeny lessened the likelihood of convergence with adjacent branches. Hypothetical colonies with either constantly high or constantly low branching angles sustained greater numbers of branch collisions and lost more budding sites as a result. Some Recent species assigned to *Stomatopora* have very similar patterns of decreasing bifurcation angle through astogeny (Lagaaij 1963b, fig. 4; Buge 1979), showing that this adaptive trait has persisted in geological time. Interestingly, colonies of the runner-like Silurian coral *Aulopora* independently evolved the same pattern of astogenetic decrease in branch bifurcation angles (Helm 1999).

A pattern found in some cheilostomes and referred to as **composite multiserial** (Silén 1987) is transitional between uniserial and conventional multiserial encrusting colonies with tight packing of the zooids. Here, a backbone of branching chains of distally budded zooids develops, with the sectors between the chains subsequently becoming infilled by zooids budded from distolateral loci (Figure 4.1C). The resulting colony may have a star-shaped outline as in the extant cheilostome *Electra pilosa* (Silén 1987, fig. 1). Much more common in both cheilostomes and cyclostomes are multiserial encrusting colonies in which tightly packed zooids are budded from a smooth, circumferential growing edge. This pattern in cheilostomes was termed **unitary multiserial** by Silén (1987).

In most of these cheilostomes the ancestrula produces one distal and two distolateral buds (e.g. Figures 3.13H, I, 3.15). Subsequent buds from the paired distolateral zooids and their progeny grow along the sides of the ancestrula, completely encircling it and meeting proximally of the ancestrula to produce the circumferential growing edge that characterizes mature colonies both of cheilostomes and stenolaemates (e.g. Figure 4.5). Circumferential growth is achieved when the

Figure 4.5 Small colonies of encrusting stenolaemate bryozoans with circumferential growing edges: (A) fixed-walled cyclostome *Mesonopora concatenata* in which calcified exterior frontal walls occupy the flat colony surface between the apertures (Jurassic, Bathonian, Bradford Clay; Bradford-on-Avon, Wiltshire, UK); (B) free-walled trepostome *Eostenopora globosa* with a domed colony surface and contiguous apertures (Devonian, Eifelian, Junkenberg Fm.; Römmersheim, Germany). Scale bars = 500 μm.

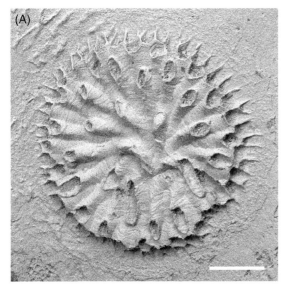

Figure 4.6 Cheilostome row bifurcation showing change in zooid width along rows. In the right-hand figure, one of the two sibling zooids following the bifurcation is a polymorph (avicularium).

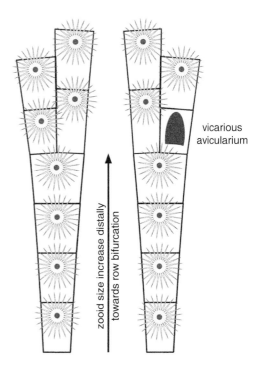

vicarious avicularium

zooid size increase distally towards row bifurcation

colony is just few weeks old in *Celleporella hyalina* (Cancino and Hughes 1988). The importance of having a circumferential growing edge for spatial competition is discussed below (Chapter 5.1).

Sheet-like encrusting cheilostomes sometimes exhibit a clearly visible arrangement of zooids in radial rows. The number of rows has to multiply as the colony grows and its circumference increases. This is achieved through **row bifurcation** in which a parent zooid produces two buds instead of one (Figure 4.6). Typically, the width of zooids increases distally along each row, culminating in a wide zooid followed by a bifurcation (e.g. Banta and Holden 1974). The two daughter zooids resulting from the bifurcation are narrow, slightly more than half of the width of their parent zooid. With the addition of further zooids, the

Figure 4.7 Spiral budding around the ancestrula (a) in the cheilostome *Biselenaria offa*; the small polymorphs are vibracula (Eocene, Lutetian, Bracklesham Group; Selsey, Sussex, UK). Scale bar = 200 μm.

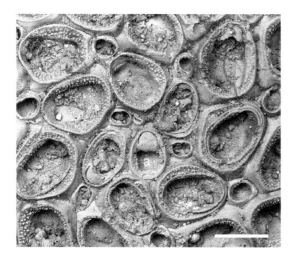

width of the new rows increase until they themselves bifurcate. The geometry of this common pattern of growth in multiserial cheilostomes introduces an inherent variability in zooid size: zooid width depends partly on the position of the zooid relative to a row bifurcation. Sometimes one of the two sibling zooids at a row bifurcation is a vicarious avicularium (Figure 4.6, right).

The overall direction of budding in sheet-like multiserial colonies is radially outwards from the ancestrula. Exceptions can be found in a few cheilostomes that have spiral budding. In *Biselenaria*, for example, the zooids are budded from the ends of two tightly wound spiral series (Figure 4.7). Autozooids at the outer ends of the spirals bud two daughters – an autozooid and a vibraculum – to extend the length of the spiral.

The genealogy of specific zooids in bryozoans can be inferred easily in some taxa, for example uniserial encrusters, but this may be less straightforward in multiserial species. In cyclostomes and other stenolaemates, zooids develop within a confluent budding zone, called the common bud by Borg (1926b) to reflect the shared parentage of the zooids in which polypides form before the cystids are complete. Parent–daughter relationships cannot be determined in these bryozoans but are more apparent in cheilostomes where the cystids are typically formed in advance of the polypides. However, genealogical complexity is introduced by the existence of **bud fusion** in many multiserial cheilostomes. Here, new zooids are formed by the coalescence of buds from two or more parental zooids.

Whereas encrusting colonies grow essentially in two dimensions, erect growth away from the surface of the substratum is three-dimensional and thus more complex. Hemispherical or massive, mound-like colonies characterize some stenolaemates, particularly trepostomes, and are formed by ontogenetic elongation of the tubular zooids such that the older ones at the centre of the colony are taller than the youngest zooids around the perimeter of the colony (Figure 4.5B). Alternatively, they can be produced by multiple layers of overgrowth. This growth pattern is also seen in cheilostomes with mound-like colonies, where successive generations of frontally budded zooids cover those beneath (e.g. Ristedt 1996; Figure 1.4B). Alveolar pores (= frontal septula) distributed around the edges of the zooids often form the loci of these frontal buds in ascophoran-grade cheilostomes (e.g. Cook 1985, pp. 40–43). Other methods for producing multilamellar colonies that occur in both cheilostome and cyclostome bryozoans include spiral patterns of overgrowth 'hinged' on pivot points (Taylor 1976; Figure 4.1D), and sporadic eruptive budding of a **pseudoancestrula** zooid (Figure 4.1E, F) onto the colony surface giving rise to a radial pattern of overgrowing zooids (e.g. Poluzzi and Sabelli 1985; Banta and Crosby 1994). In the cheilostome *Schizoporella*, the pseudoancestrula can develop from cell masses composed mainly of coelomocytes remaining after regression of older zooids that are ostensibly moribund but which in reality are dormant (Cummings 1975). Scholz and Hillmer (1995) distinguished two types of mound, or nodular, cheilostomes: 'S-nodules' comprising multilaminar sheets in which only the upper layer is alive; and 'C-nodules' formed by frontal budding and where the entire colony remains alive.

Fully erect colonies grow significantly away from the substrate surface as tree-like branching structures or plate-like masses, and usually have more restricted areas of attachment to the substrate. The most basic type of erect growth can be seen in the cheilostome *Flustra* where collisions between converging lobes at the margin of the encrusting colony base result in zooids growing back-to-back perpendicular to the colony base to form erect bifoliate fronds (Silén 1981). Alternatively, localized increases in the rates of zooidal budding and/or the rates of extension of tubular zooids can elevate parts of the colony above the level of the encrusting base. In the case of stenolaemates, enhanced growth is usually associated with the formation of endozones at the centres of erect branches (Plate 4D–E). Here, the zooids are orientated parallel to the branch axis and typically have thin walls, New zooids usually develop within the endozone exposed at distal branch tips. Zooidal budding in the centre of the endozone 'pushes' older zooids outwards towards the circumference of the branch. Their orientation thus changes from parallel to approximately perpendicular to the branch axis, linear extension rate declines, and their walls thicken into the cortical branch exozone. Continued lengthening of zooids in the exozone can be responsible for branch thickening. Thickening may require the budding of additional zooids in the exozone to keep pace with increasing branch circumference.

The exact locations and shapes of buds in stenolaemates employing this endo/exozonal mode of erect growth varies between taxa. New buds in stenolaemates either occupy the space between two or more pre-existing zooids, a pattern termed **interzooecial budding** (Plate 4I), or less commonly the space partitioned from a single zooid, which is known as **intrazooecial budding** (McKinney 1977a). Budding can be well-ordered and cyclical, or seemingly disordered. In some species, budding occurs only in the endo-

zonal axis of the branch, while in others it is concentrated at the boundary between endozone and exozone. Occasionally, the centre of the endozone is occupied by a distinct **axial bundle** of zooids that do not pass into the exozone (Figure 4.8A, B). An **axial cylinder** (Figure 4.8C) around which budding occurs is characteristic of some taxa, including the cryptostome *Rhabdomeson* (Wyse Jackson and Bancroft 1995) and the cyclostome *Entalophora* (Walter 1970). The exact loci of origin of new zooids within budding zones may rotate with growth, describing a helical trajectory in three dimensions. Sometimes this spiral budding pattern is reflected by an obvious spiral arrangement of zooids on the branch surface, as in the cyclostome *Spiropora* (Voigt and Flor 1970), but this is not always the case.

An unusual pattern of endozonal budding has been described from several trepostome genera (e.g. *Rhombotrypa*) in which the zooids are four- rather than six-sided in cross section (Boardman and McKinney 1976). Transverse sections of branches reveal a chequerboard-like pattern of quadrate zooids in the endozone. New buds develop in cycles at corners where the walls of four zooids meet and are also four-sided but oriented at 45° to the earlier zooids (Plate 4G).

All of the erect colonies described above have cylindrical branches of approximately circular cross section. Other erect colonies, both of stenolaemates and cheilostomes, have flattened branches varying from ovoidal to planar in cross section, and colonies that can be palmate or frondose. These colonies most often grow by zooidal budding from the two opposite sides of a **median lamina** (Figure 4.8D–F), as in ptilodictyine cryptostomes (e.g. Karklins 1983), and are termed **bifoliate**.

Branch bifurcation in ramose and palmate colonies is usually brought about by fission of the axial zone of budding at the tip of a branch. In ramose colonies with branches of circular cross-sectional shape, the branch and budding locus first become elliptical in cross-section and are then medially constricted before splitting to complete the bifurcation.

Figure 4.8 Axial structures visible in transverse sections of some cryptostomes: (A), (B) axial bundle in *Streblotrypa* (*Streblascopora*) *marmionensis* (Permian, Sakmarian, or Artinskian, Callytharra Fm.; Blair's Camp, Western Australia); (C) axial cylinder in *Pseudorhabdomeson kansasense* (Carboniferous, Moscovian, Horquilla Fm.; Cerros de Tule, Sonora, Mexico); (D) median budding lamina in *Astrovidictya sparsa* (Ordovician, Katian, Vormsi Stage; Vormsi Island, Estonia); (E), (F) median budding lamina in *Trigonodictya cyclostomoides*, the enlarged view showing the beading caused by the presence of styles (Ordovician, Katian, Vormsi Stage; Vormsi Island, Estonia). Images courtesy of Andrej Ernst. Scale bars: A, C = 500 μm; B, D, F = 200 μm; E = 1 mm.

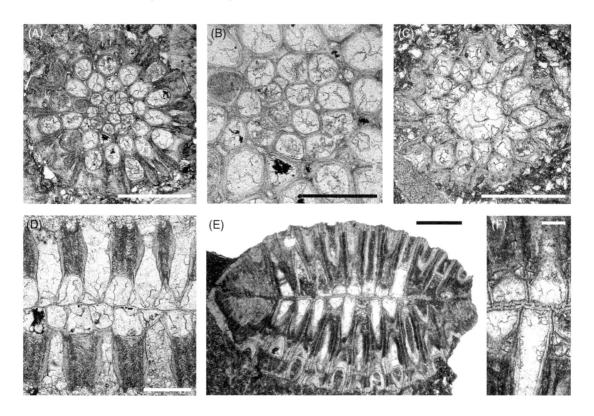

4.3 Functional Morphology of Colony-form

The functional morphology of bryozoan colony-form is important not only in the context of understanding why bryozoans are so diverse, but also in explaining the relationship between colony-form and environments, which is of particular interest when attempting to use the shapes of fossil bryozoan colonies to infer palaeoenvironments. Simple geometrical considerations provide insights into the functional significance of some bryozoan colony-forms, while observational and experimental evidence has been important in understanding function in a few cases. The main 'sculptors' of colony-form in sessile bryozoans appear to be: (i) alternative ways of competing for and utilizing substrate space; (ii) resistance to breakage in erect colonies; and (iii) generation of multizooidal currents for efficient suspension feeding. A broad division of bryozoans into encrusting, erect, and free-living is helpful when describing the functional morphology of colony-form.

4.3.1 Encrusting colonies

Encrusting bryozoans today, as well as in the geological past, utilize a wide variety of substrates, both abiogenic and biogenic. The former range in size from sand grains, through pebbles, cobbles, and boulders, to rocky platforms, submarine cliffs, and caves, while biogenic substrates mainly consist of calcareous shells (see Chapter 6.1).

Substrate space is typically a limiting resource for organisms encrusting hard (and firm) substrates, and competition between bryozoans and other sclerobionts can be intense (see Chapter 5.1). Bryozoans employ several colonial strategies for acquiring and retaining substrate space. Some of these are clearly reflected in colony-form, notably the contrast between colonies growing as uniserial runners and multiserial sheets.

Compared to the tightly packed zooids of multiserial sheets, runners have widely dispersed zooids leaving large areas of vacant substrate between their branches. New zooids are typically budded only at the distal tips of the branches, while the sides of the branches effectively consist of inert skeleton. The combination of non-encrusted substrate between branches where competitors can settle and grow, and the unprotected branch sides, makes runners particularly vulnerable to overgrowth. Indeed, observations show that runners are routinely overgrown, not just by sheet-like bryozoans but also by encrusters belonging to other phyla too. What then is the advantage of having a runner-like colony-form? Buss (1979) and Jackson (1979a) showed that runners are better able to exploit **spatial refuges** on the substrate surface where the probability of mortality from overgrowth and other processes is reduced. By dispersing their zooids over wide areas of the substrate surface, there is a greater chance that at least some of the zooids in the colony will escape overgrowth or other forms of mortality. Accordingly, large runner-like bryozoan colonies often comprise isolated groups of zooids that remain after

destruction of the intervening zooids. However, this **fugitive strategy** is evidently successful in circumventing complete mortality of the colony. Reparative growth from the broken ends of the remaining zooids is commonly observed in runners, resulting in a bipolar pair of zooids growing in opposite directions (Figure 4.1G). Restoration of physical connections between separated parts of the colony are sometimes seen in runners. For example, in the Cretaceous anascan cheilostome runner *Herpetopora*, regrowth from the broken ends of branches may occur towards open pores along the edges of other zooids (Taylor 1988a), and similar patterns of growth can be seen in the Recent ascophoran cheilostome *Hippothoa* (Figure 4.1H) and Ordovician colonies of the primitive cyclostome *Corynotrypa* (Figure 4.1I). The refuge-seeking ability of runners is enhanced by the fact that on average they have more elongate zooids than sheet-like species (McKinney and Jackson 1989).

Runners persist by spreading their zooids widely over the substrate to increase the chances that at least some will survive while 'tolerating' the loss of others. In contrast, sheets endeavour to occupy completely an area of the substrate and to defend it against competitors. Sheets usually have a growing edge that extends around the entire circumference of the colony, allowing both active defence and also offence against competitors for substrate space through growth. As described above, specific budding patterns during early astogeny are employed to establish this circumferential growing edge which is crucial to the success of the **confrontational strategy** used by sheets.

Spots are similar to sheets but, as the name suggests, have colonies that are very small in size. They show precocious sexual reproduction (Figure 4.2B), with ovicells for embryonic brooding sometimes appearing in the first generation of budded zooids in cheilostome spots. Early reproduction has been interpreted as a strategy for exploiting ephemeral substrates, or taking advantage of short-lived opportunities on more persistent substrates (Bishop 1989).

Multilayered growth is a characteristic of some encrusting bryozoans. By overgrowing their own old and moribund zooids, these colonies re-use the substrate space they had previously seques-tered. Organisms fouling the colony surface, including microbial mats (Ernst, Munnecke, and Oswald 2015), may be overgrown, while barnacles and serpulids can often be found deeply embedded within large multilayered colonies (e.g. Klicpera, Taylor, and Westphal 2013, fig. 2d).

The ascophoran cheilostome *Schizoporella errata* forms thick multilayered colonies in the Mediterranean Sea. A study of this species in the Ligurian Sea (Cocito et al. 2000) showed how cur-rent strength and biotic associates influenced the overall shape of the colonies. Colonies exposed to strong currents were found to be low and mound-like, whereas those in quieter conditions grew erect (Figure 4.9). Erect growth often occurred loosely around branching hydroids, producing irregular tubular upgrowths with flared distal ends. Similar shaped colonies, but devoid of the perished hydroids, occur in the Neogene fossil record (Moissette and Pouyet 1991). Hydroids are not the only stimulus for tubular colony-forms: growth around rhizomes of the seagrass *Thalassia* is believed to be responsible for the tubular *Schizoporella floridana* colonies found in the Pleistocene Miami Limestone of Florida (Hoffmeister, Stockmann, and Multer 1967), which is consistent

with the tubular colonies seen in many modern bryozoans growing on seagrasses (Plate 10F–H).

Enhanced strength resulting from multilay-ered growth may allow some colonies to continue growth beyond the confines of the substrate, pro-ducing free-lying domes, as in some Australasian species of the cheilostome *Celleporaria* (Taylor, Gordon, and Batson 2004; Schmidt 2007). The anachronistic bryozoan fauna of the Eocene Castle Hayne Limestone of North Carolina also contains free-lying colonies belonging to several cheilostome genera (McKinney and Taylor 2003). Externally similar hemispherical, free-lying colonies can be found among Palaeozoic stenolae-mates such as *Prasopora* (e.g. Lev, Key, and Lighthart 1993; Sanders, Geary, and Byers 2002), but here the dome-shaped colony is more often unilamellar, with thickening achieved through growth of the tubular zooids perpendicular to the colony surface rather than frontal budding, as noted earlier (p. 100).

One of the earliest known bryozoans – *Dianulites borealis* from the Tremadocian of Arctic Russia – has dome-shaped colonies that are interpreted to have lived on soft substrates experi-encing low rates of sedimentation and erosion (Ernst et al. 2014b). Some dome-shaped bryozoan

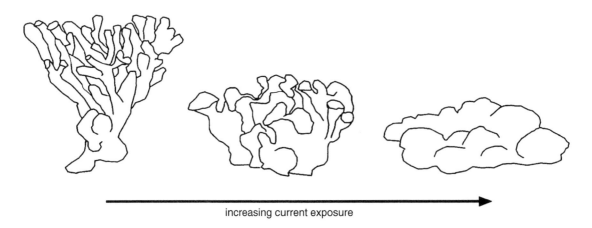

increasing current exposure

Figure 4.9 Variation in colony shape between sites experiencing different hydrodynamic exposure in the cheilostome *Schizoporella errata* from the Ligurian Sea (redrawn from Cocito et al. 2000).

colonies have conspicuously concave undersides (Plate 4B). These include Ordovician trepostomes from Sweden in which the concavities were interpreted by Spjeldnaes (2006) as resulting from growth during periods of sediment erosion. Spjeldnaes reasoned that erosion progressively lowered the sediment surface around the colony, prompting the colony to grow obliquely downwards. Conversely, he concluded that dome-shaped colonies with convex bases grew during periods of sediment deposition.

Several cyclostome genera, particularly of Cretaceous age, have massive colonies consisting of layers of stacked subcolonies (Flor and Hillmer 1970; Buge and Voigt 1972; Balson and Taylor 1982; Reguant 1993; Hara 2001; Plate 8H). New layers are formed by coordinated upward and outward growth of subcolonies that often entails budding by intrazooecial fission (e.g. Nye and Lemone 1978). Growth of subcolonies is essentially independent and each subcolony is presumed to be functionally and physiologically semi-autonomous, with the continuity of the colony being partly restored through regular fusion of subcolonies at their outer edges. Subcolonies have a polygonal shape when seen on the colony surface. In some of these cyclostomes, successive layers are closely juxtaposed, resulting in a compact colony structure. In others, new layers overarch underlying layers and the colony contains empty spaces. It is tempting to suggest that the layering of the colonies is seasonal, with each new layer forming in the spring after a period of winter dormancy. However, this has yet to be demonstrated for any of the massive cyclostomes composed of stacked subcolonies.

4.3.2 Erect colonies

Erect growth away from the surface of the substrate has several advantages. Importantly, it frees the colony from the spatial constraints of the substrate and the requirement to compete with other encrusters sharing the same substrate. Compared with two-dimensional encrusters, exploitation of three-dimensional space potentially allows a far greater colony size to be achieved, while the faster flow at higher levels above the substrate pays dividends in terms of planktonic food supply and amelioration from the threat of burial by accumulating sediment, at least for the distal parts of the colony. With respect to feeding, in an experimental study of the encrusting cheilostome *Membranipora membranacea*, colonies were made to grow on differently shaped substrates that mimicked erect sheet-like and erect tree-like colony-forms (Pratt 2005). Particle clearance rates, and hence feeding potential, were found to be greater in both erect forms than in the natural encrusting colonies. Set against these various advantages, erect colonies must cope with biomechanical demands that do not occur in encrusting colonies receiving structural support from the substrates on which they grow.

Mostly found in the order Cyclostomata, fungiform colonies (Figure 4.3A) have a relatively narrow stalk, often comprising kenozooids, attached to the substrate that supports an expanded head or capitulum containing the feeding zooids (Taylor and Grischenko 1999). Few competing encrusters seem able to grow up the stalk, while those growing around the base may have the incidental effect of bolstering the attachment of the fungiform colony to the substrate. The restricted size of the capitulum limits the number of zooids in the colony, although some fungiform cyclostomes develop radial branches and/or bud secondary mushroom-shaped subcolonies from the primary subcolony. In a few genera (e.g. *Stylodefranciopora*, see Voigt and Vávra 2006), stacked fungiform subcolonies are employed to construct a branching ramose colony.

The majority of erect bryozoans have large colonies growing significantly beyond the substrate to which they are attached at the base. In most species, erect growth originates from a single stem but some species develop multiple stems growing from a single encrusting base. As an erect colony grows, its profile becomes larger and branches extend further into faster flow regimes above the boundary layer. Accordingly, bending forces acting on the colony increase and are particularly concentrated at the base of the stem, increasing the likelihood of branch breakage and dislodgement of the colony from its substrate. Reinforcing the skeleton is one solution to this problem. This is achieved in various ways. In some cyclostomes with autozooids opening only on one side of the branch (e.g. *Erksonea*: Fig. 3.28G), kenozooidal overgrowths develop on the opposite side (Hinds 1975). Palaeostomates with long tubular zooids commonly have thick skeletal walls in the outer exozone (e.g. *Orthopora*, see Ernst and Nakrem 2015, fig. 13H), while fenestrates may thicken the extrazooidal skeleton on the obverse sides of the branches.

Intracolonial overgrowths are developed occasionally to routinely in bryozoans of most orders and all ages (e.g. Bigey 1981). In erect species these overgrowths capitalize on the supportive scaffolding provided by the erect framework of the colony and in turn contribute to the mechanical reinforcement of the branches. The Jurassic cyclostome *Terebellaria* is particularly notable in this respect (Figure 4.10). This genus has an unusual growth pattern in which a single spiral overgrowth, or a succession of annular overgrowths, originates at the distal growth tip and passes down the branch towards the colony base, each increment adding thickness and strength to the colony (Taylor 1978). Secondary attachments to the substrate develop in some erect bryozoans, augmenting the original attachment. In *Nevianipora* and a few other cyclostomes, these take the form of prop-like outgrowths on the undersides of the branches (e.g. Buge 1979, pl. 6, fig. 4). Similarly interpreted as functioning in secondary attachment (Tavener-Smith 1965) are spines at the proximal end of the basket-shaped colonies of the Carboniferous fenestrate bryozoan *Ptilofenestella*. Measuring up to 8.5 mm long, each spine can be almost as long as the rest of the colony.

Skeletal thickening in fenestrates can completely envelop older zooids close to the colony base. An extreme example of this is found in some genera with lyre-shaped colonies (McKinney 1977b). These have an arched reticulate frond with a bow-shaped thickening at the proximal end (Plate 7G). Colonies are inferred to have rested horizontally on the seafloor, the bow-shaped thickening pointing into the prevailing current, with the open end of the arch facing downstream. Not only did the weight of the thickening provide stability to the free-lying colony, but the orientation of the colony promoted water flow through the fenestrules from the convex upper side of the frond, into the cavity beneath the frond and out through the open end.

Archimedes is the most distinctive of all fossil bryozoans (Plate 7B), receiving its name from the resemblance between the thickened colony axes and the screw-shaped water pump invented by the Greek philosopher Archimedes. In terms of function this fenestrate genus used the screw axes as supports for a spiralling, typically fenestellid meshwork containing the feeding zooids. Each screw axis of *Archimedes* can exceed 50 cm in length, which would have allowed elevation of the lophophores to much higher levels above the seafloor for suspension feeding than in related fenestellids with more conventional morphologies. Both sinistral and dextral screws are found in species of *Archimedes* in equal numbers (Taylor and Sendino 2013); indeed, both chiral forms can occur in a single colony having more than one screw. The iconic colonial architecture of *Archimedes* also had implications for feeding currents (see below).

A characteristic feature of many fenestellids is the development of a skeletal superstructure above the main meshwork. This is best seen in *Hemitrypa* from the Carboniferous (Figure 4.11A). Nodes along the main meshwork support an overlying superstructure consisting of a hexagonal reticulum, with individual hexagons centred on the apertures of the autozooids in the main meshwork. While some of the more robust superstructures developed in fenestellids may have played a major role in mechanical strengthening (McKinney 1987), the delicate construction of the *Hemitrypa* superstructure suggests that strengthening the colony is not its main function (McKinney, Taylor, and Lidgard 2003). A similar superstructure occurs in the Cretaceous cheilostome *Ubaghsia* (Larwood, Voigt, and Scholz

Figure 4.10 The distinctive cyclostome *Terebellaria ramosissima* (Jurassic, Bathonian; Calvados, France) in which erect branches become thickened and strengthened by the development of a spiral overgrowth of zooids directed proximally towards the colony base: (A) external view with branch growth tip at top; (B) longitudinal thin section of a branch of similar size; (C) growing edge of an overgrowth; (D) transverse thin section showing the spiral overgrowth. Scale bars: A = 1 mm; C = 500 μm.

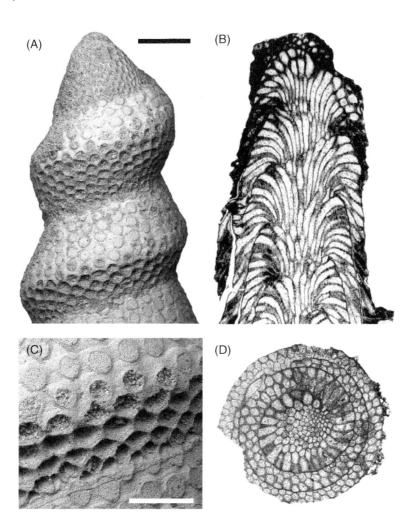

2008), which has encrusting colonies that receive structural support from the substrate on which they grow (Figure 4.11B). A more plausible functional hypothesis for the *Hemitrypa* and *Ubaghsia* superstructures is as a protective grill against predators and/or impacting objects. The first of these functions is also suggested by analogy with a living deep-sea bryozoan, *Mucropetraliella reticulata*, which has a superstructure incorporating avicularia that have a likely defensive function (Figuerola, Gordon, and Cristobo 2018).

Figure 4.11 Superstructures developed above the colony surface: (A) in the fenestrate *Hemitrypa proutana* each polygonal opening in the superstructure (formed by extrazooidal calcification) lies above a zooidal aperture which are arranged in two rows along the branches that are oriented from bottom to top in this view (Carboniferous, Warsaw Beds; Warsaw, Illinois, USA; (B) in the cheilostome *Ubaghsia* sp. the superstructure is formed by polymorphic zooids, concealing the costate frontal shields of the zooids, which are visible in the lower part of the image where parts of the superstructure are broken off (Cretaceous, Maastrichtian Chalk; Sidestrand, Norfolk, UK). Scale bars = 1 mm.

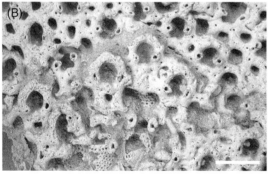

In an exceptional series of papers, Alan Cheetham, Erik Thomsen, and Lee-Ann Hayek (Cheetham and Thomsen 1981; Cheetham and Hayek 1983; Cheetham, Hayek, and Thomsen 1981; Cheetham 1986a) investigated the biomechanics of erect growth in cheilostomes with palmate (adeoniform) colonies. Limitations on the

growth of such colonies may be dependent on the ability of colonies to resist breakage from strong currents or impacts with objects in the water, as well as geometrical constraints to minimize crowding of zooids and interference between growing branches. Adeoniform cheilostomes begin erect growth from an attached base and develop through bifurcations of their palmate branches. Twisting of the planes of successive branch bifurcations results in a more or less three-dimensional and bushy colony. Both the spacing between bifurcations and the angles of bifurcation may change during growth. Rapid increase in the surface area of the colony with growth leads to a corresponding increase in the drag forces experienced by the colony in flowing water and consequently the probability of breakage, particularly close to the colony base where these forces are at their greatest. One solution to this problem is for branches to thicken at the base of the colony. This is seen in many ascophoran-grade cheilostomes in which calcification is added to the outer surface of the zooidal frontal shields, basal zooids often becoming totally enveloped in thick extrazooidal calcification and sealed (Figure 4.12). Branch thickening rate in palmate cheilostomes was found to increase through geological time, i.e. species have become better constructed to resist breakage. There is also a trend through geological time for branch bifurcation angles to increase towards the 90° value that maximizes branch spacing. As Cheetham (1971) noted, thickened frontal walls in cheilostomes first appeared in encrusting taxa where they served different functions (e.g. predator defence, see Chapter 5.2), later becoming co-opted for their role in strengthening rigidly erect colonies.

Bifoliate colonies with flattened branches or fronds vary widely in overall shape. Some are tree-like, as described above, but in others the edges of the branches are fused to give complex colonies comprising numerous box-like components (Plates 6P, 12A–C), a morphology characteristic of many frondose cheilostomes and one which presumably endows mechanical strength. In the peculiar Carboniferous cystoporate *Evactinopora*,

Figure 4.12 Extrazooidal calcification obscuring the zooids at base of a colony of the ascophoran cheilostome *Metrarabdotos moniliferum* (Pliocene, Coralline Crag Fm.; Suffolk, UK). Scale bar = 5 mm.

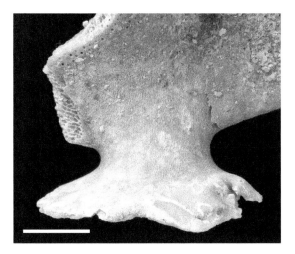

Figure 4.13 Variation in the shapes of colonies according to microhabitat in the cyclostome *Idmidronea atlantica* living in the Mediterranean near Marseille, France (redrawn from Harmelin 1973).

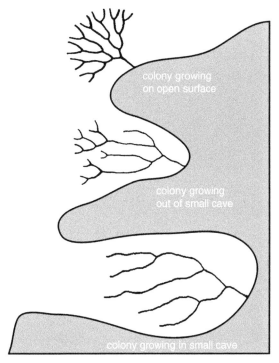

short bifoliate branches radiate from the centre of the colony in a star-like pattern (Plate 7H). Colonies of this genus having five branches are frequently mistaken for echinoderms. The functional morphology of *Evactinopora* awaits study.

Ecophenotypic variation can occur in erect colonies depending on environmental factors in different habitats. The cyclostome *Idmidronea atlantica* has planar colonies with delicate bifurcating branches. Colonies from a variety of habitats in the Mediterranean were collected by Harmelin (1973) who studied variation in their shape. Those growing in the open were found to have symmetrical colonies with short distances between branch bifurcations and thick branches, contrasting with the straggly, asymmetrical, irregularly bifurcating, and narrow-branched colonies growing in semi-obscure caves (Figure 4.13).

An alternative to resisting stresses is to flex and 'go-with-the-flow' as in flustriform cheilostomes such as *Flustra foliacea*. These have weakly mineralized skeletons which allow colonies to bend in the currents. While feeble mineralization of the skeleton in flustriform cheilostomes allows bending, it simultaneously removes the potential defensive role of the skeleton against certain kinds of predators. However, as Dyrynda (1985) has shown for *Flustra*, colonies produce alkaloids and terpenoids that are toxic to predators, suggesting a trade-off.

The elastic joints of articulated bryozoan colonies are an alternative way of providing flexibility (Figure 4.14; Plate 6E). Articulation has evolved convergently on several occasions in both stenolaemate and gymnolaemate bryozoans (e.g. Wyse Jackson, Ernst, and Suárez Andrés 2017; Taylor and Waeschenbach 2019). Sometimes the joints are formed by rings of cuticle, but in other instances they consist of cuticular strands

Figure 4.14 Articulated colonies of cyclostomes (A–C) and cheilostomes (D–F); joints are indicated with arrows, except in (F) where every zooid or pair of zooids is linked by a joint: (A) *Crisia fragosa* (Recent; Bacia de Campos, Rio de Janeiro State, Brazil); (B), (C) *Crisulipora occidentalis*, a cyclostome belonging to a clade that evolved articulation independently of *Crisia*, with detail of an articulation socket (C) (Recent; San Diego, California, USA); (D) lateral branch articulated with a parental branch in *Margaretta buski* (Recent; Pernambuco, Brazil); (E) *Adeona articulata* is a rare example of an articulated species in a genus where most species are not articulated (Recent; Sulawesi, Indonesia); (F) the catenicellid *Cribricellina cribraria* with multiple articulations (Recent; Wellington Harbour, New Zealand). Scale bars: A, F = 200 µm; B = 500 µm; C = 100 µm; D, E = 1 mm.

(Viskova 1991). In both cases, the articulations function as flexible nodes linking rigid internodes, as in the cheilostome *Cellaria* and the cyclostome *Crisia*. The joints of *Cellaria sinuosa* develop through breakage of the skeletons of zooids at points of branch ramification, and comprise bundles of chitinous tubes formed underneath the cryptocystal skeleton and which are unrelated to the outer cuticle (Schäfer, Bader, and Blaschek 2006). In crisiid cyclostomes, the joints comprise uncalcified rings of the otherwise mineralized skeleton (Borg 1926a), but their mode of formation has not been investigated. A family of Palaeozoic cryptostomes – Arthrostylidae – is characterized by articulated colonies (Blake 1979), and a single genus – *Arnaopora* – of seemingly articulated fenestrates has recently been described from the Devonian of NW Spain (Suárez Andrés and Wyse Jackson 2018).

Articulation attains a zenith in Catenicellidae, a Late Cretaceous–Recent family of cheilostomes that is particularly diverse and abundant in

Australasia. Catenicellids have articulations between each of the zooids forming the uniserial branches (Figure 4.14F; Plate 1B). Branch ramifications are preceded by asymmetrical pairs of skeletally conjoined zooids (geminate pairs), each of which forms the point of origin for a single daughter branch, and colony branching patterns follow strict rules according to species (e.g. Wass 1977). The dense bushy colonies of catenicellids have a distinctive shape resulting from their inwardly curled branches, which cause the zooids to open on the concave sides facing the centre of the colony. Colonies are able to resist strong wave and current action. Post-mortem decay of the cuticular nodes releases the individual zooids as sand-sized grains into the sediment.

In some cheilostomes the only articulations are located just above the encrusting base (e.g. *Adelascopora*, see Hayward and Winston 2011, fig. 27). Many ptilodictyine cryptostomes from the early Palaeozoic also had colonies with a single joint attaching the erect part of the colony to a basal holdfast: the holdfast contains a crater-like depression (Figure 4.15A) that accommodated the conical proximal end of the erect stem. The best example of this morphology is found in the

Silurian species *Ptilodictya lanceolata* (Plate 7F) in which colonies have a single scimitar-shaped branch. There are no exact morphofunctional analogues among Recent bryozoans. However, some species of the cheilostome genus *Melicerita* come close in having colonies comprising a scimitar-shaped branch (e.g. Brey et al. 1998, fig. 1; Smith and Lawton 2010, fig. 1; Key et al. 2018, fig. 2), although this branch is anchored into particulate sediments using rootlets (Bader and Schäfer 2004).

Often overlooked when considering the functional morphology of erect bryozoans is the role played by the substrate. Many erect (and encrusting) species grow attached to substrates which themselves are able move in the current, including fleshy macroalgae, seagrasses (Plate 10F–H), and a variety of animals with arborescent morphologies, such as sponges, hydroids, and the tubes of chaetopterid annelids. These substrates may be important not only in 'absorbing' the buffeting of waves and currents and allowing bryozoans to withstand forces greater than if they were fixed to a rigid substrate, but also in elevating colonies to higher tiers for suspension feeding. For example, some of the Silurian rugose corals described by Vinn and Toom (2016a) attached to

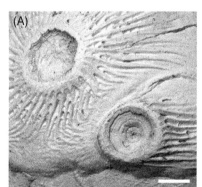 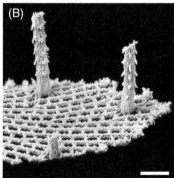 

Figure 4.15 (A) holdfast of unidentified ptilodictyine cryptostomes with two concave articulation facets (Silurian, Wenlock, Mulde Beds; Mulde Tegelbruk, Gotland, Sweden); (B) barbed spines (tips broken off) growing from the obverse surface of a fenestellid (Lower Carboniferous; County Fermanagh, Eire); (C) fenestellid (meshwork visible on left) with two stellate spinose processes which were originally considered to be hydrozoans and given the genus name *Palaeocoryne* (Lower Carboniferous; Halkyn, Clwyd, UK). Scale bars: A, B = 1 mm; C = 5 mm.

vertical stems of unknown organisms during life were encrusted by bryozoans that presumably received similar benefits from their elevated position as those inferred for the corals.

The poor fossilization potential of flexible substrates means that they are seldom preserved in the fossil record. However, they are sometimes moulded – bioimmured – by the bases of the colonies attached to them (e.g. Cheetham, Jackson, and Sanner 2001). It is not uncommon for bases of erect bryozoans found in the fossil record to comprise hollow cylinders that wrapped around a now-perished substrate (Plate 8L). The lack of identifiable substrates in many fossil bryozoan assemblages can also point to the former attachment of colonies to perished organic substrates (e.g. Cheetham 1975). Hara and Jasionowski (2012) noted the presence of cylindrical holes in small multilamellar colonies of *Celleporina* from the Miocene (Sarmatian) of Ukraine and suggested that these cheilostomes were attached to the cylindrical stems of chlorophyte algae during life. The cool-water carbonate shelf off southern Australia is rich in bryozoans, many of which are attached to living benthic animals with either no mineralized skeleton or, in the case of sponges, spicular skeletons that fall apart during fossilization (Hageman, James, and Bone 2000). These authors described criteria that enable recognition of this epizoic ecology in fossils, including bioimmured impressions of the surfaces of the hosts, unattached basal walls, and hollow ('pseudovinculariiform') colonies indicative of growth around perished substrates.

Cylindrical colony bases are also characteristic of a significant number of Late Palaeozoic fenestellids (e.g. McKinney and Burdick 2001), implying that fenestellid colonies often grew attached to the stems of arborescent organisms. By attaching to substrates that moved in the currents, fenestellids with relatively fragile meshworks may have been capable of colonizing high-energy regimes, an ecology that would not be predictable from a simple mechanical analysis of their skeletons. Relevant to this arboreal ecology is the presence in many fenestellids of solid spinose outgrowths

of the skeleton (Bancroft 1988). These range from unbranched spines (Plate 6K), some covered in backward-pointing barbs (Elias and Condra 1957, pl. 17, fig. 10; Figure 4.15B), to stellate clusters of spines once believed to be hydrozoans (*Palaeocoryne*) that attached to the surface of the bryozoan colony (Figure 4.15C). While such spines in fenestellids may have played a role in defence, it is more likely that their primary function was as grappling structures to facilitate stabilization of colonies growing among arborescent animals and plants (Taylor 1999).

Elevation above the sediment surface was achieved in some bryozoans by attachment to the stems of crinoids. The Carboniferous cystoporate *Meekoporella balladoolensis*, once mistaken for a coral, has spherical compound colonies consisting of conjoined pyramidal subcolonies (Wyse Jackson, Taylor, and Tilsley 1999). While most fossils of this species are broken and the substrate of attachment cannot be determined, some can be seen to have grown around the circumference of crinoid stems, which likely held the colonies aloft (Figure 4.16).

4.3.3 Free-living colonies

Bryozoans have evolved various kinds of free-living colonies not anchored in one place by a substrate of attachment. The most common are the **lunulitiform** cheilostomes (Plate 6L–M). These have small, cap-shaped colonies, approximately circular in plan view and with autozooids and vibracula opening only on the upper, convex side of the colony. Setae of the vibracula extend downwards from the outer edge of the colony to support the colony slightly above the sediment surface (Cook 1963). At the centre of the colony is a tiny substrate such as a shell fragment or sand grain, preferentially selected by the larva on which to settle (e.g., Driscoll, Gibson, and Mitchell 1971). This substrate is often completely enveloped by the calcified skeleton of the bryozoan. Alternatively, in some colonies a substrate may be entirely lacking when the colony is a product of regrowth from a fragment of a pre-existing colony, or was budded at the margin of a parent colony

Figure 4.16 Reconstruction of a colony of the Carboniferous cystoporate *Meekoporella balladoolensis* from the Isle of Man growing *syn-vivo* around a crinoid stem. The bryozoan has a compound colony consisting of contiguous pyramid-shaped subcolonies (redrawn from Wyse Jackson, Taylor, and Tilsley 1999, fig. 10).

Figure 4.17 Two living colonies of the lunulitiform cheilostome *Selenaria maculata* photographed in a laboratory aquarium, the colony on the left beginning to climb onto that on the right using the hair-like setae of the vibracula to locomote. The colonies are about 10–13 mm in diameter (see Cook and Chimonides 1978).

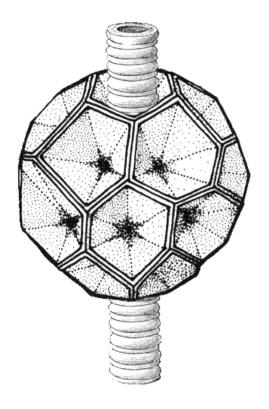

and then broke free (see below). In addition, living species of the Australian genus *Lunularia* lack a substrate, the larvae settling on the undersides of adult colonies to which the young colonies initially adhere.

Freedom from an anchoring substrate endows lunulitiform colonies with the potential for automobility (Figure 4.17) through coordinated movements of their setae (Chapter 3.11). This was first observed in living colonies of *Selenaria* by Cook and Chimonides (1978) and subsequently in the convergent cupuladrians by O'Dea (2009). As remarked by O'Dea (2009), automobility can have

only minimal benefit for dispersal or the avoidance of predators and instead may function to unbury colonies engulfed by sediment.

A secondary advantage of mobility may have evolved in some mobile *Selenaria* colonies which have male zooids with two very long tentacles, from the tips of which sperm are released by analogy with other bryozoans. Protruded for just a few seconds at a time, these tentacles are equal in length to the entire radius of the colony. Chimonides and Cook (1981) noted the frequent correlation between such male zooids and non-stationary colonies, as in *Hippopodinella* and *Hippoporidra* that encrust gastropod shells occupied by hermit crabs. Could the long tentacles of the male zooids introduce sperm into the close vicinity of other colonies when these are nearby? This idea is currently speculative (p. 65) but warrants investigation.

Colonies of very similar shape, and which are also often classified as 'lunulitiform', occur in other cheilostome genera (e.g. *Anoteropora*: see Cook and Chimonides 1994b) but lack vibracula with setae and instead are anchored into the sediment using rootlets. These species are capable of inhabiting deeper water settings than mobile lunulitiform cheilostomes which are typical of the shallow shelf.

Some non-lunulitiform bryozoans without rootlets or vibracula are also free-living or, more strictly, **free-lying**. Free-lying, dome-shaped colonies typically appreciably larger and more robust than lunulitiform colonies occur among several Cenozoic–Recent cheilostome genera (e.g. McKinney and Taylor 2003; Plate 6D). Cook (1965) described Recent examples from muddy and sandy sea bottoms off northwest Australia. These colonies grow beyond their original shell substrates, sometimes picking up and incorporating 'secondary substrates' into their basal surfaces from the sediment surface.

The peculiar Arctic ctenostome *Alcyonidium disciforme* has, as the name implies, disc-shaped colonies, each with a central hole (Kuklinski and Porter 2004). Inhabiting muddy seabeds, colonies can tolerate high rates of sedimentation and grow up to 47 mm in diameter. They are able to sequester sand particles, possibly to act as ballast to help colonies to regain the correct orientation after they have been overturned by currents.

The Ordovician trepostome *Dianulites fastigiatus* is characterized by straight to slightly curved cone-shaped colonies (Figure 4.3B). The sides of the cone comprise exterior wall, while the broad top of the cone was the active growing surface of the colony bearing the polygonal apertures of the feeding and other zooids. Colonies are inferred to have lived partly embedded in muddy sediments, much like the corals and rudists they resemble (Taylor and Wilson 1999a). Cone-shaped subcolonies developed on the upper surfaces of some parent colonies and were formed by renewed growth after temporary swamping by sediment. Examples of specimens comprising stacks of subcolonies are likely to have been the result of multiple burials and regenerations.

Bryoliths (i.e. 'bryozoan stones') are spheroidal colonies that totally enclose their substrates, indicating growth on all surfaces brought about by rolling on the seabed (Balson and Taylor 1982; Thiel, Cuffey, and Kowalczyk 1996; Kidwell and Gyllenhaal 1998; James, Foster, and O'Sullivan 2006). Little is known about the ecology of bryoliths. They resemble rhodoliths formed by calcareous algae (e.g. Amado-Filho et al. 2012), ostreoliths formed by oysters (Wilson, Ozanne, and Palmer 1998), and balanuliths by acorn barnacles (Nielsen 2009).

James, Foster, and O'Sullivan (2006) described bryoliths from the Gulf of California occurring in high densities in three habitats – rhodolith beds, sand, and cobble bottoms – all in shallow water (2–14 metres). Most of the bryoliths consisted of the branching cyclostome *Diaperoforma*, but others comprising colonies of the cheilostome *Schizomavella* were found in one rhodolith bed. The *Diaperoforma* bryoliths were up to 57 mm in maximum diameter, those from deeper water being more variable in size and shape than examples from shallower water. Further to the north along the Pacific coast of North America, a site in San Francisco Bay contains abundant colonies of the cheilostome *Schizoporella errata* showing two main colony-forms: bryoliths and tubes (Zabin et al. 2010). Molecular analysis failed to reveal any differences between the two colony-forms, showing that, at least in this example, bryolith formation reflects ecophenotypic plasticity rather than being a genetically controlled trait.

Most described bryoliths come from the Neogene, although geologically older examples have been recorded in the Ordovician (Thiel, Cuffey, and Kowalczyk 1996) and Devonian (Denayer 2018). Neogene bryoliths formed by the cheilostome *Calpensia* are particularly striking (Plate 8F–G). These colonies reach at least 8 cm in diameter and it is clear that they must have been overturned regularly during their growth. Sectioned specimens often reveal a bivalve shell substrate at the centre, surrounded by up to 100 layers of bryozoan zooids.

The Late Cretaceous cheilostome *Volviflustrellaria taverensis* forms small bryoliths that are spindle or ball-shaped, about 1 cm in width, and reminiscent of fusuline foraminifera (Figure 4.18A, B). Colony growth occurred from a single growing edge which extends from one pole of the spindle

Figure 4.18 Small free-living colonies of the cheilostome *Volviflustrellaria taverensis* are inferred to have rolled on the chalky seafloor (Cretaceous, Campanian Chalk; Norwich, Norfolk, UK): (A) almost spherical colony with the growing edge at the bottom; (B) larger, spindle-shaped colony; (C) detail showing elongate vibracula on either side of a row of autozooids, of which the one in the middle has an immersed ovicell. Scale bars: A, B = 1 mm; C = 500 μm.

to the other. Vibracula (Figure 4.18C) may have supported the colony above the sediment surface rather like a lunulitiform bryozoan.

But what causes bryoliths to roll? Water flow is one possibility, either through ambient currents or during storms, but periodic overturning by animals such as crabs and fishes offers a plausible alternative. A combination of currents and bioturbation were believed to be responsible for overturning of the *Diaperoforma* bryoliths mentioned above; at one site in the Gulf of California, echinoids and crabs were observed disturbing bryoliths. In scleractinian corals of similar spheroidal morphology (coralliths), as well as some *Schizoporella* bryoliths living off southern Brazil, Capel et al. (2012) showed that bioturbation caused by sand dollars was capable of displacing colonies. In the case of the tiny colonies

Figure 4.19 Conescharelliniform cheilostome colonies figured upside down in order that the zooids, which have the reverse polarity to the colony as a whole, are oriented with their distal ends at the top; during life, a rootlet would have emerged from the narrower ends of the colonies to anchor them into the sediment: (A) *Lacrimula kilwaensis* (Oligocene, Rupelian; Pande Peninsula, Kilwa District, Tanzania); (B) *Conescharellina* cf. *angustata* (Pleistocene; Java, Indonesia). Scale bars = 500 μm.

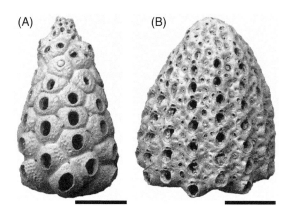

of *Volviflustrellaria taverensis* living on the Chalk seabed, continuous budding of zooids from the solitary growing edge may have shifted the balance of the colony and caused it to rotate during growth.

Finally, various cheilostomes have small, rooted colonies termed **conescharelliniform** (Figure 4.19), which includes the 'orbituliporiform' category of Cook and Lagaaij (1976). These typically inhabit muddy or sandy sea bottoms (Cook 1966), often in deeper water settings where other bryozoans are rare to lacking. Colonies are small and vary from spheroidal, conical, stellate to elongate and flattened in shape (Bock and Cook 2004b). All have rhizooids extending from the colony base into the sediment. Budding is typically reversed such that the distal ends of the zooids are closer to the proximal, rootlet-bearing end of the colony than they are to the distal, growing end of the colony (very few bryozoan genera have zooidal polarities opposite to that of the colony but the encrusting ascophoran cheilostome *Fatkullina* furnishes another exam-

ple: Grischenko, Gordon, and Taylor 1998). Conescharelliniform colonies can occur in great numbers in some environments and may be recovered during sampling for microfossils. For example, a core taken across the Eocene–Oligocene boundary in Tanzania contained abundant colonies belonging to three species of conescharelliniform cheilostomes along with one lunulitiform species (Di Martino et al. 2017). Little is known about the ecology of conescharelliniform cheilostomes apart from the outstanding study of Hirose (2011) who kept a living colony of *Conescharellina catela* alive in the laboratory for two months. He discovered that the rootlets anchoring the colony were also able to raise it above the substrate – to a height nearly five times that of the colony itself – and could be used by the colony to right itself, even though they appeared to contain no muscles.

4.4 Colony Propagation in Lunulitiform Bryozoans

Lunulitiform bryozoans are capable of forming new colonies sexually as well as asexually. They employ a variety of modes of asexual propagation (O'Dea et al. 2008). Like other bryozoans, sexual colonies result from the settlement of a sexually produced larva released by a parent colony. These colonies have an ancestrula at the centre and grow symmetrically outwards to achieve an almost perfectly circular outline shape (Figure 4.20A, B). However, in some populations, asexual colonies produced by **mechanical fragmentation** outnumber sexual colonies. Breakage of colonies can occur by strong currents or more probably through the activities of predatory and grazing animals as well as biological disturbance, factors generally correlating with higher levels of primary productivity (O'Dea et al. 2004; O'Dea and Jackson 2009). Zooidal budding can resume from the perimeters of the typically triangular wedge-shaped fragments (Figure 4.20C, D), the rate of regenerative growth being coordinated so that a colony of almost circular outline is formed. Colonies that

Figure 4.20 Comparison of sexual (larvally recruited) and asexual (clonal) colonies of the free-living bryozoan *Cupuladria biporosa* (Recent; Caribbean): (A) upper surface of a sexual colony; (B) underside of the same sexual colony; (C) more irregular upper surface of a larger asexual colony; (D) underside of the same asexual colony showing the triangular outline of the fragment from which the colony developed. Scale bars = 1 mm.

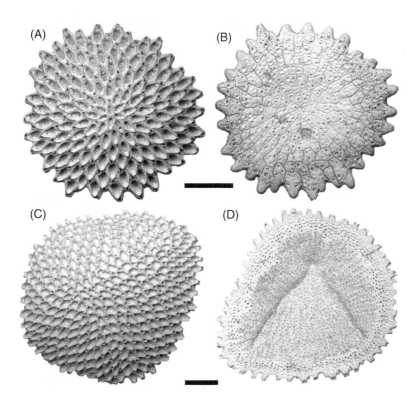

have developed in this way can be recognized in fossils and preserved recent material by the absence of an ancestrula and also from the preserved outline of the original fragment which is usually most obvious on the underside.

Autofragmentation is a second mode of clonal propagation observed in some species of *Cupuladria*. As colonies grow, lateral connections between some rows of zooids fail to develop, resulting in ever-deepening notches around the perimeter of the colony. These notches facilitate fragmentation. Survivorship of new colonies produced by autofragmentation is greater than those produced by simple mechanical fragmentation in the same

species (O'Dea 2006), underlining the value of this mode of propagation. Some lunulitiforms produce fan-shaped subcolonies around the periphery of the colony which are weakly attached and readily break off to form new colonies. This mode of asexual reproduction has been termed **colonial budding**. Subsequent growth results in a relatively symmetrical colony resembling a sexually formed colony. A fourth mode of asexual propagation – **peripheral fragmentation** – is known from a single Pliocene species of *Discoporella* (Figure 4.21). Massive skeletal thickening is developed on the undersides of the colonies near their centres, whereas the perimeter of the colony is poorly calcified and prone to

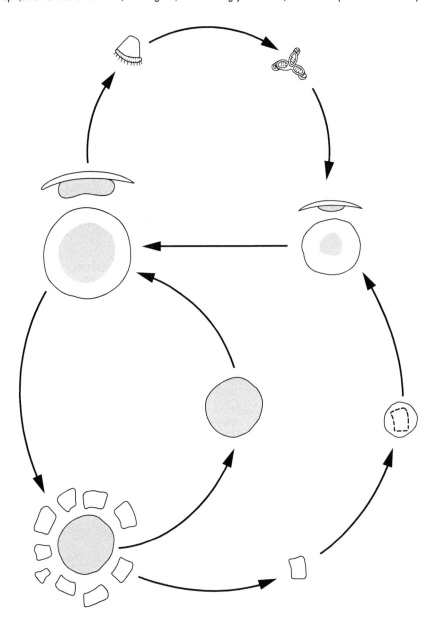

Figure 4.21 Cycle of clonal propagation by peripheral fragmentation in the free-living Upper Pliocene cheilostome *Discoporella* sp. (after O'Dea et al. 2008, text-fig. 11). The strongly calcified, thickened part of the colony is in grey.

being broken off as rectangular fragments from which new colonies can regenerate.

4.5 Multizooidal Feeding Morphologies

Many of the colony-level features observable among bryozoans are related to variations in flow patterns employed by the colony for suspension feeding. The synergistic effect of the feeding currents created by individual lophophores produces two main types of multizooidal flow patterns: (i) chimney systems in which water flows towards the colony surface, is filtered by the zooids, passed to excurrent chimneys and vented at 180° to the incurrents; and (ii) unidirectional systems in which water flows towards the colony, and after filtration passes through gaps to be vented in the same direction as the incurrents. Chimney systems occur in both encrusting and erect bryozoans whereas unidirectional systems are found only in erect bryozoans.

4.5.1 Chimneys

A landmark paper in bryozoan functional morphology was published by Banta, McKinney, and Zimmer in 1974. This provided the first convincing explanation of the function of regularly spaced hummocks – **monticules** – that cover the colony surfaces of many stenolaemate bryozoans, especially Palaeozoic trepostomes, and which were previously enigmatic in terms of function. Observations of living colonies of the cheilostome *Membranipora* by Banta, McKinney, and Zimmer (1974) revealed regularly spaced chimneys of exhalent flow between areas occupied by the lophophores of autozooids pumping water towards the colony. No lophophores were present at the centres of these chimneys and the lophophores of zooids immediately surrounding chimneys leant outwards. The distance between chimneys in *Membranipora* was shown to be very similar to monticule spacing, prompting the inference that monticules functioned as chimneys because the lophophores of autozooids on their flanks would have leant outwards as in *Membranipora*. Subsequent observations (e.g. Cook 1977a; Cook and Chimonides 1980; Winston 1978; Lidgard 1981; Taylor 1991; Grünbaum 1997; Shunatova and Ostrovsky 2002; Von Dassow 2005) have confirmed the role of monticules as loci of exhalent flow (Figure 4.22). Autozooids with obliquely truncate lophophores often surround monticules. The greater length of the tentacles on the side of the lophophore closest to these excurrent loci contributes to the channelling of exhalent currents outwards from the chimney. Modelling has demonstrated the significant feeding advantage

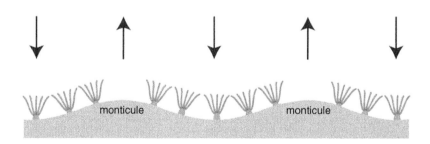

Figure 4.22 Schematic vertical section through a bryozoan with monticules. The outward-leaning lophophores on the flanks of the monticules, plus the lack of lophophores on the summits of the monticules, creates a multizooidal feeding current pattern in which incurrents are focused on intermontical areas with excurrent chimneys over the monticules themselves.

gained by bryozoans with tightly packed lopho-phores and chimneys (Eckman and Okamura 1998); notably, chimneys vent plankton-depleted water far enough away from the colony that it is less likely to be re-filtered by other zooids.

It has become clear that the old paradigm of each feeding zooid acting autonomously with respect to the generation of feeding currents is flawed. Instead, **multizooidal current systems** are very common among bryozoans, some colony-wide but most organized at the subcolony level. For instance, small mound-like cyclostome colonies may function like a single monticule, with outward-leaning zooids around the edges pumping water centripetally (Cook 1977a). After filtering by the individual zooids, exhalent water in these colonies is then passed centripetally, sometimes along specific channels between rows of autozooids, where it is expelled from the central region devoid of feeding zooids.

Monticules are usually almost circular in outline shape (Figure 4.23). However, they are elliptical with parallel long axes in colonies of the trepostome *Spatiopora* that encrusted orthoconic nautiloid shells (Plate 10A). The monticles are elongated parallel to the axis of the nautiloid, suggesting a response to an ambient flow regime in the same direction which may indicate either attachment to a swimming host (Wyse Jackson, Key, and Coakley 2014), or less likely current orientation of the shells of dead nautiloids.

It is important to note that not all monticules in living bryozoans are sites of exhalent chimneys: Shunatova and Ostrovsky (2002) observed cheilostomes from the White Sea in which incurrents were focused on the monticules and excurrents departed from intermonticular regions, the opposite of the patterns described above. In *Schizomavella lineata*, for example, monticules alternate through time as sites of inhalent and exhalent flow, depending on patterns of polypide degeneration and regeneration. When zooids occupying intermonticular spaces lack polypides but those on monticules possess polypides, directionality is reversed from that expected, with the monticules forming loci of inhalent flow.

Figure 4.23 Monticule formed of non-feeding kenozooids surrounded by radial rows of autozooids in the cyclostome bryozoan *Bimulticavea* cf. *mutabilis* (Cretaceous, Campanian, Biron Fm.; Talmont, Charente Maritime, France). The centre of the monticule can be inferred to have been the site of a chimney of exhalent current flow, venting water filtered by the autozooids that would have flowed towards it centripetally along the valleys between the autozooidal rows. Scale bar = 1 mm.

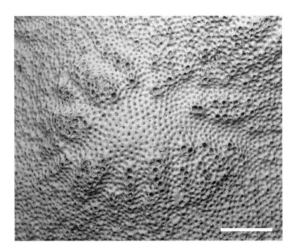

Monticules are not the only structures preserved in bryozoan skeletons that are indicative of multizooidal feeding current systems. By mapping the distributions of autozooidal apertures in Recent and fossil bryozoans it is possible to predict multizooidal flow patterns: concentrations of apertures indicate regions of inhalent flow, whereas regions of the colony devoid of apertures – and consequently lacking lophophores – probably represent sites of exhalent flow (Taylor 1979a). Thus, in cyclostomes with apertures aggregated into fascicles (Plate 8E), these concentrations can be inferred to be areas of inhalent flow, with exhalent currents departing from regions between. Likewise, flat **maculae** containing few or greatly reduced numbers of apertures of feeding zooids are regularly distributed across the surfaces of many Palaeozoic stenolaemates and can be inferred as sites of exhalent flow (Plate 10C).

The striking subcolonial organization of some fossil stenolaemates is at least in part related to the development of subcolony-based feeding current systems. This may include the Ordovician trepostomes *Eichwaldopora* (see Koromyslova 2011) and *Hexaporites* (see Pushkin and Popov 2005), the Devonian cystoporate *Botryllopora* (Ernst, Taylor, and Bohatý 2014), the Permian cystoporate *Hexagonella* (Plate 10D), the Cretaceous cyclostome *Cellulipora* (Buge and Voigt 1972), and the Neogene cyclostomes *Blumenbachium* (Balson and Taylor 1982) and *Centronea* (Hageman, McKinney, and Chandler 2008). In the Devonian cystoporate *Glyptopora* the elongate maculae developed where the bifoliate branches forming the box-like colony are joined have been reconstructed as loci of exhalent current flow by Ernst and Vachard (2017a, fig. 5). Maculae can also be stellate, as in the cystoporate *Constellaria* in which autozooidal apertures are clustered into radial series, with vesicular skeleton between the rays and at the centres of each subcolony (Plate 10B).

Using serial acetate peels, Key, Thrane, and Collins (2002) studied the development of maculae in a ramose trepostome from the Permian of Greenland. The spacing and size of the maculae increased as the branch grew in diameter and the shape of the maculae became more stellate. Intercalation of new maculae apparently brought an end to these trends. The authors interpreted this as an indication of the dynamic nature of excurrent chimneys. In the same species, Key, Wyse Jackson, and Vitiella (2011) described how exhalent water was probably channelled in branching systems towards the centres of the maculae, comparing these to stream networks emptying into closed basins with centripetal drainage.

Anstey (1987, 1989) inferred flow patterns in some Palaeozoic stenolaemates and suggested that some monticules were loci of incurrents (cf. Shunatova and Ostrovsky 2002) while others were bypassed by currents. The method he used depended on the premises that (i) cystiphragms (blister-like structures on the interior walls of the zooids; see Chapter 3.13) constrained expanded zooids; see Chapter 3.13) constrained expanded lophophores to lean towards the side of the zooid with the cystiphragm; and (ii) lunaria (hood-like structures extending over one side of the skeletal aperture; see Chapter 3.13) forced expanded lophophores to lean away from the side of the zooid with the lunarium (as was also presumed by Patzkowsky 1987). With the polarity of flow so established, Anstey was able to construct hypothetical flow lines from one aperture to the next and produce maps of inferred multizooidal feeding currents.

The orientations of the distal parts of zooidal skeletons, including tubular peristomes surrounding the apertures of some stenolaemates and cheilostome orifices, provide additional insights into multizooidal feeding currents. Peristomes direct the orientation of lophophores parallel to their axes, allowing inference of the focus of inhalent currents. Here the fundamental difference between the construction of stenolaemate and cheilostome zooids is important. The exposed part of the tentacle sheath supporting the lophophore in stenolaemate zooids is at best very short (McKinney 1988; Plate 5C), meaning that the level of the mouth seldom extends beyond the edge of the skeletal aperture and the orientation of the lophophore is determined by the enclosing skeleton. In contrast, cheilostomes generally have long exposed tentacle sheaths (Plate 5D, F) which can bend in various orientations independently of the skeleton (e.g. Cook and Chimonides 1980). The consequence of this difference is that evidence for multizooidal current patterns is often expressed in the skeletons of stenolaemates, whereas such patterns are not always reflected in cheilostome skeletons. Like *Membranipora*, many species of living cheilostomes develop regularly spaced chimneys across the colony surface but these are not always manifested in the skeleton and therefore cannot be recognized in fossils.

Multizooidal feeding current systems may be lacking only in bryozoans with widely dispersed zooids, where gaps between lophophores provide ample space for the exhalent currents of individual zooids to depart the colony surface. In narrow branched erect colonies, monticules and other maculae are lacking but as soon as branch width increases to more than about 5 mm, maculae begin to be present (McKinney 1986b).

The edges of the flattened branches of palmate colonies can serve as outlets for excurrents, as in the cheilostome '*Hippodiplosia' insculpta* (Nielsen 1981). This function is sometimes reflected in the presence of non-feeding polymorphs (kenozooids and avicularia) along branch edges. A variant of this pattern can be inferred for bryozoans with cribrate colonies (Plate 8C–D). Here the bifoliate branches divide and anastomose, leaving a planar colony pierced by lacunae. The lacunae are spaced roughly the same distance apart as are monticules and other maculae, occupying an average of 12% of the colony surface (Taylor 2012). While it might at first seem that filtered water would flow through the lacunae (cf. the fenestrate colonies described below), this is precluded by the presence of active zooids on both sides of the colony. Instead, it can be inferred that lacunae function like monticules or other maculae as sites of chimneys, venting filtered water away from the colony surface on each side of the colony (Figure 4.24).

4.5.2 Unidirectional systems

The multizooidal feeding currents described above all involve water being drawn towards the colony surface and, after filtering, turned through approximately 180° and expelled from the colony in the same direction as it came. Such systems run the risk of interference between inhalent and exhalent currents flowing in opposite directions. Unidirectional systems are more hydrodynamically efficient and are employed widely by both passive and active suspension feeding animals belonging to diverse phyla. Examples include the parabolic filtration fans of the passive suspension feeding crinoids, and the active suspension feeding sponges that pump water through their bodies.

Not surprisingly, efficient unidirectional systems have evolved convergently in many groups of bryozoans having erect colonies (Chapter 9.11). Such systems were first inferred for extinct fenestellids on purely functional morphological grounds by Cowen and Rider (1972), and shortly afterwards confirmed from observations of living phidoloporid cheilostomes with a similar colony-form (Cook 1977a). Both fenestellids and phidoloporids have erect, 'fenestrate' colonies with relatively narrow branches arranged in a plane and autozooids opening only on one side (Chapter 4.1). The

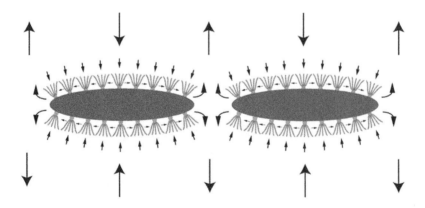

Figure 4.24 Schematic transverse section through two branches of a cribrate bryozoan. Large arrows show the prevailing multizooidal feeding currents, small arrows those created by individual lophophores. Excurrents can be inferred to have exited from the colony above gaps between branches.

Figure 4.25 Schematic transverse section through three branches of a fenestrate bryozoan. Large arrows show the prevailing multizooidal feeding currents, small arrows those created by individual lophophores. Currents pass through the fenestrules from the side of the colony bearing lophophores to the reverse side.

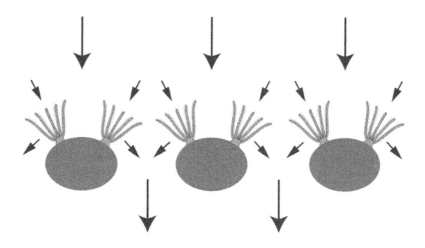

lophophores of the autozooids pull water towards the branch surface, filter food particles and pass exhalent water through the fenestrules towards the branch reverse side and away from the colony (Figure 4.25).

The overall shape of colonies employing this kind of unidirectional flow system varies (McKinney 1981b). Some have complex convoluted colonies but in others the colony takes the form of either a simple fan or an inverted cone. Colonies with a conical morphology can have the autozooids opening on the inside of the cone producing centrifugal flow through the meshwork, or more commonly, on the outside of the cone so that the flow system resembles that seen in many conical or cup-shaped sponges, with a centripetal flow of water through the meshwork into the central cavity and away through the open end of the cone. A good example of the latter are the small basket-shaped colonies of *Ptilofenestella carrickensis* described from silicified residues of the Early Carboniferous limestone at Carrick Lough, Northern Ireland (Tavener-Smith 1965).

Suárez Andrés and Wyse Jackson (2015b) reviewed thoroughly the feeding current patterns inferred for Palaeozoic bryozoans of the order Fenestrata, building on the pioneering research of F.K. McKinney. Bryozoans belonging to this order all created unidirectional currents, the combined action of the lophophores driving water from the frontal (obverse) towards the dorsal (reverse) sides of the branches. However, there are several variations on this universal pattern. The reticulate meshwork in *Archimedes* (Plate 7B) winds around the screw axis in a helicospiral. As zooids in this genus open on the upward-facing side of the meshwork, overall water flow would have been in a proximal direction towards the base of the colony. This venting of filtered water towards lower whorls of the meshwork would have put the zooids located here at a disadvantage. It is also probable that a 'dead zone' of reduced flow would have existed close to the screw axis itself, coinciding with the region where zooidal apertures are closed by calcification. The same flow pattern has been observed in spiral colonies of the living cheilostomes *Bugulina turrita* and *B. turbinata* in which a proximally expanding stagnant cone develops in colony interiors (McKinney, Listokin, and Phifer 1986).

Although far less common than *Archimedes*, another fenestrate genus, *Ikelarchimedes*, has zooids opening on the downward-facing side of the helicospiral meshwork, allowing the inference of a reverse flow direction distally towards the top of the colony (Suárez Andrés and Wyse Jackson 2015a, fig. 1G). Such a pattern was also deduced for the Eocene cyclostome *Crisidmonea archimediformis* which has bifurcating branches wound around an *Archimedes*-like screw axis (Taylor and McKinney 1996).

Extending a simulation model developed by McKinney and Raup (1982) for helical bryozoan growth, McGhee and McKinney (2000) found the morphospace occupied by the erect helical colonies of *Archimedes* to be constrained from the theoretical maximum in order that successive whorls were not so deeply nested within one another to hamper efficient suspension feeding. A subsequent study incorporated data from other bryozoans with erect helical growth forms (McKinney and McGhee 2003). This research found the branches to be held at an average of 50–60° to the colony axis, which was inferred to be the functional optimum for current flow within such helical colonies.

In lyre-shaped fenestellids (Plate 7G), which were mentioned above (p. 106), zooids open on the convex upper surface of the arched meshwork. The closed end of the colony with the skeletal thickening pointed towards the direction of ambient current flow. Lophophores of zooids around the fenestrules would have vented filtered water into the barrel-like area beneath the meshwork, after which it would have flowed out through the open end of the arched colony (McKinney 1977b).

The most extreme modification among genera of Fenestrata can be seen in the peculiar Devonian genus *Ernstipora*, which has encrusting colonies made up of dome-shaped subcolonies each comprising short branches with zooids opening on the surface but concealed beneath a superstructure (Suárez Andres and Wyse Jackson 2015a). The tops of the subcolonies are open and were interpreted to have functioned like monticules as exhalent chimneys.

Colonies with deep bifoliate branches dividing and anastomosing in a plane perpendicular to the median lamina to give a reticulate structure (Plates 8L, 10E) were particularly characteristic of some Mesozoic cyclostome genera (e.g. *Reticulipora* and *Reticrisina*), but can also be found in a few Palaeozoic cystoporates (e.g. *Ramipora*). This colony-form has been termed reticuliporiform or eccentric bilaminate (McKinney 1986b, p. 136). Like the fenestrate colonies described above, it is likely that water flowed through the gaps between the branches, with the distal-facing orientation of the zooids implying a direction from the distal growing edges of the branches proximally towards their bases (Taylor 1979a).

While in most bryozoans with colonies forming erect planar fans, water flows towards the frontal sides of the colony, passing between the branches and away from the dorsal side, this is not always the case. McKinney (1991) observed feeding currents in the cyclostome *Idmidronea*, finding the reverse of this flow direction, i.e. from abfrontal ('dorsal') to frontal. This correlates with the presence of zooids with larger more powerful lophophores near the branch crests on the frontal surface which channel water towards the branch crests for expulsion. The effect is amplified by the lophophores of zooids flanking the branch crests being obliquely truncate, with the longest tentacles closest to the crest. These bryozoans thus conserve the direction of excurrent flow that can be inferred in their encrusting antecedents (Figure 4.26), where the equivalent of the dorsal side of the branch is cemented to the substrate and excurrent flow occurs over the branch crests (e.g. *Idmonea*; see Taylor 1979a). Indeed, *Idmidronea* and its relatives often have extensive encrusting bases comprising bifurcating, ribbon-like branches in which the flow pattern present in the radially asymmetrical erect branches is likely already established before erect growth commences. In contrast, bryozoans with erect planar fans employing the more conventional frontal

Figure 4.26 Inferred multizooidal feeding currents in cross sections of the encrusting cyclostome *Idmonea* (Plate 6C) and its erect relatives such as *Idmidronea*. The larger, obliquely truncate lophophores suggest that excurrents depart from above the branch crest in *Idmonea*. The excurrent flow direction has been observed to be replicated in the erect branches of *Idmidronea*, where multizooidal current flow is from reverse (dorsal) to obverse, i.e. the opposite direction to that seen in most fenestrate bryozoans (cf. Figure 4.25).

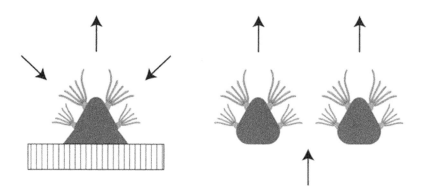

to abfrontal current flow pattern seemingly evolved from taxa with erect, radially symmetrical branches with autozooids opening around the entire circumference subsequently becoming restricted to one side only of the branches (e.g. Taylor 1987). Thus, the pattern of multizooidal current flow is phylogenetically constrained. By extension, observations of flow direction in living bryozoans with planar fan systems may provide insights into the origin of these bryozoans from either encrusting or erect progenitors. One such bryozoan is the cyclostome *Fenestulipora* (Taylor and Gordon 1997). This deep-water genus, recorded in water 270–1901 metres in New Zealand and Indonesia, has never been observed alive but there is circumstantial evidence that water flow was from abfrontal to frontal: peristomes of the tiny zooids of the cyclostome *Stomatopora* encrusting the frontal surfaces of *Fenestulipora* branches were observed to bend into the fenestrules towards the abfrontal side of the fouled colony (Taylor and Gordon 1997, fig. 5e), suggesting that the lophophores of the *Stomatopora* zooids were directed towards an inhalent flow created by the host colony which was in an abfrontal to frontal direction.

4.6 Life Histories

Bryozoans vary enormously in life history patterns, especially timing of sexual reproduction and relative allocation to colony growth through budding vs. sexual reproduction including the production of larvae. Contrasting life history strategies were described in a selection of living species by McKinney and Jackson (1989). Variations in life history strategies can occur both between and within habitats. An example of the latter was described by Herrera and Jackson (1996) who studied three encrusting cheilostome species from Caribbean coral reefs in Panama: *Parasmittina areolata*, *Stylopoma spongites*, and *S. projecta*. Whereas *Parasmittina* grew at about one-third the rate of the two *Stylopoma* species, it produced 12–17 times more larvae.

While colony growth rate is difficult to estimate in fossil bryozoans, it is possible to gain insight into allocations and timing of female sexual activity in species that have skeletal larval brood chambers (i.e. gonozooids in cyclostomes and ovicells in cheilostomes), and also of male sexual activity in the few cheilostomes with distinctive male zooids. However, this has been an under-researched area of palaeobryozoology. A rare exception is a study of 15 species of Mesozoic encrusting cyclostomes forming small discoidal colonies (McKinney and Taylor 1997). Significant variations between species were found in the proportions of fertile colonies possessing gonozooids, values ranging from 2.5–88.2%. Many of the fertile colonies had multiple gonozooids budded simultaneously and therefore arranged in a ring parallel to the distal growing edge. In most colonies, growth ceased after formation of the gonozooid (or gonozooids), and presumably liberation of the larvae, which implies a **semelparous** life history strategy. A few colonies belonging to some of the species had more than one generation of gonozooids and were thus **iteroparous**.

Aspects of the life histories of free-living, lunulitiform bryozoans have been described previously (Chapter 4.4).

4.7 Colonial Integration

Colonial integration expresses the degree to which the colony functions as a single entity rather than a collection of individual zooids. Integration involves resource sharing among zooids and elevates animal colonies above the level of simple aggregates of clonal individuals (Mackie 1986). Much has been written about colonial integration in bryozoans (e.g. Boardman and Cheetham 1973; Cook 1979) and other colonial animals (e.g. Oren et al. 2001; Cartwright 2003), but quantifying integration is hampered by our poor knowledge of the exact ways in which the zooids within colonies interact physiologically, through metabolite exchange, nervous coordination, or other means (see Burgess et al. 2017). Various proxies for colonial integration have been suggested, such as highly determinate colony growth patterns, zooidal polymorphism reflecting division of labour, and multizooidal feeding currents indicating cooperative acquisition of planktonic food, but the levels of these traits may conflict within particular species. For example, the Cretaceous bryozoan *Herpetopora* has very irregular, straggly, runner-like colonies with no avicularia, implying poor integration, yet the proclivity of branches to restore connections through active growth towards pores in other branches (as noted above on p. 103), as well as the strong astogenetic gradients within colonies, could be taken to indicate a high level of integration.

Some kind of soft tissue connectivity between the zooids in a colony exists in all living bryozoans. Similar connectivity can be inferred in fossil bryozoans from the presence of pores in the skeletal walls (Chapter 2.2), or through evidence for hypostegal tissues connecting zooids over the colony surface above the distal ends of the skeletal walls, as in all palaeostomates and some cyclostomes. The independence of wall porosity and polymorphism was shown by Bone and Keough's (2010) comparisons between two sympatric ascophoran cheilostomes, identified as *Watersipora subtorquata* and *Mucropetraliella ellerii*, from the Recent of Victoria, Australia. Contrary to predictions, *Watersipora subtorquata* with only autozooids has more pores per zooid than does the *Mucropetraliella ellerii* which has avicularian polymorphs requiring trophic support.

Each bryozoan zooid has its own nervous system and this connects with those of the other zooids in the colony through the pores in the skeletal walls to form a colonial nervous system, as detailed in the comprehensive review written by Gruhl and Schwaha (2016). Nervous pulses are able to pass from each zooid to its neighbours (Thorpe, Shelton, and Laverack 1975). This explains the simultaneous retraction of the lophophores of multiple zooids when a single zooid is stimulated, as well as the coordinated movements of setae to bring about locomotion in free-living colonies of *Selenaria*.

Partitioning of functions between different types of zooids demands transfer of metabolites. This is particularly relevant to bryozoans in which brooding zooids nurturing incipient larvae are unable to feed, but is also vital for the nutritional support of newly budded zooids at growing edges. Little is known about this aspect of colonial integration. However, Best and Thorpe (1985) fed ^{14}C labelled plankton to colonies of the cheilostome *Membranipora*, tracing the movement of the isotope using autoradiography. They found the isotope to be transported predominantly towards active growing edges at an average speed of 1.49 mm per hour. In a subsequent study (Best and Thorpe 2002), the same authors presented evidence to show that transport occurred using the funicular system which links zooids through pores in the skeletal walls (see also Lutaud 1985). Transport predominantly towards the growing edge was also found by Miles et al. (1995) in separate experiments on *Membranipora* using ^{14}C and ^{35}S tracers. These findings are consistent with the fact that funicular connections between zooids occur through pore plates containing specialized cells that have a particular polarity, favouring lipid movement in a proximal to distal direction (Bobin 1977).

'Sacrifice' of zooids to the overall benefit of the colony is another measure of integration. For instance, autozooids at the centres of some monticules degenerate so that an excurrent chimney can form above them. The common overgrowth of proximal zooids near the bases of erect colonies (Figure 4.12) to strengthen the colony provides another example of individual forfeit for the sake of the colony as a whole.

Embryonic brooding in bryozoans can furnish further evidence of colonial integration. In cheilostomes, the calcified roofs of the most primitive ovicells are formed by the zooid distal to that supplying the embryo (Figures 3.15B, 3.16). In essence, the distal neighbour constructs the 'crib' for the embryo of the maternal zooid (e.g. Nielsen 1985). In some cheilostomes (e.g. *Bugula neritina*: Woollacott and Zimmer 1972) three zooids are apparently involved in ovicell formation: the maternal and distal autozooids plus a kenozooid. Cyclostome gonozooids often overgrow incipient autozooids around them, preventing them from supporting a polypide, an indication that the colony is placing greater value on sexual reproduction than it is on individual feeding zooids.

When cultured under stress in low pH conditions to simulate ocean acidification, colonies of the cheilostome *Cryptosula pallasiana* allocate proportionally more resources to the budding of new zooids at the colony growing edge at the expense of maintaining existing zooids in which polypide recycling (degeneration-regeneration) is greatly reduced (Lombardi et al. 2017). This too can be viewed as dominance of the colony over individual zooids.

It is easy to assume that traits associated with higher levels of colonial integration will increase during the evolution of colonial animal groups. McShea and Venit (2002) tested this for cyclostome bryozoans by mapping the states of three criteria of coloniality (connectedness between zooids, differentiation among zooids, and presence of subcolonies) onto a published skeletal morphological tree (Taylor and Weedon 2000). Their results found no clear tendency for colonial integration to either increase or decrease. However, this study was undertaken before the availability of molecular trees and the method used by McShea and Venit would be worth applying to a molecular tree now that we know of the problems associated with using skeletal morphology to construct phylogenetic trees in cyclostomes (Chapter 8).

4.8 Endolithic and Etching Bryozoans

Some gymnolaemate bryozoans are capable of boring or etching into calcareous substrates (Figure 4.27), especially mollusc shells but occasionally also into sedimentary hardgrounds or other bryozoans possessing calcareous skeletons (Winston and Hayward 1994) as well as the parchment-like tubes of polychaetes (Pröts, Wanninger, and Schwaha 2019). Although traditionally all deep boring bryozoans are assigned to the order Ctenostomata, in *Penetrantia* the presence of ovicells, a calcareous operculum, and even a calcareous lining have led to the suggestion that this genus should instead be classified in the order Cheilostomata (Soule and Soule 1969; Smyth 1988). Among indisputable Ctenostomata, anatomical differences between boring genera imply multiple origins of an endolithic lifestyle.

Pohowsky (1978) published the most complete study of boring bryozoans, summarizing their taxonomy and biology. All species have runner-like colonies which are very easily overlooked as the cavities occupied by the zooids are slender and their openings on the surface of the substrate through which the lophophores protrude are extremely small (Figure 4.27A). Wetting substrates can help to reveal bryozoan borings but in many cases the borings only become obvious after abrasion of the substrate exposes the branching pattern of the colony lying a little way beneath its surface. The best way of studying fossil borings, which have been recorded

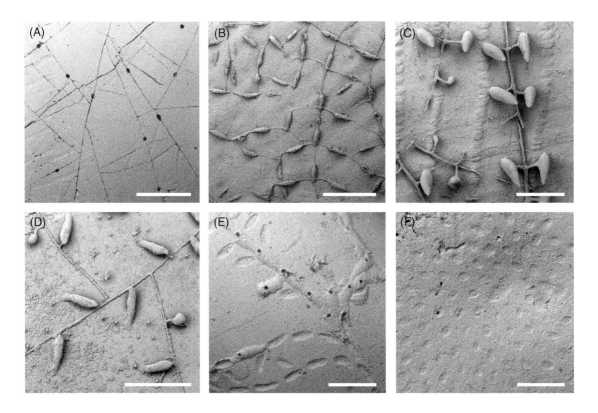

Figure 4.27 Ctenostome borings (A–D) and cheilostome etchings (E–F); (A) partly exumed branches and apertures of *Ropalonaria? arachne* on the surface of a shell (Jurassic, Oxfordian, Ampthill Clay; Waterstock, Oxfordshire, UK); (B) resin cast of *Ropalonaria venosa* with zooids (Ordovician, Katian, Waynesville Fm.; Oxford, Ohio, USA); (C) resin cast of *Haimeina michelini* with almost vertical zooids linked by stolons (Jurassic, Pliensbachian or Toarcian, Marlstone; Stroud, Gloucestershire, UK); (D) resin cast of *Spathipora brevicauda* with zooids attached to stolons (Middle Miocene; Manthelan, Indre-et-Loire, France); (E) ichnofossil *Finichnus dromeus* revealed beneath exfoliated zooids of *Tecatia robusta* encrusting a belemnite guard (Cretaceous, Maastrichtian; Trimingham, Norfolk, UK); (F) ichnofossil *Finichnus peristoma* on the surface of a bivalve shell once covered by a cheilostome, each pit etched by a zooid (Pleistocene, Nukumaru Limestone; Waiinu Beach, New Zealand). Scale bars: A, B, E = 1 mm; C, D, F = 500 μm.

back to the Ordovician, is to impregnate them with resin and then dissolve the shell to reveal the resulting cast of the colony (Figure 4.27B–D).

There has been disagreement about whether bryozoan taxa established on the basis of cavities in shells should be treated as ichnotaxa when fossil (e.g. Botquelen and Mayoral 2005), or incipient ichnotaxa when based on Recent material. The fact that the borings represent a precise natural mould of the colony and its zooids has been used by some to justify treating fossil bryozoan borings as body fossils rather than trace fossils (Pohowsky 1974). This distinction is important as trace fossils are governed by separate rules in the International Code of Zoological Nomenclature. However, as Bromley (2004) remarked, taxonomic misunderstanding can be eliminated as long as the same name is used by bryozoologists as by ichnologists.

Very little is known about the biology of endolithic bryozoans. They are presumed to bore into calcareous substrates using chemical means. This must occur soon after larval settlement on the substrate surface as the ancestrula is fully immersed. Some colonies inhabit living molluscs but the small size of the borings compared, for example, to those of sponges, suggests that they may not significantly weaken the shells of their hosts (Smyth 1988). Indeed, subsequent growth of the mollusc shell can result in the openings of the zooids being covered, and the bryozoan zooids are also vulnerable to overgrowth by calcareous algae (Smyth 1988), whether on living molluscs or gastropod shells tenanted by hermit crabs, which are commonly colonized by endolithic bryozoans (Walker 1988). Remarkably well-preserved boring ctenostomes from the Devonian of Ukraine have phosphatized soft tissues including setigerous collars (Olempska 2012, 2015).

Contrasting with the deep borings made by ctenostomes, shallow etchings into calcareous substrates can be produced by cheilostome bryozoans belonging to several genera in which the basal walls of the zooids have an uncalcified window allowing soft tissue access to the substrate beneath (Taylor, Wilson, and Bromley 1999; Rosso 2008). When colonies are detached from the substrate a distinctive pattern of elliptical pits can be seen corresponding to the arrangement of the zooids in the colony, with uniserial and multiserial encrusters showing clear differences (Figure 4.27E, F). This trace fossil was originally called *Leptichnus* but this name was replaced by *Finichnus* when *Leptichnus* was found to be preoccupied. Recorded as far back as the Maastrichtian, *Finichnus* can often be found on fossil shells that lack the calcareous skeletons of the tracemakers, indicating taphonomic loss of the body fossils of the bryozoans.

5 | Biotic Interactions

Aside from capturing plankton for feeding, which has been discussed previously, bryozoans exhibit biotic interactions of various kinds with other organisms. Interactions are, of course, better understood in living bryozoans than in fossils. Nevertheless, the mineralized skeletons of fossil bryozoans can be copious sources of information about certain kinds of biotic interactions. The most notable of these is competition for substrate space which can be studied using skeletal overgrowths between colonies encrusting shells or other hard substrates. Direct evidence for predation is rarer in bryozoans and mainly consists of tiny boreholes in the skeletons of some zooids, but interactions with symbionts are commonly recorded as embedment fossils and bioclaustrations in fossil bryozoan skeletons.

5.1 Competition

The sessile habit characterizing the great majority of bryozoans inevitably brings them into conflict with neighbouring organisms (see Taylor 2016). Unlike mobile animals, bryozoans do not have the option of moving away from their competitors, which places a premium on developing effective adaptations to cope with competition. Bryozoans growing on hard substrates routinely confront other sclerobionts belonging to the same or different species. Encounters result in interference competition for living space. The usual outcome is for the dominant competitor to overgrow the subordinate competitor. With regard to competition with species belonging to other phyla, bryozoans tend to be overgrown by algae, ascidians, and sponges, but are typically able to non-lethally overgrow the exoskeletons of serpulid polychaetes and barnacles. This pattern of dominance may be reflected in ecological succession,

Bryozoan Paleobiology, First Edition. Paul D. Taylor.
© 2020 Natural History Museum. Published 2020 by John Wiley & Sons Ltd.

with bryozoans occurring on hard substrates during early stages of colonization but being replaced by sponges and ascidians through time.

5.1.1 Patterns of competition for substrate space

Space is acquired at the expense of competitors either by **marginal overgrowth** or by **fouling** (Figure 5.1). Marginal overgrowth occurs when one encruster encounters another on the surface of a substrate, whereas fouling results from the settlement of a larva on the surface of an existing sclerobiont followed by growth over that surface.

Marginal encounters may have one of three outcomes: **simple overgrowth**, **reciprocal overgrowth**,

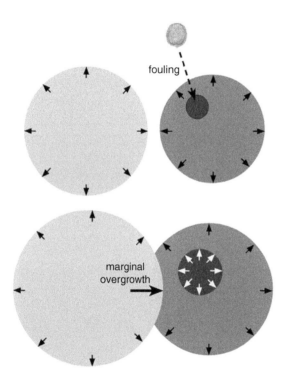

Figure 5.1 Marginal overgrowth and fouling. Arrows show centrifugal growth directions of colonies in plan-view represented as grey discs. In the case of fouling, a larva settles on the surface of an established colony, overgrowing the fouled colony as it develops. Marginal overgrowth occurs when the growing edge of one colony succeeds in budding zooids onto the surface of another colony.

and **stand-off** (Figure 5.2). In simple overgrowths between competing bryozoans, one colony overgrows the other along the entirety of the contact between them. Overgrown zooids die, resulting in partial mortality of the colony which may be a precursor to total colony mortality if all of the colony is overgrown. Todd and Turner (1988), however, recorded instances of ascidians completely overgrowing bryozoan colonies in which the overgrown colonies survived and rejuvenated when the ascidian was removed. Reciprocal overgrowth between bryozoans occurs when one colony overgrows the second along part of their contact but the polarity of overgrowth is reversed elsewhere. Stand-off is when the growth of both competitors is halted at their contact. This is most often observed in interactions between bryozoan colonies belonging to the same species (e.g. Cancino and Hughes 1988; Lopez Gappa 1989; Harvell, Caswell, and Simpson 1990) and is less common in interspecific encounters. Some intraspecific stand-offs between bryozoan colonies result in fusion (homosyndrome) between the two colonies (Chaney 1983; Craig 1994). In this case, newly budded zooids at the contact are formed partly by one and partly by the other colony and are genetic chimaeras. In the cheilostome *Thalamoporella gothica*, complete fusion (histocompatibility) with the formation of specialized pores between the zooids was observed only between parts of the same colony or sibling colonies but not between non-sibling colonies (Chaney 1983).

The outcome of overgrowth interactions between given pairs of bryozoan species may be completely predictable (**transitive**) or variable (**intransitive**). Intransitivity is often related to the angle at which the competing bryozoan colonies encounter one another (Jackson 1979b; Rubin 1982): a dominant species that wins all frontal, head-on encounters may be overgrown by a subordinate species when that species encroaches the flank or the rear of the dominant. Another factor responsible for intransitivity is the condition of the competitors, as has been shown in overgrowth interactions between the cheilostome *Watersipora* and the coral *Oculina* (Fine and Loya 2003). *Oculina* normally overgrows *Watersipora* but after periods of coral bleaching dominance is reversed and *Watersipora* overgrows *Oculina*.

Figure 5.2 The three possible outcomes – overgrowth, stand-off, and reciprocal overgrowth – when two encrusting bryozoan colonies (represented by grey discs) encounter one another during growth along the surface of a substrate.

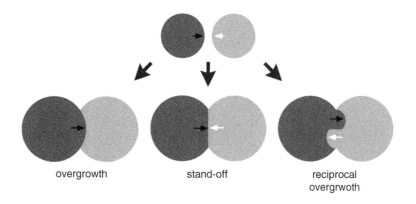

overgrowth   stand-off   reciprocal
                                         overgrwoth

In communities containing more than two encrusting species, competitors may form either a **competitive hierarchy** or a **competitive network.** Competitive hierarchies show simple patterns of species dominance; for example, species A overgrows species B, B overgrows C, and A also overgrows C. In contrast, in competitive networks species A overgrows B, B overgrows C, but C overgrows A, meaning that there is no single dominant species (Buss and Jackson 1979). These two different structures have implications for the development and diversity of encrusting communities. In competitive hierarchies, the top-ranked competitor in the absence of disturbance or other confounding factors will eventually overgrow all of the others and monopolize substrate space. In contrast, competitive networks result in greater variability in the development of encrusting communities (Karlson and Buss 1984) and competitive exclusion may not occur, especially when there are early settling species present that can pre-empt substrate space (Edwards and Schreiber 2010). According to Barnes (2002), there is a latitudinal gradient in transitivity among encrusting bryozoans with a greater proportion of competitive hierarchies compared to networks at higher latitudes where poor spatial competitors are overgrown more consistently than at high latitudes.

5.1.2 Adaptations for keeping and gaining substrate space

The importance of competition for substrate space in encrusting bryozoans is reflected by numerous adaptations for defending space already occupied and for winning new space from competitors. The circumferential growing edges characteristic of most encrusting bryozoans are important in providing the colony with an active perimeter for defence and offence. Zooidal budding patterns in the early astogeny of stenolaemate and cheilostome bryozoans ensure that this encompassing growing edge (Figure 4.5) is formed at an early stage of colony development. The edges of the branches of runner- and ribbon-like encrusters are more vulnerable to overgrowth as they are generally inert and cannot respond through zooidal budding. Instead, and as described above (Chapter 4.3.1), runners and ribbons employ a fugitive strategy for which defending substrate space is of less consequence. Some weak competitors may partly compensate for the likelihood of being overgrown by recruiting onto hard substrates before competitively dominant bryozoans appear (Bowden 2005).

Large size correlates with competitive success in most organisms (Maynard Smith and Brown

1986). Size is manifested at the level of both the zooid and the colony in bryozoans. With respect to colony size, Buss (1980) found a strong positive correlation between colony area and overgrowth ability in three species of cheilostomes encrusting rocks. However, López Gappa (1989) found colony size to be only important as a determinant of overgrowth success between species with otherwise similar competitive abilities. As for zooid size, there is little comparative data for living bryozoans but in a Japanese Pleistocene bryozoan community 105 of 173 interactions (61%) were won by species with larger zooids (Taylor 2016), while studies of overgrowth competition among cheilostomes from the New Zealand Pleistocene reported below have yielded similar results.

Large zooids not only present higher obstacles to be surmounted by competitors trying to overgrow them, but they may create stronger feeding currents capable of depleting the planktonic food available to competitors (Buss 1981). The long-term competitive superiority of cheilostome over cyclostome bryozoans (Chapter 9.15) correlates not only with the typically larger size of cheilostome zooids but also with the contrasting patterns of water flow at the margins of the colonies. In cyclostomes, exhalent water depleted of plankton tends to be channelled towards chimneys at the centres of colonies, whereas in cheilostomes it is passed outwards towards any encroaching competitors, including cyclostomes, thereby placing them at a trophic disadvantage (McKinney 1992).

Lidgard (1985) distinguished different budding processes in cheilostomes that are correlated with success in overgrowth competition: (i) **intrazooidal budding** (Figure 5.3A) in which new buds at the growing edge are formed through a discontinuous process of bud expansion from the pore chambers of parent zooids; (ii) **zooidal budding** (Figure 5.3B) where new buds are partitioned by growth of interior walls more or less continuously; and (iii) **multizooidal budding** (Figure 5.3C) resembling zooidal budding but in which new buds extend more than two zooids in length so that the growing edge of the colony forms a so-called 'giant bud' (Silén 1982). To some extent

these three categories are end members, and not all cheilostome species can be readily classified into one or the other. Nevertheless, there is evidence that species with zooidal budding perform better in overgrowth competition than those with intrazooidal budding, while the switchover from zooidal to multizooidal budding may apparently be induced by the presence of a competitor in species that normally show zooidal budding (Lidgard and Jackson 1989). The greater proportion of species with zooidal budding found encrusting relatively stable substrates, where competition is likely to be more important than physical disturbance, corroborates the notion that these budding processes have an adaptive value in competition for substrate space. In evolutionary time, Lidgard (1986) was able to show a significant decline in the proportion of species with intrazooidal budding from the Late Cretaceous to Holocene in North American cheilostome assemblages.

Multilayered growth is another trait believed to enhance the ability to compete for space (Lidgard and Jackson 1989; Taylor 2016). This has several advantages: it allows organisms that have previously overgrown or fouled the colony to be themselves overgrown, as well as increasing the effective thickness of the colony making it more of an obstacle to overgrowth. Buss (1981) reported competitive interactions between two cheilostomes, *Antropora tincta* with small zooids and *Onychocella alula* with larger zooids. Despite the size disadvantage of *Antropora*, this bryozoan was able to halt the growth of *Onychocella* by interlocking its zooids with gaps in the growing edge of the *Onychocella* colonies and subsequently producing a new layer of zooids that could overgrow these colonies.

Spines of various types furnish obstructions against organisms attempting to overgrow bryozoans. Zooids with long spines at the edges of colonies of the cheilostome *Electra pilosa* (which is known as the 'hairy sea-mat' on account of its spines) appear to block, or at least delay, overgrowth by ctenostome bryozoans (Stebbing 1973). However, this is not the sole function of *Electra* spines as they also protect zooidal lophophores in high-energy,

Figure 5.3 Encrusting cheilostomes with different budding processes recognized by Lidgard (1985): (A) intrazooidal budding in an unidentified cribrimorph showing the stepped growing edge (Cretaceous, Coniacian or Santonian Chalk; Northfleet, Kent, UK); (B) zooidal budding in *Rhagasostoma inelegans* with smooth growing edge (Cretaceous, Campanian Chalk; Norwich, Norfolk, UK); (C) multizooidal budding in *Parasmittina winstonae* showing a 'giant bud' not yet subdivided into zooids (Recent; Langkawi, Malaysia). Scale bars = 500 μm.

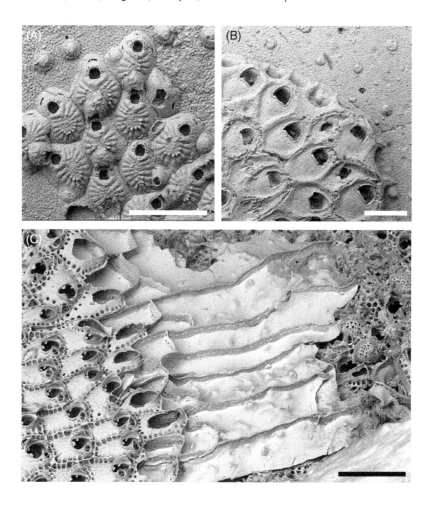

abrasive environments (Bayer et al. 1997). The antler-like spines of *Doryporella alcicornis* (see Grischenko, Mawatari, and Taylor 2000, fig. 3C), project outwardly from the colony growing edge, like the spears of a Macedonian phalanx. They are thus oriented towards potential spatial competitors and can presumably interfere with their growth. **Tower cells** are spine-like structures up to 8.4 mm long produced by columnar outgrowths of the frontal cuticles of zooids in the cheilostome bryozoan *Membranipora* (Cook and Chimonides 1980). Concentrations of tower cells at the boundaries between colonies of *Membranipora* prompted Xing and Qian (1999) to suggest that their function was at least in part related to defence against overgrowth.

Another kind of cuticular expansion – **stolons** – are produced at the growing edges of encrusting colonies in some cheilostomes. Stolons interfere with the growth of spatial competitors (Osborne 1984). In the case of intraspecific competition between colonies of the cheilostome *Membranipora membranacea*, ephemeral stolons reduce the size of the newly budded zooids in the colony that is being competed against (Harvell and Padilla 1990), as well as slowing their growth rates (Padilla et al. 1996). Several species of cheilostome bryozoan colonies live symbiotically with hydroid cnidarians. The hydrorhizae of these hydroids ramify across the surfaces of the host bryozoan colonies and sometimes become encased in bryozoan skeleton (Puce et al. 2007). Bryozoans harbouring hydroid symbionts perform better in competition for space than those lacking symbionts (Osman and Haugsness 1981; Ristedt and Schuhmacher 1985).

Elevating the growing edge off the substrate to make it harder to surmount is a common strategy employed in spatial competition by bryozoans, particularly cyclostomes that often grow as small cup-like colonies (Figure 5.4). Stebbing (1973) found that whereas isolated colonies of the lichenoporid *Disporella hispida* seldom developed elevated edges, colonies close to other organisms invariably raised their margins, seemingly as a response to the threat of being overgrown (Mariani 2003, fig. 3). The trade-off for bryozoans adopting this 'rampart strategy' is that they forego the potential for continued growth along the surface of the substrate. The rampart strategy is thus more strictly defensive than some other adaptations which combine defence with offence.

Chemical defences are employed by many marine animals, particularly non-motile species unable to flee from predators (Dyrynda 1986). These bioactive **natural products**, or **secondary metabolites** (Christophersen 1985; Figuerola and Avila 2019), are potentially potent against predators as well as competitors for substrate space. However, their efficacy in spatial competition has yet to be well established in bryozoans.

5.1.3 Spatial competition in fossil bryozoans

Direct evidence for competition of any sort is seldom seen in the fossil record. A rare exception is provided by competition among sclerobionts, particularly bryozoans: when carefully interpreted, skeletal overgrowths (Figure 5.5) provide the best opportunities for investigating competition in deep geological time. An interpretive caveat is that skeletal overgrowths can not only occur

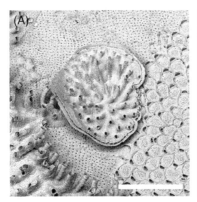 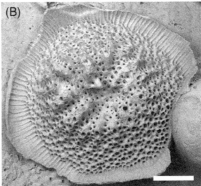

Figure 5.4 Some encrusting cyclostomes are able to elevate their growing edges, providing an obstacle to overgrowth: (A) small colony of *Microeciella* with margins raised above competing cyclostome (left) and cheilostome (right) colonies (Recent; Otago Shelf, New Zealand); (B) fossil *Disporella* with raised edges (Pleistocene, Setana Fm.; Kuromatsunai, Hokkaido, Japan). Scale bars = 1 mm.

Figure 5.5 Examples of overgrowths involving bryozoans in fossil communities: (A) spirorbid polychaete fouling the cheilostome *Porella* (Pleistocene, Setana Fm.; Kuromatsunai, Hokkaido, Japan); (B) cyclostome '*Berenicea*' (left) overgrowing cheilostome *Wilbertopora* (Cretaceous, Campanian, Aubeterre Fm.; Royan, Charente Maritime, France); (C) cyclostome *Hyporosopora* (top left) overgrowing another cyclostome, *Reptomultisparsa cobra* (lower right) (Jurassic, Bathonian, Hampen Marly Beds; Enstone, Oxfordshire, UK); (D) cheilostome *Escharoides* overgrowing branch flank of oligoserial cyclostome *Platonea* (Pleistocene, Nukumaru Limestone; Waiinu Beach, New Zealand); (E) standoff between two colonies of the cheilostome *Antarctothoa tongima*, one growing from the top left, the other from the bottom right (Pleistocene, Nukumaru Limestone; Waiinu Beach, New Zealand); (F) overgrowth interaction involving three cheilostome colonies, a colony of *Puellina* (lower right) overgrowing two colonies of *Haywardipora*, that on the upper left overgrowing the colony on the lower left with ovicellate zooids (Pliocene; San Clemente Island, Los Angeles California, USA); (G) reciprocal overgrowth between two cheilostomes, the colony of *Reginella* (lower left) overgrowing that of *Microporella* (upper right) to the left of the asterisk but being overgrown by it to the right (Pleistocene, Setana Fm.; Kuromatsunai, Hokkaido, Japan); (H) colony of the runner-like cyclostome *Corynotrypa* being overgrown by a sheet-like trepostome (Ordovician, Sandbian, Bromide Fm.; Arbuckle Mountains, Oklahoma, USA). Scale bars: A, C, E = 500 µm; B, D, F, G, H = 1 mm.

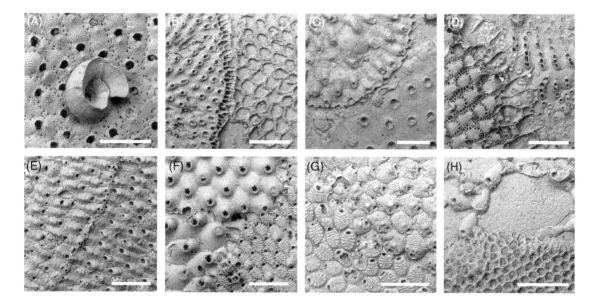

during the life of the overgrown sclerobiont (i.e. *syn vivo*), but also after its death (i.e. post mortem). As the proportion of post-mortem overgrowths is likely to be high on long-lasting hard substrates (such as the undersides of coral heads or cobbles) where multiple generations of encrusters may be present, these are less suitable targets for studying competition than smaller, more ephemeral substrates like bivalve or brachiopod shells. With respect to the latter, a study of encrusted shells of the brachiopod *Bouchardia*

rosea collected from the Brazilian shelf suggested that colonization by bryozoans and other epibionts occurred only during a brief interval of time after host death, individual shells offering ecological snapshots even though the shells in the studied assemblage ranged from 0–3000 years old (Rodland et al. 2006, 2014).

Confidence that interactions observed in fossils are *syn vivo* is increased when the sclerobiont community contains: (i) examples of reciprocal overgrowth, providing unequivocal evidence that the two interacting encrusters were alive at the same time (Figure 5.5G); (ii) intraspecific encounters resulting in stand-offs (Figure 5.5E), which are likely outcomes when both competitors were alive at the same time but not in post-mortem overgrowths; and (iii) reaction effects, such as raised edges in encrusters that are overgrown. Furthermore, McKinney's (1995b) analysis of *synvivo* and post-mortem overgrowths in a living bryozoan community showed that post-mortem effects dilute the signal of dominance between encrusters but do not reverse it. This finding favours a probabilistic approach to the study of fossil overgrowths and underlines the need for large samples if competitive rankings in bryozoan communities are to be estimated with confidence.

In spite of the unprecedented potential for documenting competition in the geological past, relatively few studies have been published on overgrowth interactions in sclerobiont communities containing bryozoans (reviewed by Taylor 2016). Many have concerned small collections from single localities and stratigraphical horizons (e.g. Taylor 1979b, 1984a; Liddell and Brett 1982; Wilson 1985; Alvarez and Taylor 1987; Lescinsky 1997; Nebelsick, Schmid, and Stachowitsch 1997; Rosso and Sanfillipo 2005). These studies have found that subordinate competitors for space, somewhat paradoxically, are often the most abundant species in sclerobiont communities, which is also the case for some modern communities (Barnes and Clarke 1998; Centurión and López Gappa 2001; Clark, Stark, and Johnston 2017). Fouling in the fossil communities tends to involve mostly the settlement of small organisms, especially the spirally coiled tubes of spirorbid polychaetes (Figure 5.5A) and convergent microconchids, onto the surfaces of bryozoan colonies, while, as predicted, runner-like bryozoans generally lose encounters with sheet-like species (Figure 5.5H). Some particularly striking examples of stand-offs have been recorded between sheet-like bryozoans and edrioasteroids, the bryozoans encircling the sedentary echinoderms without overgrowing them (Müller, Hahn, and Bohaty 2013; Sprinkle and Rodgers 2010).

New research led by Lee Hsiang Liow has employed much larger samples than previously used to address the dynamics of interactions over geological time and to identify traits that correlate with success in spatial competition. This work has focused on bryozoan-encrusted molluscs collected from a sequence of five Pleistocene shell beds in the Wanganui Basin of New Zealand. Including a Recent sample from the adjacent Cook Strait, Liow et al. (2016) examined 7088 cheilostome–cheilostome interactions on 751 bivalve shells. It was found that most of those species which were statistically winners remained so throughout the 2-myr period bracketed by the shell beds. Likewise, losers remained losers. Thus, the prevailing picture is of competitive stasis. Competitive ability showed no correlation with abundance, in accordance with the studies mentioned above. A similar data set was employed by Liow et al. (2017) to show that zooid size correlated strongly with competitive ability, species with large zooids routinely overgrowing those with smaller zooids. Interestingly, some species showed a coordinated increase in zooid size through time, followed by a decrease, which is believed to have been driven by environmental changes. Finally, a study by Liow et al. (2019) looked at how well combinations of numerous traits were at predicting overgrowth outcomes. Large zooid size was confirmed as being generally advantageous for success in overgrowth, although its importance varied through time, but bryozoans with heavily calcified zooids, spines, and avicularia seemed to be at a disadvantage in overgrowth battles.

The importance of competition for space among bryozoans and between them and competitors belonging to other taxonomic groups remains to be demonstrated in terms of both ecological dominance and evolutionary success. Many overgrowths may be merely 'border skirmishes'. As described more fully below (Chapter 9.15), however, the study of McKinney (1995a) showed that about two-thirds of overgrowth interactions between encrusting cyclostome and cheilostome bryozoans over the last 100 million years resulted in wins for cheilostomes, a long-term dominance that may be relevant to the progressive replacement of cyclostomes by cheilostomes through evolutionary time.

5.1.4 Competition for other resources

While living space is generally believed to be the primary limiting resource for sessile organisms inhabiting hard substrates, trophic competition has often been deemed of minor importance because planktonic food is superabundant and need not be competed for in most of the habitats colonized by suspension feeding sclerobionts. However, evidence that plankton can be depleted in at least some bryozoan habitats opens up the possibility of competition for food among bryozoans and between bryozoans and other sessile suspension feeders. In addition, at microscale levels, competition for plankton may mediate interactions between bryozoan colonies for substrate space, as already mentioned above.

Competition for oxygen has also been shown to be more important than was previously believed. Ferguson, White, and Marshall (2013) found oxygen availability to be highly variable in some habitats colonized by bryozoans and other sessile suspension feeders. Encrusting cheilostomes belonging to *Hippopodina*, *Schizoporella*, and *Watersipora* were found in the Ferguson et al. study to deplete oxygen in their vicinity much more than did erect species of *Bugula*, raising the intriguing possibility that oxygen rather than space may be the main limiting resource to bryozoans living in some habitats.

5.2 Predation

Relatively little is known about predation on bryozoans. It was formerly considered that living bryozoans contain so little tissue that they scarcely warrant the attention of predators. This view has since been challenged with the accumulation of an increasing number of records of a wide diversity of marine animals preying on bryozoans – the literature search of Lidgard (2008) found

almost 400 predator species. In addition, many features of bryozoan skeletons, including cheilostome frontal shields and some types of spines, as well as the presence of polymorphs equipped with jaw-like mandibles, are best interpreted as functioning in defence against predators.

Predation on bryozoans can occur at the level of individual zooids, groups of zooids or even the entire colony. Because it typically results in the demise of one or a few zooids rather than death of the entire colony, predation bears some similarity to parasitism on unitary animals: in principle, consumption of zooids may be compensated by the budding of new zooids, and there is the potential for an equilibrium to be reached between the predator and its prey. However, research is needed to assess the extent to which this ever actually occurs in living bryozoan colonies.

5.2.1 Predators of Recent bryozoans

A wide variety of marine animal groups include bryozoans in their diets. Prominent among these are pycnogonids (sea spiders) and nudibranchs (sea slugs). Also known to consume bryozoans are platyhelminths, nematodes, and amphipod, isopod, copepod, and decapod arthropods, as well as chiton and gastropod molluscs, echinoid, holothurian, asteroid, and ophiuroid echinoderms, and fishes (McKinney, Taylor, and Lidgard 2003, table 1). Some of this predation may be considered to be incidental (e.g. when bryozoans growing among algae are eaten by largely herbivorous grazers: Gordon 1972), or may involve planktivores consuming bryozoan larvae. In other instances, predation is highly specific, with some predators preying on only one bryozoan species (Cook 1985, p. 18).

Perhaps the most complete account of single-zooid predation on bryozoans concerns the pycnogonid *Achelia echinata* eating zooids of the cheilostome *Flustra foliacea* (Wyer and King 1973). These pycnogonids are usually found in the distal parts of *Flustra* fronds but seldom at the growing edge. They use the two distolateral spines of the bryozoan zooids for orientation and position themselves just distally of the zooid with their proboscis above the orifice. When the operculum begins to open the tip of the proboscis is thrust into the zooid and small pieces of the zooid are bitten off by the jaws at the end of the proboscis. Feeding takes about 10 minutes. Additional examples of pycnogonids preying on bryozoans can be found in the review of pycnogonid feeding by Dietz et al. (2018).

Nudibranchs graze both erect and encrusting bryozoans. The nudibranch *Polycera hedgpethi* consumes entire branches of the erect cheilostome *Bugula neritina* (Bone and Keough 2005). Experimental studies revealed no response by the bryozoan in terms of alterations to its growth or sexual reproduction. Thompson (1958) described the feeding behaviour of the nudibranch *Adalaria proxima* when attacking zooids of its favoured bryozoan prey *Electra pilosa*. The radula is employed in short bursts of 3–8 strokes until the membranous covering of the bryozoan zooid is breached, after which the lips of the nudibranch are applied and the contents of the zooid are sucked out by means of dilations of the buccal pump. It is unclear whether the bryozoan skeleton is damaged during this process. In the epiphytic cheilostome *Membranipora membranacea*, the presence of predatory nudibranchs induces the formation of sharp, cuticular spines in newly budded zooids (Harvell 1986). These are effective in protecting the zooids against predation but only at the cost of depressing rates of colony growth and sexual reproduction (Yoshioka 1982a). Grünbaum (1997) showed that the reduction in growth rate suffered by colonies with spines was because the spines interfered hydrodynamically with the feeding currents of the colony. The nudibranch *Okenia hiroi* is a specialist predator on the cheilostome *Integripelta acanthus* which, as the trivial name implies, possesses thorny spines, uniquely for the entire family Eurystomellidae (Gordon and Rudman 2006). The spines of *I. acanthus* are clearly ineffective in preventing predation and it remains to be demonstrated if they have any effect on the feeding rate of the nudibranch.

Interestingly, two nudibranch predators (*Corambe pacifica* and *Doridella steinbergae*) have dorsal surfaces patterned to resemble the zooids of the *Membranipora* colonies on which they graze (Yoshioka 1986), presumably serving as camouflage against their own predators. Harvell (1984) studied the behaviour of three nudibranch predators of the erect cheilostome *Dendrobeania lichenoides*, finding all three species preferentially to consume the younger, outer parts of colonies, which is consistent with greater palatability here. It is worth emphasizing that not all spines in cheilostomes are direct responses to predation: as previously mentioned (Chapter 5.1), the long spines of *Electra pilosa* can be induced by wave-related abrasion, buffering the colony against damage, as well as delaying overgrowth by competing ctenostomes, but nevertheless may have an incidental effect against nudibranch predators.

Juvenile asteroids have been observed to feed selectively on cyclostome bryozoans (Day and Osman 1981), everting their stomachs over the colonies and ingesting the tissues.

An important function of frontal walls in bryozoans is almost certainly protection of the retracted polypide from predators capable of breaking into the cystid and consuming the contents within. Accordingly, the ability of frontal walls to resist puncture and compressive stresses is likely to be important. Best and Winston (1984) determined experimentally the stress required to break the frontal walls of nine species of recent encrusting cheilostomes from Jamaica. Particularly in puncture tests, frontal walls of the tested anascans were more easily broken than those of the ascophorans, while zooids at the edges of colonies were weaker than those nearer to the centres. These results corroborate the general expectation that anascans are more vulnerable to these predators than are ascophorans, as are young zooids near the growing edge.

Figuerola et al. (2017a) found that extracts from two Antarctic cheilostomes were repellent to a potential amphipod predator. Colonies of the ctenostome bryozoan *Alcyonidium nodosum* that encrust the shells of living individuals of the gastropod *Burnupena papyracea* in South Africa produce secondary metabolites that protect the host gastropod from predation by the rock lobster *Jasus lalandii*, thus improving their own survival (Gray, McQuaid, and Davies-Coleman 2005).

Cook (1985) described an example from Ghana of a large colony of the cheilostome *Trematooecia magnifica* supporting a population of at least 30 predatory nudibranchs. Conservative estimates of bryozoan growth rate and the zooid consumption rate by the predators suggested that the formation of new zooids greatly exceeded the rate at which zooids were lost to these predators.

5.2.2 Predation in fossil bryozoans

While the skeletons of fossil bryozoans occasionally archive attacks made by predators, the interpretation of the resulting structures can be difficult, not least because almost nothing is known about the skeletal signatures of predation in living bryozoans.

5.2.3 Predator stomach contents

The gut contents of an individual of the Permian shark *Janassa* were found to contain pieces of the fenestrate bryozoan *Acanthocladia* (Schaumberg 1979). This is perhaps the only example in the fossil record of bryozoan remains in either guts or coprolites.

5.2.4 Putative predator drillholes

There is a voluminous literature on predator borings in the fossil record. Small round holes in shells have been recorded from the Ediacaran onwards, increasing in prevalence in the late Mesozoic and especially the Cenozoic (Kowalewski, Dulai, and Fürsich 1998). Borings are particularly common in mollusc shells, but are also recorded in brachiopods, barnacles, echinoids, and serpulid tubes (see Klompmaker et al. 2015). Most are attributable to predatory gastropods, with a smaller number made by octopods. The trace fossil represented by the boring – *Oichnus* – is typically a millimetre or more in diameter. However, similar borings on a micronscale described in foraminiferan tests (Nielsen and Nielsen 2001) as well as some bryozoans (see below) seem to have been made by tiny predators

of unknown identities but perhaps juvenile or very small species of gastropods.

Circular and oval borings 40–90 μm in diameter are present in some Middle Jurassic to Paleocene cyclostome bryozoans. Although not all of these borings need have been made by predators (as, for example, when multiple borings penetrate a single zooid, or terminal diaphragms of senesced autozooids are bored), in some cases a predatory origin can be inferred with greater confidence. Putative predatory borings (Figure 5.6A) have been described in species of Late Cretaceous Eleidae (Taylor 1982), the cyclostome family unique in having calcified opercula closing autozooidal apertures and developing mandibulate heterozooids (eleozooids). Borings in one fragment of the eleid *Meliceritites durobrivensis* (Gregory) from the British Chalk demonstrated a clear pattern, with 58% of autozooids that retained their opercula having been bored through the operculum, whereas only one of these was bored through the frontal wall. In contrast, only 16% of eleozooids were bored. Among living cyclostomes possessing small heterozooids (e.g. nanozooids), the polypide is much smaller than that of the autozooids, and the same can be assumed in *M. durobrivensis* in which the eleozooids are much smaller than the autozooids. The preferential occurrence of borings in autozooids is therefore consistent with predation on the larger, more nutritious, autozooids, while penetration through the opercula indicates a locus of attack via the aperture consistent with that described by many predators of modern bryozoans.

Very similar borings are known in some Recent cheilostomes. Even here, however, the identity of the borer is unknown. In large frondose colonies of the cheilostome *Microporella hyadesi* from the South Atlantic, a high proportion of zooids are penetrated by borings 60–65 μm in diameter (Figure 5.6B). These normally pass through the frontal shield, and are sometimes sealed internally by the frontal shield of an intramural bud (see below) that developed within the empty cystid of the host zooid (Wilson and Taylor 2006). Similar borings in a species of cheilostome

Figure 5.6 Putative predatory drillholes: (A) eleid cyclostome *Meliceritites durobrivensis* with circular borings in the calcareous opercula of several autozooids (Cretaceous, Coniacian or Santonian Chalk; Chatham, Kent, UK); (B) cheilostome *Microporella hyadesi* with boring (lower left) in an autozooid and two additional borings (upper left and centre right) beneath which the frontal shield has apparently been repaired (Recent; South Atlantic). Scale bars = 200 μm.

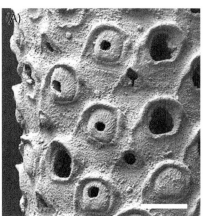
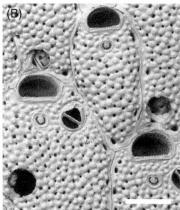

mentioned by Gordon and d'Hondt (1991) may pierce the ovicells, suggesting that the predator was consuming the yolky embryos within. Gastropods are the most likely culprits – Franz (1971) commented that very young individuals of the boring predatory snail *Urosalpinx cinerea* attacked encrusting bryozoans such as *Cryptosula* but did not mention whether they bored into the bryozoan skeleton. Countersunk borings reported from the recent encrusting cheilostome *Integripelta acanthus* were thought most likely to have been made by a marginelliform gastropod (Gordon and Rudman 2006).

5.2.5 Intramural budding

Intramural buds are new zooids budded into the empty cystids of dead zooids. Common in many cheilostomes (Taylor 1988a; Berning 2008), and also found in eleid cyclostomes (Taylor 1994c), they can be recognized by the presence of an additional mural or apertural rim located immediately within the original rim (Figure 5.7). The outer rim belongs to the original zooid while the inner rim defines the aperture or orifice of the new zooid. Repeated intramural budding in one zooid can

lead to a sequence of successively smaller zooids arranged like Russian dolls (e.g. Poluzzi, 1980, pl. 4, fig. 6; Figure 5.7C). Intramural buds can occur both in autozooids and heterozooids, such as avicularia (e.g. Koromyslova 2014b), and it is not uncommon for an autozooid to host an intramurally budded avicularium (Figure 5.7E, F) or kenozooid in cheilostomes, or an eleozooid in eleid cyclostomes (Figure 5.7H). It may be that the reduced space available inside the host autozooid precludes the intramural budding of another autozooid and forces formation of a smaller polymorph. Very occasionally, a single zooid is occupied by more than one intramural bud formed simultaneously. Nothing is known about the process of intramural budding but it is assumed that the new zooids are budded from neighbouring zooids through pores in the vertical skeletal walls. Usually the bud has the same polarity as the host zooid but this is not always the case and reverse polarity intramural buds can occur (Figure 5.7D).

A firm link has yet to be established between intramural budding and predation, although it is reasonable to suppose that at least some intramural buds developed after attacks by zooid-level

Figure 5.7 Intramural buds, possibly indicating reparative growth after predation, in cheilostomes (A–F) and cyclostomes (G–H): (A) normal autozooid with paired distolateral avicularia (lower right) and an intramurally budded autozooid also with associated avicularia (upper left) in *Antropora* aff. *japonica* (Pleistocene, Setana Fm.; Kuromatsunai, Hokkaido, Japan); (B) intramural bud (lower right) visible within the damaged walls of an autozooid in *Fenestrulina* sp. (Pleistocene, Setana Fm.; Kuromatsunai, Hokkaido, Japan); (C) two generations of intramural buds within a host zooid of *Cauloramphus disjunctus* (Pleistocene, Setana Fm.; Kuromatsunai, Hokkaido, Japan); (D) reversed polarity intramural bud (centre) in *Porella* sp. (Pleistocene, Setana Fm.; Kuromatsunai, Hokkaido, Japan); (E) intramurally budded avicularium in an autozooid of *Corbulella fossa* (Pleistocene, Lower Castlecliff Shellbed; Wanganui, New Zealand); (F) intramurally budded avicularia filling the orifices of two autozooids (arrows) in *Exochella longirostris* (Recent; Antarctica); (G) intramural bud in a damaged zooid of *Corynotrypa* sp. (Ordovician, Sandbian, Bromide Fm.; Sulphur, Oklahoma, USA); (H) intramurally budded eleozoid with small semi-circular aperture set within the terminal diaphragm of an autozooid of *Atagma sp.* (Cretaceous, Santonian; Vesterival, Seine Inférieux, France). Scale bars: A, B, C, E, F, G = 200 μm; D = 500 μm; H = 100 μm.

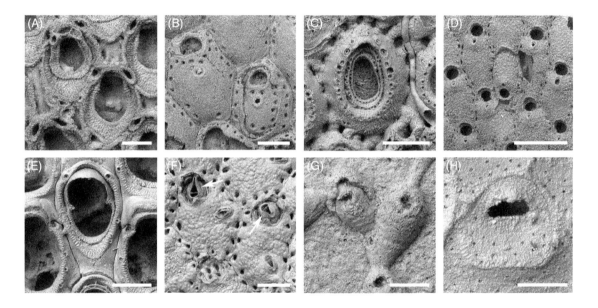

predators. Indeed, the prevalence of intramural buds in bored zooids of *Microporella hyadesi* is suggestive of a boring predator triggering the formation of the intramural buds.

Menon (1973) reported free-living nematodes (*Pelagonema obtusicauda*) living inside zooids of the cheilostome *Electra pilosa*, noting that they occupied most of the chamber and destroyed the zooid before migrating to the next zooid after enlarging the pore plate in the skeletal wall between. By moving between zooids in this way, the nematodes do not rupture the frontal membranes. Similar animals could, potentially, be responsible for zooid mortality leading to intramural budding.

Should the link between intramural budding and predation be firmly established through studies on living bryozoans, this would open the way for employing the spatial and temporal distributions of intramural buds in fossil bryozoans as indicators of predation intensity (McKinney, Taylor, and Lidgard 2003).

5.3 Symbioses

Interest in symbioses involving living bryozoans has mostly concerned associations that have not been recognized in the fossil record, notably symbiotic bacteria in marine bryozoans and cnidarian parasites in freshwater bryozoans. The living cheilostome *Bugula neritina* hosts a symbiotic bacterium called 'Candidatus Endobugula sertula' that synthesizes **bryostatin** (Haygood and Davidson 1997). A particular variety of this bioactive molecule – bryostatin 1 – is a modulator of the enzyme protein kinase C and has been trialled as a drug against tumours, HIV/AIDS, and Alzheimer's disease in humans (Kollár et al. 2014; Wender et al. 2017). Its function in bryozoans seems to be to make the larvae of *Bugula neritina* distasteful to fish predators (Lindquist and Hay 1996; Lopanik, Lindquist, and Targett 2004; Linneman et al. 2014). Molecular analysis of the host bryozoan *Bugula neritina* has shown it to be a complex of cryptic species having differing thermal tolerances, with only the Type D cryptic species containing the bacterium that synthesizes bryostatin 1 (Fehlauer-Ale et al. 2014).

Aside from the bryostatin-producing bacterium, a wide range of other bacteria have been found in association with bryozoans. Heindl et al. (2010) were able to culture no fewer than 340 bacteria in just 14 bryozoan species they collected from habitats in the Baltic and Mediterranean seas. Almost one-third of these bacteria showed antibacterial activity. As most were found to be Gram-negative bacteria, they could potentially hinder the growth of Gram-positive bacteria on the bryozoan surface, possibly to the benefit of the host bryozoans.

Freshwater phylactolaemate bryozoans often harbour myxozoan parasites. These aberrant cnidarians are disadvantageous to the host bryozoans in various ways – including reducing colony growth rates and bringing about malformations of the statoblasts used in propagation (Hartikainen, Fontes, and Okamura 2013) – and cause economically damaging diseases such as Proliferative Kidney Disease when passed on to salmonid fishes.

Marine bryozoan colonies are able to accommodate small symbionts such as spionid polychaetes, sometimes developing tubular skeletons around them (Laubier 1959). Tanaidacean crustaceans can also stimulate the formation of similar structures in host bryozoans, but in neither these nor the spionids is there any indication of an obligatory symbiosis (Taylor 1991). In contrast, Lörz, Myers, and Gordon (2014) described an unusual association between a new taxon of amphipods, *Bryoconversor tutus*, and the cheilostome bryozoan *Onchoporoides moseleyi* in New Zealand. The tiny amphipods appear to live entirely within the basal coelom on the abfrontal sides of the bryozoan branches where autozooids are lacking, seemingly surviving on nutrients provided by the bryozoan host.

The sessile nature of bryozoans means that they frequently form 'incidental symbioses' with a wide variety of other organisms in the broad sense of two different species living together permanently or semi-permanently. These include brachyuran crabs (Key et al. 1999; Savoie, Miron, and Biron 2007), isopods (Key and Barnes 1999), limulids (Key et al. 1996, 2000), pycnogonids (Key, Knauff, and Barnes 2012), and sea snakes (Key, Jeffries, and Voris 1995). More intimate symbioses also occur in which one or both of the symbionts show apparent adaptations to the other. One or both of the species can be obligatory symbionts (i.e. always found in association with the other), while the degree to which the symbioses are species specific can also vary.

A non-obligatory symbiosis involves the sponge *Halisarca* overgrowing the palmate cheilostome *Smittina cervicornis*, leaving only the orifices of the bryozoan zooids uncovered (Harmelin, Boury-Esnault, and Vacelet 1994). Bryozoan colonies with sponge symbionts have longer zooids and higher peristomes than those without, but their growth seems to be unaffected and it is possible that the symbiosis is mutualistic, the two participants cooperating in the production of feeding currents.

5.3.1 Intergrowths, embedments, and bioclaustrations

Sessile suspension feeders often intergrow, typically with a larger symbiont hosting a smaller one that is embedded in its skeleton. Colonial organisms such as bryozoans are particularly common in intergrowth symbioses because of their inherent morphological plasticity and the ability to accommodate smaller symbionts without fatal consequences to the colony as a whole (Taylor 2015). A subcategory of intergrowths in which the embedded symbiont is soft-bodied and is represented only by a mould in the skeleton of the host is termed **bioclaustration** (Palmer and Wilson 1988). Most bioclaustrations involving bryozoans and other colonial animals (corals and stromatoporoid sponges) have been described from the Palaeozoic and these unmineralized 'endosymbionts' apparently peaked in diversity in the Middle Devonian (Tapanila 2005).

Vinn et al. (2019) described conulariids (scyphozoan cnidarians) embedded in colonies of nine different species of trepostome bryozoans from the Upper Ordovician of Estonia. The symbionts are smaller than usual for conulariids and are oriented approximately perpendicular to the surfaces of the bryozoan colonies, having apparently recruited onto the living bryozoans and maintained their quadrate apertures slightly above the surface of the growing colony. Cystoporate bryozoans belonging to two genera, *Fistuliporidra* and *Altshedata*, hosted rugose corals during the Early Devonian in northern Spain (Sendino, Suárez Andrés, and Wilson 2019), forming symbiotic associations that developed in a similar way to those from the Estonian Ordovician.

A skeletal intergrowth common in the Neogene of Europe and North Africa comprises thick-branched colonies of the cheilostome *Celleporaria palmata* hosting the scleractinian coral *Culicia parasita* (Cadée and McKinney 1994; Chaix and Cahuzac 2005). Corallites of *Culicia* are distributed across the surfaces of *Celleporaria* colonies and kept pace with the outward growth through frontal budding of these bryozoans, becoming ever more deeply embedded in the host (Figure 5.8A, B;

Plate 7C). Whereas the host bryozoan has a calcitic skeleton, the skeleton of the coral is aragonitic and its loss may leave regularly spaced perforations in the bryozoan colony which are not always recognized as the former locations of diagenetically leached corallites (e.g. El-Sorogy 2015). A third symbiont, a pyrgomatid barnacle, was occasionally involved in *Celleporaria/Culicia* symbioses from the Pliocene Coralline Crag Formation of Suffolk, settling on the tops of the corallites (Tilbrook 1997).

An Early Devonian intergrowth between a tabulate coral identified as *Aulopora* and the trepostome *Leioclema* was interpreted as mutualistic (McKinney, Broadhead, and Gibson 1990). Both coral and bryozoan normally have encrusting colonies that are limited in size by the size of their substrates. The association allowed three-dimensional erect growth and the formation of much larger colonies in the two symbionts than is typical.

Caupokeras is a bioclaustration found in the superstructures of Devonian and Carboniferous fenestrate bryozoans and interpreted as the mould of the hydrorhizal system of a hydroid overgrown by the host bryozoan (McKinney 2009; Suárez Andrés 2014). Hydroids commonly occur as symbionts of Recent bryozoans, some species spreading across the surface of the host bryozoans and having the stoloniferous basal parts overgrown by bryozoan skeleton while the polyps extend freely above (McKinney 2009, fig. 1E and F). While there is evidence that hosting a hydroid symbiont benefits the bryozoan in terms of protection against predators and increasing overgrowth abilities, there are also indications of food theft by the hydroid and even of consumption of bryozoan zooids (Boero, Bouillon, and Gravili 2000). Thus, the symbiosis may in some instances be one of parasitism.

The ichnotaxon *Anoigmaichnus odinsholmensis* was introduced for 'vermiform' bioclaustrations in Ordovician colonies of the trepostome *Mesotrypa* from Estonia (Vinn, Wilson, and Motus 2014). *Chaeotosalpinx tapanilai* (Figure 5.9A) is a similar bioclaustration found in the cystoporate

Figure 5.8 Bryozoan symbionts: (A) scleractinian coral *Culicia parasitica* on the surface of the cheilostome *Celleporaria palmata* (Pliocene, Coralline Crag Fm.; Ramsholt Cliff, Suffolk, UK); (B) in another example of the same symbiosis, a fractured colony of *Celleporaria palmata* contains the mould remaining after dissolution of a corallite of *Culicia parasitica*, showing the surrounding bryozoan zooids and the bryozoan skeletal wall lining the mould (Pliocene, Kolymbia Fm.; Cape Vagia, Rhodes); (C) three funnel-shaped bioclaustrations of a symbiont that lived in the cyclostome *Actinopora* (Cretaceous, Cenomanian; Mülheim, Westfalia, Germany). Scale bars: A = 1 mm; B, C = 500 μm.

Figure 5.9 Pseudoborings (bioclaustrations) and borings in palaeostomates: (A) longitudinal section of the bioclaustration *Chaeotosalpinx tapanilai* in the cystoporate bryozoan *Stellatoides muellertchensis*, partly infilled with sediment; note the smooth lining consisting of bryozoan skeleton (Devonian, Givetian, Ahbach Fm.; Eifel, Rhenish Massif, Germany); (B) sediment-filled *Trypanites* boring in a trepostome; note the ragged edges of the boring with penetration of sediment into the broken zooids (Ordovician, Sandbian, Bromide Fm.; Arbuckle Mountains, Oklahoma, USA). Scale bars = 1 mm.

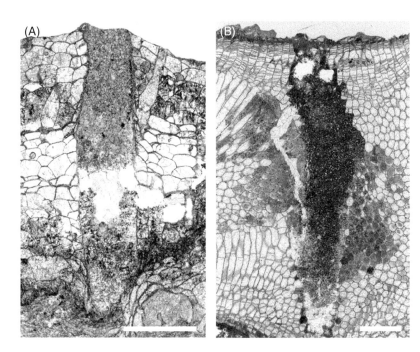

Stellatoides from the Devonian of Germany (Ernst, Taylor, and Bohatý 2014). Both of these bioclaustrations resemble tubes made by Recent cyclostomes around spionid polychaete worms and occasionally tanaidacean crustaceans (Taylor 1991).

A bioclaustration in trepostomes from the Ordovician of the Cincinnati region, USA, was named *Catellocaula vallata* (Palmer and Wilson 1988). Consisting of pits linked by narrower tunnels, it was evidently formed by the partial overgrowth of a soft-bodied organism that settled on the surface of the living bryozoan colonies. In being oriented parallel to the colony surface, *Catellocaula* differs from both *Anoigmaichnus* and *Chaeotosalpinx* which are perpendicular to the surface of the host bryozoan.

Un-named funnel-shaped bioclaustrations (Figure 5.8C) have been found in some mid-Cretaceous European cyclostomes (Taylor and Voigt 2006). Multiple funnels can be found within each colony and four species of bryozoans have been identified as hosts. The symbionts apparently invaded individual zooids at the growing edges of these encrusting colonies, causing the colony to respond by secreting a gall-like, calcified wall around them but without totally encapsulating the symbionts. As the inner surface of the gall adjacent to the symbiont was lined by bryozoan epithelial tissues, the association with the symbiont seems to have been more intimate than examples in which cuticle forms this lining (e.g. the tubes formed around spionids mentioned above).

Similar funnel-shaped structures occupied by an unidentified symbiont have been discovered recently in some living hornerid cyclostomes from New Zealand (P.S. Batson pers comm. August 2018).

5.3.2 Hermit crab symbioses

Gastropod shells inhabited by hermit crabs are habitats for a rich diversity of symbionts, among which are bryozoans (e.g. Kuklinski, Barnes, and Wlodarska-Kowalczuk 2008). Patterns of encrustation in fossil shells lacking their original occupants can be used to infer the former presence of a hermit crab (Walker 1992). The most striking hermit crab symbionts are cyclostomes and cheilostomes constructing thick, generally multilamellar encrustations over the external surfaces of the hermit-occupied shells (Taylor 1994b) (Plate 11A–G). Putative fossil examples of these symbioses always lack the occupants, which is consistent with the feeble calcification of hermit crabs, chelae excepted, and their tendency to vacate shells prior to burial.

Hermit crabs (superfamily Paguroidea) have a fossil record extending back to the Early Jurassic (Collins 2011). The oldest inferred bryozoan symbionts of paguroids date from the Middle Jurassic (Buge and Fischer 1970; Palmer and Hancock 1973) (Plate 11C; Figure 5.10). In contrast to shells that are encrusted while the gastropods are alive, or after death and empty, the presence of 'pagurid facets' and the pattern of encrustation are indicative of occupancy by hermit crabs. Pagurid facets are worn patches near the base of the shell caused by the hermit crab dragging its home along the seabed. That they were created when the bryozoan was alive is evident from the gradual thinning of the layers of bryozoan zooids towards the facet, with kenozooids progressively replacing autozooids as space became restricted.

Symbiotic bryozoans cover the external surface of the gastropod shells more or less evenly, which would not be the case had the shells rested unoccupied on the seabed with their undersides against the sediment. Bryozoan growth also extends onto the inner surfaces of the shells, which are covered by the mantle tissues of the gastropods when

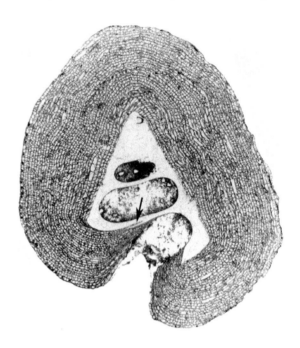

Figure 5.10 Thin section of the inferred paguroid symbiont *Reptomultisparsa incrustans* showing the clear calcite of the trochiform gastropod shell, the >30 layers of cyclostome zooids, and a layer of zooids immured (arrow) between whorls of the gastropod (Jurassic, Bathonian; Calvados, France). See Plate 11C for a similar, unsectioned colony.

alive. In many instances, the bryozoan colonies grow outwards from the aperture, forming an extension of the helicospiral chamber of the gastropod that crudely mimics a gastropod shell (e.g. Taylor 1991; Carter and Gordon 2007; Plate 11B). These bryozoans puzzled early naturalists who were unaware that the specimens represented bryozoan-encrusted gastropod shells occupied by paguroids. Looking into the aperture, no trace could be found of the gastropod shell, leading to the supposition that the shell must have been dissolved, either by the gastropod itself to compensate for the heavy bryozoan encrustation, or by the bryozoan which was then considered to be a shell-dissolving parasite. In fact, when such specimens are sectioned or broken open, a small gastropod shell can be found deeply embedded within the bryozoan colony. The gastropod shell forms a nucleus from which the extended helicospiral whorl created by the bryozoan grows. Kirkpatrick and Metzelaar (1922) made an early study of one of these symbiotic bryozoans, named by them *Conopeum commensale*, which forms 'potato-like' colonies with more than 50 layers of zooids around its hermit crab tenant in the waters off West Africa (see Klicpera, Taylor, and Westphal 2013).

Some bryozoan species living with paguroids are obligate symbionts of the crabs, as in most species of the ascophoran cheilostome *Hippoporidra* (Taylor and Schindler 2004). However, others may be found without a paguroid symbiont. Of the 13 tube-building bryozoan symbionts of hermit crabs living today on the Otago Shelf of New Zealand, at most only three are obligate paguroid symbionts (Taylor, Schembri, and Cook 1989). The symbiotic bryozoans here live with any one of six species of paguroids and there is no species specificity.

As most gastropods have shells of aragonite readily dissolved diagenetically, all that is left in many fossil symbioses is a mould of the shell. Crude replication of the gastropod shell shape by the bryozoan symbionts is presumed to be a physical response to the asymmetrical shape of the hermit crab that is adapted to living in a helicospiral chamber, plus the constraining effect on symbiont growth by the back and forth movements of the crab's body through the aperture. There is no evidence for any genetic control of the bryozoan colony growth pattern. In symbioses from the Otago Shelf, the diameter of the bryozoan tube correlates with the height of the manus of the paguroid occupant (Taylor 1991, text-fig. 1A). Coiling sometimes becomes open or evolute, endowing the colony with a curved tubular shape which, if the nucleus and embedded gastropod shell are broken off, can render their origin as paguroid symbionts obscure. Helicospiral extensions analogous to those developed in bryozoan–hermit crab symbioses occur in various sponges and cnidarians living symbiotically with hermit crabs (e.g. Cairns and Barnard 1984; Sandford and Brown 1997).

Examples of modern bryozoan–hermit crab symbioses from the Otago Shelf of New Zealand are remarkable in showing that the helicospiral tube may be started by one bryozoan species and continued by a different species following the same growth trajectory (Taylor 1991). Very occasionally colonies of two different species growing around the perimeter of the aperture together build the tube. Darrell and Taylor (1989) described fossils from the Pliocene of Florida in which helicospiral tube growth was begun by the cheilostome bryozoan *Hippoporidra* before being continued by the scleractinian coral *Septastrea* along the same pathway.

Advantages and disadvantages to the two participants in bryozoan/paguroid symbioses are poorly understood. However, the bryozoan gains from having a long-lived substrate which it may be able to monopolize, while the powerful feeding and respiratory currents of the hermit crabs create a high flux of particulate food for the bryozoan, especially in the vicinity of the tube aperture where the growth rate of the bryozoan is greatest. The hermit crab may benefit from having a domicile that grows with it, alleviating the need for it to find a larger shell, although crabs kept in the laboratory were found to switch homes (Taylor 1991). As hermit crabs are unable to maintain or repair the gastropod shells they occupy, it is possible that the symbiotic bryozoans may be important to the crabs in excluding destructive boring organisms as well as manufacturing a thick domicile difficult for shell-crushing predators of the crabs to break. Some of the symbiotic bryozoans have branching outgrowths and pigmented colony surfaces (Plate 11D, F), potentially camouflaging the crabs. On the other hand, branched and excessively thick bryozoan colonies may prove cumbersome for their crab tenants, limiting their mobility and manoeuvrability.

The fossil record of putative bryozoan/paguroid symbioses is sparse in the Jurassic and Cretaceous. The diversity of bryozoan symbionts increased steeply in the Oligo-Miocene, with at least 15 known species by the early Miocene (Taylor 1994b; Pérez et al. 2015). Both cyclostome and cheilostomes participate in this symbiosis. A few of the fossil bryozoan species were possibly obligate symbionts of hermit crabs, but others can also be found on other substrates. The ascophoran cheilostome genus *Hippoporidra* is one of the most widely distributed hermit crab symbionts at the present day. However, the aragonitic mineralogy of all but the earliest known species of this genus, *H. portelli* from the late Eocene of Florida (Taylor and Schindler 2004), biases against a more widespread occurrence of *Hippoporidra* in the fossil record. As noted above (Chapter 3.5), male zooids with long tentacles can be found in *Hippoporidra*, possibly allowing sperm to be delivered to other colonies when they are brought together by their hermit crab symbionts.

5.3.3 Other conchicoles

There are several records of bryozoans resembling those that are symbionts of hermit crabs but significantly antedating the origin of hermit crabs in the Mesozoic. Most come from the Early Palaeozoic (Plate 11H, I). The unknown **conchicoles** (Vermeij 1987) inhabiting these bryozoan-encrusted gastropod shells may have been non-paguroid arthropods. Amphipods may inhabit and carry gastropod shells (Carter 1982), although these are not an ancient group of arthropods. Alternative conchicoles may have included 'worms', such as sipunculans known to inhabit vacant gastropod shells today (e.g. Hylleberg 1970) which may be overgrown by corals (e.g. Igawa, Hata, and Kato 2017).

The most complete description of such a Palaeozoic association comes from the Middle Devonian of New York State (Morris, Linsley, and Cottrell 1991). High-spired gastropods belonging to the genus *Palaeozyglopleura* are here encrusted by the trepostome *Leptotrypella*. The pattern of encrustation is very like that seen in modern bryozoan–hermit crab symbioses, with *Leptotrypella* covering the entire outer surface of the gastropod shell and extending down into the aperture. McNamara (1978) inferred a symbiotic association between gastropods and the trepostome *Diplotrypa* in the Late Ordovician of north-west England but it is possible that the bryozoans in

the decalcified material available to McNamara were symbionts of conchicoles rather than living gastropods.

5.4 Bryozoans as Habitat Providers

Large bryozoan colonies form habitats for a wide variety of different animals at the present day (Cocito 2004; Wood et al. 2012). This was surely also true in the geological past, even though many of the associates will have been lost to the fossil record. For example, Bradstock and Gordon (1983) showed how large colonies of two cheilostomes – *Celleporaria agglutinans* and *Hippomenella vellicata* – were important habitat providers in Tasman Bay, New Zealand. One colony of *Celleporaria* with a wet weight of 6.4 kg yielded more than 51 polychaete individuals as well as 27 bivalves, one gastropod, three chitons, nine decapod crustaceans, four ascidians, and a small octopus. A fossil analogue of this bryozoan community, presumably once hosting a similar array of associated animals, was subsequently identified at a nearby Miocene locality in north-west Nelson (Gordon, Stuart, and Collen 1994).

Wood et al. (2012) reviewed the present-day distribution of habitat-forming bryozoans defined as heavily calcified, three-dimensional species regularly attaining sizes over 50 mm in three dimensions and contributing significantly to benthic habitat structure. They found such bryozoans to occur from *c.* 59°N to 77°S, most commonly on stable substrates in temperate shelf environments with consistently rapid currents. Other bryozoans, molluscs, annelids, arthropods, cnidarians, sponges, echinoderms, and macroalgae benefit from the habitat space provided by these bryozoans. Both erect and encrusting cheilostome bryozoans are habitat-forming bioconstructors in the Mediterranean today (Lombardi, Taylor, and Cocito 2014). Many occur in channels or at headlands where water flow is accelerated.

The interstices of large erect colonies are habitats for many animals seeking cryptic and hidden locations. Rao and Ganapati (1980) compared the animals associated with two bryozoans – the cheilostome *Thalamoporella* and the ctenostome *Pherusella* – in the Bay of Bengal. Colonies of *Thalamoporella*, for example, contained on average 443 individuals of sessile epizoans and 2170 of mobile animals per 100g. Accumulation of sediment in the interstices of the calcified frondose *Thalamoporella* colonies was greater than in the soft-bodied branching *Pherusella* colonies, and this was mirrored by the larger number of deposit feeders associated with *Thalamoporella* than *Pherusella*.

Meadows formed by the articulated cheilostome *Cellaria* at 35-metre depth in the northern Adriatic Sea contain a diverse associated biota of animals and plants (McKinney and Jaklin 2000). The identity of this biota varies according to depth within the bryozoan colonies: shallow, distal branches contain mainly foraminifers, rhodophyte algae, and sponges; intermediate branches are dominated by hydroids, ctenostome bryozoans, annelids, and ascidians; and deep, proximal branches by cheilostome and cyclostome bryozoans.

'Reefs' constructed by the cheilostome *Schizoporella errata* in south-eastern Brazil provide habitats for no fewer than 31 species of crabs (Alves et al. 2013) and 70 polychaete species (Morgado and Tanaka 2001). With respect to the crabs, the bryozoan colonies provide refuges for the early developmental stages of some species, whereas other species apparently use them as shelters or food sources throughout their lives. Finally, a small goby was recently described from Indonesia as *Sueviota bryophila*, the species name reflecting its association with bryozoans (Allen, Erdmann, and Dita Cahyani 2016). The fish nestles in folds of reticulate colonies of the cheilostome *Triphyllozoon*, pigmented spots on the body camouflaging it against the background pattern of the bryozoan fenestrules.

5.4.1 Endoliths

The skeletons of stenolaemate and cheilostome bryozoans can be bored by various non-predatory organisms seeking domiciles (Figure 5.9B). Boring sometimes occurred entirely during the life of the bryozoan but in other instances it began *syn vivo* and continued after death, or else was entirely post mortem.

Erickson and Bouchard (2003) introduced the ichnofossil *Sanctum laurentiensis* for borings in ramose trepostomes and cystoporates from the Late Ordovician of the Cincinnati region. These borings enter through an opening on the colony surface about 1–3 mm wide, and extend into the endozone where they expand into elongate or saccate chambers of variable size. Continued growth of the bryozoan around the opening of a boring indicative of a *syn-vivo* relationship was observed in just one example; the timing (*syn vivo* or post mortem) of other examples could not be established. It was speculated that the Cincinnatian borer was a small crustacean. *Sanctum* has since been recognized in ramose trepostomes from the Middle and Upper Ordovician of Estonia by Wyse Jackson and Key (2007) who suggested a polychaete trace maker. These authors also described the presence of another boring ichnogenus, *Trypanites*, in co-occurring hemispherical trepostome colonies, considering that both *Sanctum* and *Trypanites* were made by the same borer, their different morphologies depending on the colony-form (ramose vs. dome-shaped) of the host bryozoan.

An older boring found in some Sandbian bryozoans from Estonia – *Osprioneides kampto* – was interpreted as a post-mortem excavation as it elicited no response from the host bryozoan in terms of deflection of growth laminae (Vinn, Wilson, and Motus 2014). Other boring ichnotaxa found in fossil bryozoans are *Rogerella* (pip-shaped borings made by acrothoracican barnacles) and *Gastrochaenolites* (crypt-like borings generally made by bivalve molluscs).

Microborings are commonly observed when living and fossil bryozoans are examined using SEM (Figure 5.11A, B). They seem never to have been studied in detail and little is known about the identities of the microborers, or how they interacted with the host bryozoans.

5.4.2 Microbial mats

Bacteria are ubiquitous foulers of firm and hard surfaces in the sea (Wahl 1989), including not only

Figure 5.11 Bryozoan-associated microendoliths and microbial threads: (A) microborings and pseudopores visible on the underside of a frontal shield of the cheilostome *Pentapora foliacea* (Recent; La Spezia, Italy); (B) microborings in the frontal shield of the cheilostome *Cellarinella* (Holocene; Antarctica); (C), (D) pyritized ?cyanobacteria overgrowing zooids of the cyclostome *Corynotrypa* (Ordovician, Katian; Cincinnati, Ohio, USA). Scale bars: A, B = 50 μm; C = 200 μm; D = 100 μm.

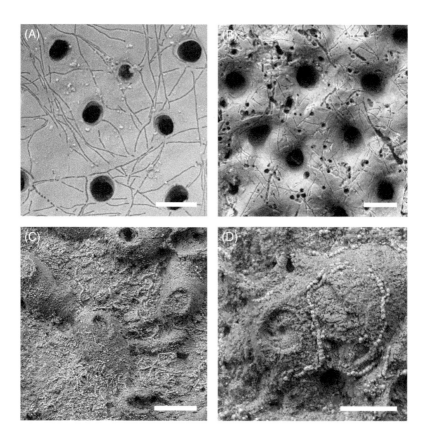

substrates encrusted by bryozoans but also the surfaces of the bryozoan colonies themselves (e.g. Scholz and Krumbein 2006; Heindl et al. 2010, fig. 1). Microbial biomats comprising communities of autotrophic and heterotrophic bacteria, as well as fungi and diatoms, potentially interact with bryozoans in various ways but empirical data detailing these interactions is almost entirely lacking. Nevertheless, Scholz and Levit (2003) have suggested

that the feeding currents of bryozoans favour the development of microbial biomats on colony surfaces, while the biomats protect the host bryozoans against predators and help to delay or halt overgrowth by other, faster-growing bryozoan colonies. They also indicated that cheilostome species with calcified exterior frontal walls (gymnocysts) are better suited to the growth of biomats. However, a study (Kittelmann and

Harder 2005) of four sympatric cheilostome species in the North Sea found only one (*Conopeum reticulum*) to be fouled by bacteria, and this was the sole species lacking a gymnocyst. As noted above (Chapter 5.3), Heindl et al. (2010) suggested that some of the bacteria on the surfaces of bryozoan colonies may hinder the growth of other bacteria.

With regard to the influence of microbial mats on settlement, Ochi Agostini et al. (2017) concluded that bryozoans and other sclerobionts were facilitated by higher biofilm bacterial densities on the surfaces of dead mollusc shells. Brancato and Woollacott (1982) showed experimentally that larvae of three species of the cheilostome *Bugula* (all now placed in other genera of Buguilidae) preferentially settled on surfaces covered by microbial films, corroborating the earlier findings of Dobretsova and Qian (2006) using *Bugula neritina*. On the other hand, Wieczorek and Todd (1997) found *Bugulina flabellata* larvae to be inhibited from settlement by biofilms. These studies underline the difficulty in generalizing making generalizations.

Sediment trapping and/or biomineralization can preserve microbial mats as various types of **microbialites**, such as stromatolites and oncolites (see Riding 1991 for microbialite classification). Some remarkable structures colloquially referred to as 'growing stones' have been known for centuries in the shallow brackish inland waterways of south-eastern Holland (Bijma and Boekschoten 1985). Often capped by stromatolitic crusts, most growing stones are built by the cheilostome *Einhornia crustulenta* and comprise multilayered spheroidal bryozoan colonies, or brittle bioherms that can reach up to one metre in thickness.

Intergrowths of bryozoans and microbial mats in the Coorong Lagoon of South Australia may be the result of seasonal changes in salinity and water level favouring either bryozoan growth or microbial mat growth (Palinska et al. 1999). The result is the formation of rigid 'bryostromatolites'. According to Kaselowsky et al. (2005), the growth

of bryostromatolites is induced by the presence of microbial biofilms that alternate with the layers of bryozoan zooids. An interesting environmental parallel to the Coorong Lagoon is the Albufeira coastal lagoon of Portugal where, during February 1990, a population explosion occurred in a cheilostome bryozoan identified as '*Membranipora savarti*' along with the serpulid *Ficopomatus enigmaticus* (Freitas et al. 1994). This event coincided with a high level of water in the lagoon due to a fluvial influx after very heavy winter rainfall which depressed the salinity down to about 6.5–9 ppt. Somewhat counterintuitively, therefore, low salinity favoured cheilostome colony growth in this lagoon despite the fact that cheilostomes as a group are more usually associated with fully marine conditions and their diversity typically declines as salinity decreases (e.g. Occhipinti Ambrogi 1985). It should, however, be noted that some cheilostome species prefer – or at least tolerate – brackish water of <10 ppt, such as *Einhornia crustulenta* in the Baltic Sea (Grabowska, Grezelak, and Kuklinski 2015).

The fossil record provides examples of bryozoan–microbial associations. A Silurian microbial-dominated reef in Gotland, Sweden, contains several bryozoan species, including thin encrusting colonies intercalated with calcimicrobes such as the probable calcified cyanobacterium *Rothpletzella* (Ernst, Munnecke, and Oswald 2015). Late Miocene (Messinian) associations between bryozoans and microbialites have been described from Sicily and Sardinia (Moissette et al. 2002). Twenty-four species of bryozoans were recorded in these shallow-water associations, including encrusting and nodular celleporid cheilostomes, as well as encrusting cheilostomes and cyclostomes growing within the microbial bioherms.

Ferruginous oncoliths, known locally as 'snuff-boxes', are found in the Bajocian (Middle Jurassic) of southern England and northern France. The microbial laminae of the oncoliths are interlayered with an encrusting biota that includes bryozoans. Palmer and Wilson (1990) argued that this biota was cryptic, growing on the undersides of the cobble-like oncoliths simultaneously with growth of the microbial laminae. Elsewhere in the Middle Jurassic, calcareous oncoliths can be densely encrusted by cyclostome bryozoans (Plate 8J).

A very different style of microbial mat preservation was recently described from the Late Ordovician of Cincinnati, USA (Wilson and Taylor 2017). Filamentous microencrusters – interpreted as cyanobacteria – form pyritized mats extending over the external surfaces of brachiopod shells and overgrowing zooids of the cyclostome bryozoan *Corynotrypa* living on the same shells (Figure 5.11C, D). It is unknown whether the mats developed after the death of the bryozoans or contemporaneously as no evidence could be found for reciprocal overgrowth.

6 Ecology and Palaeoecology

As major components of hard substrate communities since the Ordovician, bryozoans have ecological and palaeoecological importance. Some aspects of bryozoan ecology and palaeoecology, notably functional morphology and biotic interactions, have been covered already and this chapter concentrates on the habitats colonized by bryozoans as well as their broader facies associations. The ecology of living bryozoans must guide and inform our interpretations of the bryozoan palaeoecology and the nature of the ancient communities in which they lived.

The limited value in stratigraphy of the typically long-ranging and often patchily distributed bryozoans is widely acknowledged, notwithstanding a few regional applications (e.g. Mesentseva 2008). Therefore, the practical use of bryozoans in geology has focused on their value as 'facies fossils' to provide insights into palaeoenvironments. Colony-form has been widely employed for this purpose, despite our imperfect knowledge of the distributions of different bryozoan colony-forms across the spectrum of environments inhabited by these animals. More recently, zooid size variability has been used as a proxy for temperature seasonality (MART analysis). This, along with ecophenotypic variations in colony-form within species, may be more useful as palaeoenvironmental proxies.

6.1 Bryozoan Habitats

Bryozoans typically flourish in habitats with plentiful hard or firm substrates, low rates of sedimentation (or protection from smothering by sediment), and currents strong enough to replenish planktonic food resources. Areas of high productivity, such as shelf breaks, may support the

Bryozoan Paleobiology, First Edition. Paul D. Taylor.
© 2020 Natural History Museum. Published 2020 by John Wiley & Sons Ltd.

most diverse bryozoan faunas present in a given region (e.g. López Gappa 2000). In the Adriatic Sea, the highest diversities of bryozoans are found along submarine escarpments and sloping surfaces with strong and constant currents (Novosel 2005).

With few exceptions, the availability of a hard or firm substrate is essential for bryozoans. In their study of recent bryozoans from Svalbard, Kuklinski and Barnes (2005) distinguished between 'flexible' and 'stable' substrates, the former including algae and the latter shells and stones. Flexible substrates are generally ephemeral but bryozoans colonizing them are less likely to face competition from others. All the erect rigid colonies in their study were attached to stable substrates, whereas encrusting and non-rigid erect colonies occurred also on flexible substrates.

The continental shelf off Venezuela and Guyana is characterized by active deposition of silty clay in the inner shelf where bryozoans are virtually absent but no deposition over the outer shelf where a diversity of free-living and other bryozoans are present (Rucker 1967). Coarse sand and gravel seabeds are more usually colonized by bryozoans than are muds, silts, and fine sands. For example, Madurell et al. (2013) undertook a multivariate analysis of bryozoan distribution relative to environmental variables in an area of the north-western Mediterranean, finding higher diversities, abundances, and biomasses where sediments were coarser.

Most of the bryozoan species present in many coastal habitats occur underneath or on the sides of boulders and rocks, or in crevices, rather than on exposed surfaces (e.g. Cook 1985). According to Harmelin (2000), a large part of the bryozoan fauna occurring along coasts at the present day occupies cryptic habitats ranging from small cavities to large submarine caves. Spjeldnaes (2000) described cryptic bryozoan habitats in the littoral zone of Gambia, in some of which the bryozoan colonies were living in direct contact with sand grains. Likewise, the undersides of stones in Florida were found to be encrusted by a diverse bryozoan fauna, even when the stones were lying directly on sand (Winston 1982, p. 160). The ctenostome *Cryptoarachnidium* was found to be the dominant fouling organism on the undersides of artificial 'reefs' off southern California, the bryozoan

colonies fixing large amounts of fine sand between their zooids (Carter et al. 1985).

A tendency to encrust cryptic surfaces was found by Bowden et al. (2006) who deployed settlement panels at 8- and 20-metre depths over a period of three years off the Antarctic Peninsula. Spirorbid polychaetes and bryozoans were the dominant encrusters colonizing these panels. Total areal cover was <10% on all upward-facing panel surfaces, with coralline algae often the only taxa present by the end of the study. In contrast, downward-facing surfaces had up to 100% coverage. Ice scour in this environment adds to sedimentation and competition from phototrophic algae to inhibit the presence of bryozoans on upper surfaces.

Encrustation of undersides is a feature of some living organic substrates too. López Gappa and Landoni (2009) studied patterns of bryozoan encrustation on shells of the scallop *Psychrochlamys patagonica*, finding that significantly more bryozoan species and colonies encrusted the lower than the upper valves.

Not all bryozoan species react in the same way to sediment. The Antarctic cheilostome *Inversiula nutrix*, which is a poor competitor for substrate space, is nevertheless the spatially dominant species where it occurs, apparently because it has a greater tolerance to sedimentation than the other species present in the same communities (Clark, Stark, and Johnston 2017).

Probert and Batham (1979) considered the low sedimentation rate over the mid- and outer shelf off the Otago Peninsula in New Zealand to be a major factor determining the high abundance of bryozoans living here. Accumulating skeletal remains, along with pebbles, provide hard substrates for the bryozoans, while the fast-flowing Southland Current enhances nutrient supply. Large bryozoan colonies are the dominant benthic animals over an area of roughly 110 kilometres^2 from 70-metre depth to the shelf break, particularly over the mid-shelf (Wood and Probert 2013). They include thickets of the cyclostome *Cinctipora elegans*. This ramose species also forms linear reef-like structures 0.5–1 metres high in eastern Foveaux Strait where aggregate mounds contoured by tidal currents can reach more than 10 kilometres long by half a kilometre wide (Gordon, Taylor, and Bigey 2009).

Proximity of source populations can also be important, at least in the short term. Long and Rucker (1970) noted the slow rate of bryozoan fouling of settlement panels in deep-water arrays they had deployed because of the large distance from inshore source populations and consequently the sparsity of bryozoan larvae, which are short-lived in the majority of species.

6.1.1 Intertidal boulders

Boulders in intertidal zones, particularly in the deep intertidal, are commonly colonized today by bryozoans (Gordon 1972; Dick and Ross 1985; Dick, Grischenko, and Mawatari 2005). Most colonies occur on boulder undersurfaces sheltered from sunlight and which remain damp when exposed. In Hawaii, boulders of lava at four intertidal sites yielded 32 encrusting species, dominated by ascophoran cheilostomes, including a few species found only growing over calcareous algae that had colonized the same volcanic substrates (Dick, Tilbrook, and Mawatari 2007). The undersides of pieces of reef-flat rubble in Western Samoa are colonized by 10 bryozoan species of mostly small jointed colonies, which are concentrated at the centres of the cobbles and surrounded by sponges and ascidians (Gordon 1989). Locally, the number of species and individuals can be positively correlated with boulder size (e.g. Grzelak and Kuklinski 2010).

Relatively few rocky shoreline communities have been described in the geological record (Johnson 1992), but some contain bryozoans. They include a Late Cretaceous community from Baja California in which bryozoans encrust oyster shells and rocks (Lescinsky, Ledesma-Vázquez, and Johnson 1991). Mergl (2004) described an Ordovician (Arenig) community from Bohemia containing decalcified erect and encrusting bryozoans, as well as sponges and brachiopods, which were interpreted to have lived attached to the sides of crevices in boulders and cliffs at a high palaeolatitude.

6.1.2 Subtidal rocky bottoms, cobbles, and pebbles

Bryozoans are the most diverse and abundant phylum associated with pebbles in the Dover Strait at the present day (Foveau et al. 2008). Muddy seafloors in the Arctic that are unsuitable for most bryozoans but contain dropstones released from the ice are 'oases' for bryozoans adapted for life on hard substrates (e.g. Oschmann 1990; Kuklinski et al. 2006; Kuklinski and Bader 2007). Lithic clasts, mostly of dropstone origin, in the Greenland Sea support rich bryozoan assemblages (Kuklinski 2009). Their diversity, as well as the disparity of morphotypes present, tends to increase with depth, correlating with reduced levels of disturbance from icebergs and wave action. Iceberg scour in the Antarctic has a profound effect on epibenthic communities: few bryozoans are found in recently disturbed areas, but as recovery proceeds there is a transition to communities rich in erect cheilostomes including *Cellarinella*, *Melicerita*, and *Cellaria* (Bader and Schäfer 2005). Subtidal cobbles north of Phuket, Thailand, are colonized by bryozoans on both upper and lower surfaces but coverage is greater on lower surfaces, particularly of encrusting species, with erect forms such as phidoloporid cheilostomes having fenestrate colony-forms occurring mainly on upper exposed surfaces (Sanfilippo et al. 2011).

Turning to the fossil record, Wilson (1986) described cyclostome bryozoans and other sclerobionts associated with cobbles in the Early Cretaceous (Aptian) of Faringdon, Oxfordshire, where sponges and phosphatic pebbles are other common substrates (Figure 6.1A). The cobbles provided two main habitats for encrusting bryozoans: exposed outer surfaces, and the cryptic interiors of vacant borings made by bivalves. Bryozoan diversity on outer surfaces is low, the main taxon present being *Reptoclausa* which has thickly calcified colonies apparently able to tolerate abrasion due to the cobbles being rolled and sand-blasted. In contrast, the walls of the borings provided a more benign habitat for several taxa, including delicate runners and ribbons. A late Eocene 'mobile rockround fauna' from Oamaru, New Zealand, was described by Lee, Scholz, and Gordon (1997). The pebbles and cobbles are encrusted by 70 species of cyclostome and cheilostome bryozoans, along with serpulids, coralline algae, and foraminifera. This warm-water biota was inferred

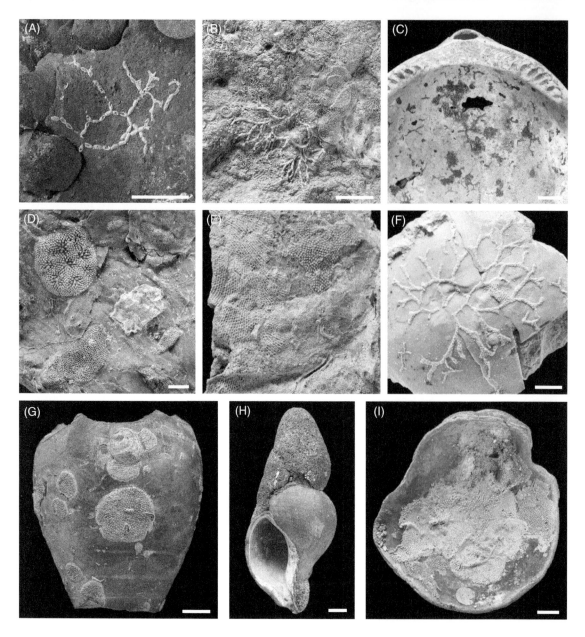

Figure 6.1 Abiotic (A–B) and biotic (C–I) fossil substrates colonized by bryozoans: (A) phosphatic pebble encrusted by the cyclostome *Voigtopora* (Cretaceous, Aptian, Faringdon Sponge Gravel; Faringdon, Oxfordshire, UK); (B) sedimentary hardground (probably the roof of an undercut) encrusted by numerous cyclostome bryozoans including runner-like colonies of *Stomatopora* (Jurassic, Bathonian, Forest Marble; Moonfleet, Dorset, UK); (C) interior of a shell of the bivalve *Glycymeris* encrusted by small bryozoan colonies (darker patches) (Pleistocene, Red Crag Fm.; Brightwell, Suffolk, UK); (D) underside of a stromatoporoid with an encrusting community of runner-, ribbon-, and sheet-like bryozoans (Silurian, Llandovery, Visby Beds; Lickershamn, Gotland, Sweden); (E) underside of a scleractinian coral encrusted by large, sheet-like cheilostomes (Miocene, Serravallian; Samarinda, East Kalimantan, Indonesia); (F) ribbon-like cyclostome encrusting the test of the irregular echinoid *Echinocorys* (Cretaceous, Coniacian Chalk; Chatham, Kent, UK); (G) upper stem and cup of the crinoid *Apiocrinus elegans* hosting several discoidal cyclostome colonies (Jurassic, Bathonian, Bradford Clay; Wiltshire, UK); (H) sinistral gastropod *Neptunea angulata* with a large colony of the cheilostome *Turbicellepora* encrusting the apex, probably *syn vivo* (Pliocene, Red Crag Fm.; Walton-on-the-Naze, Essex, UK); (I) interior of a bivalve shell encrusted by several colonies of cheilostomes, including a large, ovicellate colony of *Powellitheca waipukurensis* (Pleistocene, Nukumaru Limestone; Waiinu Beach, New Zealand). Scale bars = 5 mm.

to have accumulated in a current-swept channel between volcanic islands at 25–50-metres depth.

Cobbles and pebbles in the Pleistocene of Calabria, Italy, are encrusted by bryozoans and serpulids, along with subordinate foraminifera, molluscs, and calcareous algae (Di Geronimo, Rosso, and Sanfilippo 1995). Perfect preservation of the encrusters, such as bryozoans with intact oral spines, points to the fact that the clasts were stable at the time they were encrusted. The bryozoans encrusting the cobbles and pebbles at another Pleistocene locality, Kuromatsunai in northern Japan (Plate 8I), are so abundant and diverse that the locality has been dubbed 'Kokemushi Paradise', *kokemushi* being the Japanese vernacular name for bryozoans. At least 120 bryozoan species have been identified here, and it is not unusual to find more than 20 species growing on a single clast (Taylor et al. 2012). Bryozoans encrust all sides of the clasts, occasionally totally enveloping small-sized pebbles. This marine gravel within the Setana Formation is interpreted to represent the infill of a submarine channel, like the deposits described above from Faringdon and Oamaru. It is possible that tsunamis were responsible for the periodic transportation of clasts into the channel from shallower waters. The clasts became encrusted by bryozoans that benefited from the strong, plankton-laden currents passing along the channel between periods of disturbance and burial.

6.1.3 Sedimentary hardgrounds

Hardgrounds are "synsedimentarily lithified carbonate seafloors that became hardened *in situ* by the precipitation of a carbonate cement in the primary pore spaces" (Wilson and Palmer 1992, p. 3). Environments containing hardgrounds and mechanisms of their formation were reviewed by Christ et al. (2015). Hardgrounds furnish a variety of microhabitats that can be colonized by bryozoans (Figure 6.1B)

The upper surfaces of Palaeozoic hardgrounds are occasionally colonized by encrusting or erect palaeostomate bryozoans but densities tend to be low (e.g. Palmer and Palmer 1977; Brett and Liddell 1978; Vinn and Toom 2016b). There is no indication that the bryozoans living on hardgrounds differ taxonomically from those recruiting onto other hard substrates. Undercut hardgrounds provide cryptic habitats for bryozoans. A good example is a Late Ordovician hardground from Kentucky in which the roofs of the cavities are encrusted by the trepostome *Stigmatella personata* (Buttler and Wilson 2018). The same bryozoan also encrusts the exposed upper surfaces of the hardground, underlining the lack of polarization in the Ordovician between taxa utilizing these two contrasting habitats.

Encrusting cyclostome bryozoans are a major component of the fauna of a Middle Jurassic (Bathonian) hardground at Bradford-on-Avon, Wiltshire, colonizing the roofs of crevices formed by burrowing animals before lithification (Palmer and Fürsich 1974). A few erect bryozoans also occur on the exposed upper surface of the hardground, often detached before burial of the hardground and found loose in the overlying Bradford Clay. Bryozoans can also be found encrusting the cryptic habitats furnished by the hardened walls of *Thallasinoides* burrows formed before lithification in hardgrounds from the Bathonian White Limestone Formation of southern England (Fürsich and Palmer 1975).

The surfaces of *in-situ* hardgrounds in the Late Cretaceous Qahlah Formation of Arabia are seldom colonized by bryozoans (Wilson and Taylor 2001). However, the Qahlah hardgrounds that were broken up into cobbles, as well as the walls of burrows formed in the hardgrounds prior to their lithification, support a moderately diverse bryozoan assemblage of 11 genera, along with representatives of other sclerobiont groups. These encrusting bryozoans lived cryptically, as in the Jurassic examples described above.

Authigenic mineralization of limestone substrates at between 65 and 510 metres depth in the Tyrrhenian Sea, Italy, has produced ferromanganese crustgrounds (Toscano and Raspini 2005). As with many hardgrounds, the encrusting biota is polarized, with bryozoans found on the more irregular lower surfaces. These include colonies of the cheilostome *Puellina* showing active precipitation of botryoidal Fe-Mn oxides over the skeletal surfaces.

Hiatus concretions – concretions that have been reworked and exhumed – are occasionally encrusted by bryozoans. Numbering six genera, cyclostomes are the most taxonomically diverse group found on hiatus concretions in the Middle Jurassic of the Polish Jura where carbonate hiatus concretions in mudstones formed mobile hard substrates repeatedly overturned by storm currents (Zaton et al. 2011).

6.1.4 Submarine caves and cavities

The walls of submarine caves provide habitats for many bryozoans (e.g. Scholz and Hillmer 1995; Novosel 2005). A good example described by Harmelin (1997) is a 120-metre-long dark cave in the French Mediterranean with an entrance at a depth of about 15 metres. Fifty-three bryozoan species were found in this dark cave. The inner parts of the cave, more than 20 metres from the entrance, show a clear pattern of decreasing bryozoan diversity with distance from the entrance – from 40 to 10 species – with the proportion of cyclostomes to cheilostomes increasing over this distance from roughly 20% to 50%. Dark caves in the same region contain a lower proportion of erect species compared to encrusting species than do better-illuminated habitats (Harmelin 2000). Similar patterns were found in three Sicilian caves (Rosso et al. 2012), with species diversity declining, erect species becoming less common, and the proportion of cyclostomes increasing with distance into the caves. Such changes may be due to the decrease in plankton that has been quantified in some other Mediterranean caves (Palau et al. 1991).

Subreefal palaeocaves of Late Jurassic (Kimmeridgian) age have been exhumed on the coast near La Rochelle in south-west France where they form subaerial caves in the modern coastal cliffs (Taylor and Palmer 1994). The palaeocaves, each of which can be over one metre high and five metres deep, have walls formed by thrombolitic microbialites encrusted by a range of Jurassic organisms, including cyclostome bryozoans.

Small cavities between 6 and 15 cm high are developed in a condensed sequence of Jurassic pelagic carbonates in Slovakia (Schlögl et al. 2008).

The roofs of these Tithonian micro-caves are densely encrusted by serpulid worm tubes among which are cyclostome bryozoans that interrupt two phases of serpulid bioconstructions.

Caves that developed between large boulders following a Pliocene marine transgression in Rhodes have walls encrusted by sheet-like cheilostomes, dominated by the anascans *Parellisina* and *Onychocella* (Steinthorsdottir and Håkansson 2017). Although the sediments infilling the caves contain numerous fragments of erect bryozoans, there are no indications of broken bases of these bryozoans on the cave walls and it is unclear whether they represent cave-dwellers or were transported into the caves from elsewhere.

6.1.5 Shellgrounds

The shells of molluscs are frequently encrusted by bryozoans, and often form concentrations represented by shellgrounds today and by shellbeds in the fossil record (Figure 6.1C). For example, in the fjords of northern Norway, shells of the pectinid *Chlamys islandica* at 20–70 metres are heavily fouled by bryozoans associated with other sclerobionts such as sponges, serpulid worms, and barnacles (Bader and Schäfer 2005). The right valves of the bysally attached bivalves form cryptic habitats for encrusting bryozoans and serpulids, whereas the upward-facing left valves support more complex and tiered communities that also include flexible erect bryozoans.

Shellgrounds can support high diversities of encrusting bryozoans, the majority of which have sheet-like colonies. At least 60 species of cheilostomes occur commonly on disarticulated valves of four species of bivalves at 35 metres depth in the Irish Sea south of the Isle of Man (Ward and Thorpe 1989). Particular species present in this community vary in their abundances according to whether the surfaces of the shells are rough or smooth. These variations, plus differences in life histories (e.g. timing of recruitment) between species, may explain the ability of so many bryozoan species to co-exist in a single community.

Most biotic substrates encountered in the fossil record that are encrusted by bryozoans are calcareous shells. They include other bryozoans (Thomsen 1977; McKinney and Taylor 2006), foraminifera (Berning et al. 2009), stromatoporoids (Kershaw 1980; Vinn 2012; Figure 6.1D), corals (e.g. Sanders and Baron-Szabo 2008; Figure 6.1E), echinoids (Nebelsick, Schmid, and Stachowitsch 1997; Zamora et al. 2008; Borszcz, Kuklinski, and Zatoń 2013; Figure 6.1F), crinoids (Liddell and Brett 1982; Figure 6.1G), hyoliths (Galle and Parsley 2005), trilobites (Kátcha and Saric 2009; Key et al. 2010), brachiopods (e.g. Hoare and Steller 1967; Kesling, Hoare, and Sparks 1980; Sparks, Hoare, and Kesling 1980; Alvarez and Taylor 1987; Alexander and Scharf 1990; Brice and Mistiaen 1992; Lescinsky 1997; Zaton and Borszcz 2013; Plates 6B–C, 8E), crabs (Key et al. 2017), and molluscs, especially bivalves (e.g. Taylor 1979b; Bishop 1988; Ward and Thorpe 1989; Parrass and Casadio 2006; Zuchsin and Baal 2007; Zaton, Wilson, and Zavar 2011; Breton 2017; Plates 6A, 10I), but sometimes ammonites (Wilson and Taylor 2012; Luci, Cichowolski, and Aguirre-Urreta 2016), nautiloids (Luci and Cichowolski 2014; Wyse Jackson and Key 2014; Pohle et al. 2019; Plate 10A), belemnites (e.g. Pugaczewska 1965; Koromyslova et al. 2018) or gastropods (e.g. Hladilová, Zágorsek, and Ziegler 2004; Figure 6.1H). Each post-mortally encrusted shell represented a separate 'benthic island' on the seafloor, making their individual study justifiable.

Several criteria can be used to infer whether shell substrates were encrusted by bryozoans while the host was alive (*syn vivo*) or after it had died (post mortem). These vary in their potency, from unequivocal to merely suggestive (Taylor and Wilson 2003, table 4). For example, finding bryozoans encrusting the internal surfaces of shells that would have been covered by soft tissues when the host was alive can be taken to indicate a post-mortem association. Another good post-mortem indicator is the growth of a bryozoan colony across the commissure of a bivalved host from one valve to the other. Conversely, evidence of a response in the growth of the host skeleton to the presence of an encrusting bryozoan implies a *syn-vivo* association. Occasionally, bryozoan

growth halts abruptly at a prominent growth line on the host. As argued by Ager (1961) for brachiopod hosts, this points to the perturbation that caused the growth line also resulting in death of the bryozoans and can therefore be used to infer a *syn-vivo* association.

Bishop (1988) studied the spatial distributions of the small cheilostome colonies encrusting the concave interior surfaces of disarticulated bivalve shells in the Early Pleistocene Red Crag Formation of eastern England (Figure 6.1C). He found colonies to be concentrated on what would have been the highest points of the shell interiors, reflecting larval behaviour during recruitment that positioned colonies furthest from the sediment on which the shells rested.

The Wanganui Basin of New Zealand contains a shallow marine sequence spanning the entire Quaternary (Pillans 2017). There are numerous shellbeds, each 0.1–4.5 metres thick and consisting predominantly of bivalves and gastropods, that were deposited during intervals of reduced sedimentation above transgressive surfaces (Abbott 1997). The shells constituting these mid-cycle condensed shellbeds are encrusted by highly diverse communities of cheilostome and cyclostome bryozoans (Figure 6.1I), which Rust and Gordon (2011) recognized as a distinct biofacies (Biofacies 2).

6.1.6 Seeps and vents

Although not often noted, cold seeps can be colonized by bryozoans. Off south-eastern New Zealand today, carbonate chimneys up to 90 cm in height associated with cold seeps at depths of *c.* 220 metres are encrusted by bryozoans (Orpin 1991). A low-temperature hydrothermal vent described from the Early Carboniferous of Newfoundland has trepostome bryozoans intercalated with microbial carbonates (Morris et al. 2002). Records of bryozoans living in hot vents are sparse. A possible exception is provided by the ctenostome *Parachnoidea rowdeni* found associated with a bathymodiolin mussel at a sulphur-rich hydrothermal spring in the Kermadec volcanic arc (Gordon 2012). However, the ambient temperature was perhaps only 10°C at the exact site of bryozoan collection.

6.1.7 Algal epiphytes

Many modern bryozoans are epiphytes of laminarian and other marine algae, some obligatory. For instance, Rogick and Croasdale (1949) reported 29 species of bryozoans growing on mostly brown algae in the Woods Hole region of Massachusetts, either attached to blades or holdfasts. The relationships between the bryozoans and their living algal substrates can be complex. In some cases, the bryozoans diminish the amount of light reaching the algal tissues beneath (e.g. Oswald et al. 1984), and they may also decrease the output of algal spores (Saier and Chapman 2004). However, an example has been reported of a cyclostome epiphyte potentially benefitting its host by deterring the grazing of herbivorous gastropods (Durante and Fu-Shiang 1991), while ammonia excreted by some epiphytic bryozoans is a source of nitrogen for the alga *Macrocystis integrifolia* (Hurd et al. 1994). Additionally, in the calcareous red alga *Gelidium sesquipedale*, the presence of the epiphytic bryozoan *Electra pilosa* was shown to increase the availability of CO_2 to the alga, with potential photosynthetic benefits (Mercado, Carmona, and Niell 1998).

Given the poor preservation potential of non-calcified marine algae, especially in high-energy nearshore settings, it is not surprising that reports of unequivocal fossilized bryozoan epiphytes of algae seem to be lacking.

6.1.8 Seagrasses

Seagrasses today are a habitat for many bryozoans which can attach to the blades or to the rhizomes (e.g. Koçak, Balduzzi, and Benli 2002; Novosel 2005; Kouchi, Nakaoka, and Mukai 2006). In a review of living and fossil bryozoans associated with seagrasses, Di Martino and Taylor (2014) listed a total of 281 species, comprising 67 cyclostomes, 11 ctenostomes, and 203 cheilostomes. The majority of these come from the present-day Mediterranean where 152 species have been recorded. Very few Recent bryozoan species are

obligate seagrass epiphytes, although *Electra posidoniae* encrusts only seagrass blades in the Mediterranean.

Keough (1986) studied the cheilostome *Bugula neritina* on the seagrass *Thalassia testudinum* in the Gulf of Mexico. He found most colonies to be attached to the older, distal parts of the blades (like terrestrial grasses, seagrasses have a basal meristem), where colonies were able to grow faster. The anascan cheilostome *Calpensia nobilis* commonly grows around the rhizomes of seagrasses in the Mediterranean today (Cigliano, Cocito, and Gambi 2007). While being associated with *C. nobilis* might seem to be disadvantageous to the plant, fewer boring polychaete worms are found in the bryozoan-encrusted rhizomes, suggesting a possible benefit.

Seagrasses in South Australia are encrusted by a variety of bryozoans (Plate 10F–H). Among these are the cyclostome *Densipora* (see Di Martino and Taylor 2014, fig. 4E) and the cheilostome *Celleporaria cristata* (see Pouyet 1978) that have remarkably similar colonies, not only through their tubular morphology resulting from growth around the seagrass stems, but also the pattern of ridges on the colony surface oriented approximately perpendicular to the seagrass.

Several examples of ancient seagrass associations have been identified in the fossil record. The oldest are from the Maastrichtian of the type area in the Netherlands where Voigt (1981b) described silicified roots and stems of the seagrass *Thalassocharis bosqueti* overgrown by approximately 50 cyclostome and cheilostome bryozoan species. At least two of the bryozoan species – the cheilostomes *Kunradina bicincta* and *Onychocella spinifera* – seem to be restricted to seagrass habitats. Unlike the roots and stems, the leaves of these Maastrichtian seagrasses are not silicified but may be preserved as bioimmured impressions on the undersides of bryozoan colonies.

Epiphytes associated with fossils of the seagrasses *Thalassodendron* and *Cymodocea* in the middle Eocene of Florida include blade-encrusting, sheet-like colonies provisionally identified as the cheilostomes *Membranipora* and *Thalamoporella*, as well as small lichenoporid cyclostome colonies (Ivany, Portell, and Jones 1990). Cheetham, Jackson, and Sanner (2001, fig. 4.5) depicted the mould of seagrass cells on the basal attachment surface of a *Metrarabdotos* colony from the Miocene of the Dominican Republic. Two Miocene (Tortonian) species of the erect cheilostome *Vincularia* from East Kalimantan in Indonesia are inferred to have been inhabitants of seagrass beds on the basis of the occurrence of the seagrass-indicative gastropod *Smaragdia* in the same deposits (Di Martino and Taylor 2014).

On the Greek island of Rhodes, the late Pliocene Kritika Member of the Rhodes Formation contains leaves and rhizomes of the seagrass *Posidonia oceanica* along with a bryozoan community totalling 25 species (Moissette et al. 2007b; Moissette 2012). Some of the bryozoans are associated with leaf moulds, while others are found in the surrounding sediment. The former includes four species known to be obligate seagrass encrusters at the present day.

6.1.9 Mangroves

Coastal mangroves may host bryozoans, mainly encrusting the submerged roots of these trees (e.g. Fransen 1986; Creary 2003). One species, however, grows on mangrove leaves. Originally described from northern Australia, the gymnolaemate *Amphibiobeania* is able to survive periods of desiccation between tidal cycles due to the high humidity of the habitat and a cohesive coating of silt that covers the zooids (Metcalf, Gordon, and Hayward 2007). In Singapore, colonies of this 'amphibious' bryozoan can occur subaerially on leaves at heights more than one metre above low tide datum (Tilbrook and Gordon 2015).

There is scope for recognizing mangrove-associated bryozoans in the fossil record given that co-existing oysters may carry bioimmurations of mangrove stems on their attachment areas, as described by Plaziat (1970) from the Eocene of France.

6.1.10 Woodgrounds

Floating wood is well known as a substrate for gooseneck barnacles at the present day and stalked crinoids in the geological past, but the literature contains few mentions of bryozoans on either modern or fossil wood (Jebram 1969; Nikulina and Schäfer 2006). Evans and Todd (1997) described a bioimmured bryozoan on the attachment surface of the oyster *Deltoideum* from the Late Jurassic Kimmeridge Clay of Wiltshire, England, that had become broken off a piece of coalified fossil wood. The bryozoan, an arachnidiid ctenostome, evidently encrusted the wood along with an entoproct (*Barentsia*) and a brachiopod (*Discinisca*). The wood is interpreted to have been colonized after it sank to the seabed. In contrast, Kaneko (1993) described three cheilostome species encrusting a piece of wood from the Japanese Miocene, suggesting that the wood was colonized while floating, with implications for aiding bryozoan dispersal.

6.1.11 Deep sea

Modern bryozoans found in the deep sea are mostly non-mineralized ctenostomes or weakly calcified anascan cheilostomes (e.g. Hayward 1981; Gordon 1987). Cyclostomes are rare in the deep sea, although gracile articulated colonies of Crisiidae, which are more typical of the shallow subtidal, have been recorded down to 1175 metres in the Atlantic (Harmelin 1990) and 2267 metres in the Pacific (Grischenko and Chernyshev 2018). Shallow-water cheilostomes are occasionally recovered from considerable depths, notably *Flustra foliacea* at 1150 metres and *Bugula neritina* at 4060 metres, but these undoubtedly represent transported material (Hayward 1981). Grischenko and Chernyshev (2018) recorded four intertidal to shallow subtidal cheilostome species on dropstones at stations in excess of 3000 metres deep in the Kuril Basin.

Some deep-sea ctenostomes live in association with komokiaceans, becoming intergrown with the tubules of these foraminifera (Gooday and Cook 1984). Others have particularly striking colonies, such as *Pseudoalcyonidium* in which a small number of autozooids form a capitulum supported on a tall stalk with rootlets at the basal end (d'Hondt 1976). Cyclostomes living at depth on seamounts may also have small fungiform colonies, as in *Discantenna* and *Dartevellopora* (Figure 4.3A), two genera described from >900 metres depth from seamounts on the Chatham Rise, New Zealand (Gordon and Taylor 2010).

A remarkable bryozoan fauna has been discovered recently associated with polymetallic nodules in the Clarion-Clipperton Fracture Zone (CCFZ) in the eastern Pacific Ocean. Only the cyclostomes and ctenostomes have been formally described but the total number of species so far sampled is 50 in a small part of the CCFZ, making bryozoans the most diverse phylum of sessile animals on the nodules (Grischenko, Gordon, and Melnik 2018). The bryozoans are highly endemic and tend to be small, with many of the cyclostomes having stalked colonies elevating the feeding zooids above the substrate surface similar to those mentioned above from the Chatham Rise.

Uncalcified or feebly calcified bryozoans have a low fossilization potential, made worse in the deep sea by their occurrence below carbonate compensation depths at which any carbonate skeleton is likely to be dissolved. Accordingly, the fossil record of bryozoans thought to have lived below 1000 metres (i.e. at bathyal, abyssal, and hadal depths) is sparse (David and Pouyet 1984). Bryozoans from palaeodepths estimated to be more than a few hundred metres have seldom been identified explicitly in the fossil record, although there are examples from the Eocene of Tonga (Cheetham 1972), the Eocene–Oligocene of Tanzania (Di Martino et al. 2017), the Miocene of the Czech Republic (Zágoršek, Vávra, and Holcová 2007), and the Plio-Peistocene of Rhodes (Moissette and Spjeldnaes 1995). Most of these Cenozoic species have narrow branched erect colonies, some of which are articulated, and/or attached by basal rootlets. For example, the Eocene–Oligocene Pande Formation of Tanzania, a hemipelagic clay deposited at about 300–500 metres based on evidence from foraminifera, contains four cheilostome species adapted to living on soft sediments (Di Martino et al. 2017). One is a free-living, lunulitiform species, the other three are rooted conescharelliniform species.

6.1.12 Fine-grained mobile sediments

Sand- and silt-grade sediments are not normally considered to be important bryozoan habitats: the grains themselves are too small to serve as substrates for most species, moving grains can damage the delicate bryozoan lophophores (Best and Thorpe 1996), and the risk of colony burial tends to be high. Nevertheless, some bryozoans are capable of flourishing in fine-grained mobile sediments. They include free-living lunulitiform colonies, as well as recently discovered interstitial bryozoan communities. Lunulitiform cheilostomes, described in more detail elsewhere (Chapter 4.3.3), often occur on soft bottoms of silt and silty sand devoid of other kinds of bryozoans. Such environments are typically correlated with higher turbidity and sedimentation rates than those more often associated with bryozoans (Di Geronimo, Rosso, and Sanfilippo 1992).

The first interstitial bryozoan fauna was described by Håkansson and Winston (1985; see also Winston and Håkansson 1986) from Capron Shoal off the eastern coast of Florida. Ranging from 6 to 40 metres in depth, the sediment forming the shoal is a medium to coarse biogenic sand. At least 37 bryozoan species were found on grains less than 4.75 mm in size, with the majority occurring on grains under 2 mm. Many of the colonies encrust single grains but some extend from one grain to another. Precocious sexual reproduction characterizes the bryozoan species in this community, ovicells first appearing in cheilostome colonies that contain only three to six zooids. This compares, for example, with between 30 and about 2700 zooids before female sexual reproduction occurs in five common cheilostome species living on reefs in Jamaica (Jackson and Wertheimer 1985). Surface sediment on Capron Shoal is highly mobile and the bryozoans apparently occupy the sheltered habitat provided by the interstices of buried sediment grains. Among the bryozoan species recorded at Capron Shoal is the lunulitiform cheilostome *Cupuladria doma*, which is able periodically to moult the cuticle on the outer surface of the colony, thereby ridding itself of fouling algae and other epibionts to expose a new cuticular layer beneath (Winston and Håkansson 1989). Two similar communities of sand grain-encrusting bryozoan communities have since been described from south-eastern Brazil (Winston and Migotto 2005; Winston and Vieira 2013), but fossil examples have yet to be reported.

6.2 Bryozoans in Reefs and Mounds

If reefs are defined in the broadest sense as multi-individual concentrations of organisms forming topographical relief on the seabed, then bryozoans have been significant reef components since very soon after their first appearance in the Early Ordovician. The 'roles' played by bryozoans in reefs are several: frame builders, sediment bafflers, sediment trappers, binders, and cryptic encrusters. Cuffey's (2006) review of 'bryozoan-built reef mounds' distinguished three main kinds in the geological record: (i) crust-mounds, (ii) frame-thickets, and (iii) mud-mounds. Crust-mounds are small structures, composed of layered encrusting to foliaceous sheets, or tabular to globular bryozoans. Frame-thickets are small- to moderate-sized reefs consisting of erect colonies. Mud-mounds are variably sized reefs of carbonate mud containing a scattering of mostly delicate, reticulate bryozoans. The potential of bryozoans to facilitate the accumulation of mud was shown by an experimental study of living colonies of a bugulid cheilostome (McKinney, McKinney, and Listokin 1987). Suspension feeding colonies of this bugulid proved very effective in baffling, resulting in considerably more sediment accumulating downstream of living than dead colonies.

6.2.1 Bryozoans in modern reefs

Reef-dwelling bryozoans at the present day are mostly species that can also be found in non-reefal tropical habitats (Jackson, Winston, and Coates 1985), while the few reef-constructing modern bryozoans are not specialists and have wide habitat distributions. The fringing reefs off Bonaire were found to contain 73 cheilostome and 2 cyclostome species occupying cryptic habitats (Kobluk et al. 1988). Most have encrusting colonies. Bryozoan diversity tends to increase with depth on the Bonaire reef front, at least down to 60 metres.

Abrolhos Bank is located where the Brazilian continental shelf broadens and extends 200 km from the shoreline. Hosting the largest reef complex in the South Atlantic, Bastos et al. (2018) showed that the frameworks of Abrolhos mid-shelf reefs differ from typical coralgal reefs in being dominated by bryozoans interlayered with crustose coralline algae and minor corals. Twenty bryozoan species have been described from these reefs (Ramalho et al. 2018). Some have small encrusting colonies but others, such as *Celleporaria atlantica*, have large multilaminar colonies suggesting a more significant role in reef construction.

A study using settlement panels on a coral reef in Tobago, West Indies, found cheilostomes, which typically grew on cryptic, sediment-free and dimly lit surfaces, to calcify at average rates of 20.1 grams per square metre per year (Mallela 2013). It was concluded that these bryozoans are important secondary contributors to reef construction through accretion and overgrowth of cryptic reef substrates.

Little is known about the vulnerability of reef-dwelling bryozoans to environmental changes. However, bryozoans living on shallow coral reefs (10–40 metres) in Northern Bahia, Brazil, showed a significant reduction in both density and diversity following the 1997–1998 El Niño event when seawater temperatures increased (Kelmo et al. 2004).

Bryozoans are common not only in many shallow-water reefs developing within the photic zone, but can also be found in deep-water coral reefs. The deep-water *Oculina* reefs off the Atlantic coast of Florida at 70–90 metres depth host more than 40 bryozoan species mostly encrusting coral rubble (Winston 2016), while 36 species have been recorded growing on the deep-water corals *Lophelia* and *Madrepora* in the north-west Mediterranean (Zabala, Maluquer, and Harmelin 1993).

6.2.2 Ordovician

The oldest known bryozoan reefs are found in the Early Ordovician (Tremadocian) of the Yangtze Platform in central China. Because of their antiquity as metazoan reefs, they have attracted special interest (Adachi, Ezaki, and Liu 2011, 2012; Adachi, Liu, and Ezaki 2013; Cuffey et al. 2012). The reefs at Huaghuachang consist of numerous closely spaced, bifurcating columns formed by stacked domes of the esthonioporate bryozoan *Nekhorosheviella semisphaerica*, individual domes being about 10–15 mm in diameter and 5–10 mm high (Plate 2D). In some instances, continuity between the zooecial chambers in successive domes shows that they represent subcolonies belonging to the same colony, while in other cases they may be new colonies that fouled the surfaces of older moribund or dead colonies. Cuffey et al. (2012) envisaged the reefs to have grown upwards from the bioclastic sediments on the seafloor to a height of about 1–3 metres and diameter of 3–10 metres. A less conventional interpretation was proposed by Adachi, Ezaki, and Liu (2011, 2012) and Adachi, Liu, and Ezaki (2013). These authors believed that the bryozoan columns were attached to lithistid sponges, or occasionally pelmatozoan echinoderms, and grew preferentially downwards and laterally. The lithistid sponges had siliceous spicules encased in micritic or peloidal carbonate sediment after decay of the collagenous tissues.

Lamellar encrusting bryozoans, together with stromatoporoids and tabulate corals sharing similar morphologies, formed a reef-building 'consortium' that emerged during the Middle Ordovician as part of the Great Ordovician Biodiversification Event (Kröger, Descrochers, and Ernst 2017). These groups diversified almost synchronously in the late Darriwilian, expanding from cryptic and very shallow marine hard substrates onto soft substrates.

Darriwilian reefs at Duwibong in South Korea comprise intergrowths between the bryozoan *Nicholsonella*, the stromatoporoid *Cystostroma*, and minor amounts of algae and siliceous sponges (Hong et al. 2014). The reefs are small, about 80 cm wide by 30 cm high, but more compact than the older Huaghuachang reefs described above.

Cuffey et al. (2000) reviewed Ordovician bryozoan reefs in Virginia, most of which were constructed by sheet-like or massive colonies of the trepostome *Batostoma chazyensis*. The *Batostoma* reefs are small patch reefs up to 3 metres wide by 2.7 metres high, and comprise multiple colonies growing on top of one another.

Carbonate mud mounds up to 100 metre thick in the Late Ordovician Jifarah Formation of Libya contain nine species of bryozoans dominated by trepostomes (Buttler, Cherns, and Massa 2007). The mounds lack an identifiable framework or microbial fabrics, accumulated in a carbonate belt at a palaeolatitude of 60–70°S, and were likened to the Quaternary mounds of the Great Australian Bight described by James et al. (2000) (see below).

On Anticosti Island, Canada, a latest Ordovician (Hirnantian) patch reef, 20–30 metres in diameter, 2–3 metres thick, and dominated by corals, contains common encrusting bryozoans in peloidal microbialites (Ernst and Munnecke 2009).

6.2.3 Silurian

The famous Silurian reefs of Gotland, Sweden, have corals and stromatoporoids as the main frame builders but bryozoans can play secondary frame-building as well as sediment-trapping roles (Brood 1976a). One Gotland reef containing thin encrusting bryozoan colonies intercalated with calcimicrobes has already been mentioned above (Ernst, Munnecke, and Oswald 2015).

The Wenlock 'ballstone' reefs of Wenlock Edge in Shropshire, England, contain bryozoans (e.g. *Hallopora*) as secondary frame-builders to tabulate corals. Bryozoans here have more important roles as reef binders (e.g. *Fistulipora*) and as cryptic encrusters on the undersides of the frame builders (Scoffin 1971). The reefs were interpreted as patch reefs sanding 0.5–3 metres above the surrounding seafloor and on average about 12 metres in diameter.

Very different reefs containing bryozoans occur in the Lower Silurian of Anticosti Island, Canada (Desrochers, Bourque, and Neuweiler 2007). Here, mud-rich fenestrate bryozoan-sponge build-ups are up to 8 metres thick and a few tens of metres in diameter. Marine cement precipitation and organogenic lithification are interpreted to have been important in the accretion of these reefs but not frame building or sediment trapping.

6.2.4 Devonian

Mud mounds less than 2 metres in thickness in the Early Devonian of north-west Spain contain 16 species of bryozoans (Ernst et al. 2012a). Most are encrusters or ramose erect forms, the former functioning as binders in these microbial reefs. The Koneprusy Limestone (Pragian) of the Czech Republic comprises several different facies, with bryozoans peaking in diversity in the reef core facies where 13 species – seven encrusting, two ramose, and four reticulate – have been recorded (Ernst and May 2009). The bryozoans exhibit greater diversity and abundance than is usual for Devonian reefs, possibly because encrusting species functionally replaced stromatoporoids in the Koneprusy reefs.

The Middle Devonian Sabkhat Lafayrina reefs of Western Sahara have a stromatoporoid framework where bryozoans are mostly associated with the basal parts of the reef structures (Ernst and Königshof 2008). Although not frame builders, these bryozoans are interpreted to have been important in sediment stabilization and baffling, facilitating the initial phases of reef growth.

6.2.5 Carboniferous

The Carboniferous is particularly well-known for bryozoans in carbonate build-ups at a time when more typical frame-building taxa are lacking and, conversely, 'mud-mounds' peaked in abundance (Krause et al. 2004). Waulsortian mounds, first named for their occurrence near Waulsort in Belgium but since recorded elsewhere in the world, are distinctive carbonate mounds of Tournasian to early Visean age. The mounds comprise a crinoidal-bryozoan-microbial core facies surrounded by a flank facies of steeply dipping crinoidal limestones. Lees (1964) noted that in-situ fenestellid bryozoans are often common in Waulsortian 'reefs' and regarded them as important as bafflers trapping fine sediment but not frame builders, while Wyse Jackson (2006) believed

that even this contribution was relatively minor. Bryozoan diversity is low in Waulsortian mounds and the species present are not restricted to these structures. The build-ups are usually interpreted as in-situ structures that developed in moderately deep water (>100 m) on a carbonate ramp. Giles (1998), however, offered an alternative interpretation of mounds in the Sacramento Mountains of New Mexico as allochthonous structures formed by gravity-driven, downslope sediment movement, which were then colonized by bryozoans, sponges, and microbes. Ahr and Stanton (1994) believed the Sacramento mounds to be autochthonous, formed by localized accumulation of mud of probable microbial origin, with the fenestellid bryozoans simply representing reef dwellers, rather than constructors or bafflers, better preserved in the mounds than the surrounding sediments.

From Turkey, Denayer and Aretz (2012) described an Early Carboniferous microbial-sponge-bryozoan-coral bioherm. This 50-metre-thick reef contains a core facies with erect fenestellids and narrow-branched cryptostomes, as well as massive and encrusting trepostomes and cystoporates which may have multilayered colonies. The bryozoans were interpreted both as frame builders and stabilizers in this 'Cracoan-type' reef that developed along a carbonate platform margin.

6.2.6 Permian

The Zechstein Sea covering parts of northern Europe in the Late Permian contained reefs in which bryozoans were a major component. According to Hollingworth (1991), the Ford Formation reef in County Durham, north-east England, developed from a basal brachiopod-dominated coquina which became colonized by bryozoans, particularly the fenestrate genera *Rectifenestella*, *Synocladia*, *Acanthocladia*, and *Kingopora*, but also the ramose trepostome *Dyscritella*. The fenestrates served as nuclei for the growth of aragonite cements neomorphosed to botryoidal fibrous calcite. Precipitation of these marine cements, along with the binding and baffling action of the bryozoans, is believed to have been important in the establishment and growth of the main part of the reef.

An Early Permian reef at Tratau in the Ural Mountains of Russia has a core comprising massive bryozoan bafflestones with fenestellids and encrusting bryozoans preserved in life position (Vennin 2007). Bryozoans also occur in floatstones on the reef flanks and within fissures ('neptunian dykes') in the reef.

Small bryozoan reefs categorized by Zimmerman and Cuffey (1987) as 'frame-thickets' developed on the slopes of the Permian Basin in the Glass Mountains of West Texas. A high diversity of bryozoans – 83 species – has been recorded from these reefs. As the colonies are silicified, acid dissolution of the limestone matrix is able to reveal the dense tangles of mainly pinnate and recticulate fenestrate bryozoans that constructed the reefs (Plate 10J).

6.2.7 Triassic

Given their scarcity in the Triassic, bryozoans do not play significant roles in Triassic reefs. However, Sánchez-Beristain and Reitner (2017) identified the trepostome *Reptonoditrypa cautica* as an important component of some Triassic reefs, particularly in the St Cassian Formation of northern Italy where colonies functioned as binders.

6.2.8 Jurassic

Small, inverted conical sponge reefs in the Middle Jurassic (Bathonian) of Calvados, France, have abundant encrusting cyclostomes on the cryptic undersides of the platy, reef-building lithistid sponges (Palmer and Fürsich 1981). The narrow-branched ramose cyclostome *Entalophora* formed meadows adjacent to the individual sponge reefs, which are believed to have had only minor relief on the seabed. Build-ups in the Upper Jurassic (Oxfordian–Kimmeridgian) of south-eastern Spain also contain cyclostome bryozoans encrusting the cryptic undersides of cup-shaped siliceous sponges (Reolid 2007). In this case, the bryozoans and other encrusters are interpreted to have colonized the sponges after they had died but were still in life position.

The Tithonian Portland Limestone Formation of Dorset in England contains small patch reefs with various main frame builders (Fürsich, Palmer, and Goodyear 1994). Most of the reefs are formed by the coralline alga *Neosolenopora* or bivalve molluscs but some are constructed of multilayered masses (Plate 10K) of a cyclostome bryozoan provisionally identified as *Hyporosopora*. The bryozoan reefs can be up to a metre in diameter and several tens of centimetres thick, and often contain bivalves that became overgrown and embedded.

Elsewhere in the Jurassic bryozoans are occasionally reported from coral patch reefs, mostly in minor quantities as cryptic encrusters (e.g. Helm and Schülke 2006).

6.2.9 Cretaceous

The Maastrichtian (and Danian) bryozoan mounds of Denmark have attracted considerable attention. At Stevns Klint, late Maastrichtian bryozoan mounds are asymmetrical structures characterized by steep south-west faces, measure 30–35 metres in width but have a topographical relief of only about one metre (Larsen and Håkansson 2000). The high dominance of bryozoans led these authors to conclude that bryozoans were essential for mound formation. Bryozoan-rich chalks in the Turonian–Santonian sediments exposed in the cliffs around Fécamp in Normandy, France, show depositional relief once thought to be due to the presence of mounds similar to those in the younger Cretaceous deposits of Denmark. However, the detailed study of Quine and Bosence (1991) reinterpreted this sequence as a complex of channels separated by ridges.

6.2.10 Palaeogene

Bryozoan mounds are extensively developed in the Danian of Scandinavia. Cheetham (1971) recorded the distribution of bryozoans and other fossils in a middle Danian mound in southern Sweden. He considered that the mound probably formed at the shelf-edge, and recognized three biofacies: (i) flanks dominated by bryozoans; (ii) a core rich in octocorals with some colonial scleractinian corals and bryozoans; and (iii) transitional areas dominated by octocorals but with abundant bryozoans. The relative abundances of cheilostome species – though not necessarily their presence or absence – differed between these three biofacies. The Early Danian reefs at Karlby Klint in Denmark consist on average of about 30% bryozoan fragments (Thomsen 1976). Coarse bryozoan fragments occur on the south-eastern flanks of the mounds which faced into the prevailing current direction, whereas the north-western flanks and basins between the mounds are characterized by finer bryozoan fragments less able to bind the sediment. Thus, the more rapid growth of the bryozoans on the south-eastern flanks determined the overall structure and development of the mounds. Elsewhere in Denmark at Stevns Klint, mounds of similar age to those at Karlby, are bryozoan thickets that overlap and interfinger to form complexes separated by hardgrounds and erosional surfaces (Surlyk 1997; Surlyk, Damholt, and Bjerager 2006). Lines of *Thallasinoides* burrow flints pick out the asymmetrical internal geometry of the mounds (Plate 2E), which are believed to have prograded to the south-south-west parallel to the ancient coastline and towards nutrient-carrying currents.

Mounds along the edge of the Great Australian Bight, previously believed to be volcanic in origin on the basis of seismic data, were shown to be huge bryozoan mounds when well cuttings became available (Sharples et al. 2014). These middle Eocene bryozoan mounds form complexes extending for more than 500 kilometres parallel to the shelf margin. The size of individual reef mound complexes – 60–150 kilometres long, 15 kilometres wide, and up to 200 metres thick – is remarkable. The mounds have depositional dips on their basinward flanks of 8–30°, and 5–15° on landward sides. Well cuttings are dominated by fragments of erect bryozoans (Plate 10L). These await detailed study but include both cheilostomes (*Porina*, *Cellaria*, *Puellina*, *Chondriovelum*, *Foveolaria*, ?*Reteporella*, *Nudicella*, and *Exechonella*) and cyclostomes (*Patinella*, *Nevianipora*, *Hornera*, '*Entalophora*', *Idmidronea*, *Crisia*, *Platonea*, and *Diaperoecia*). The bryozoans apparently flourished as a result of decreased siliciclastic sedimentation

coupled with a rich supply of nutrients on the shelf edge. Of additional importance for mound growth was the accommodation space provided by accelerated tectonic subsidence along the continental margin associated with an increase in the rate of separation between Australia and Antarctica.

Also of middle Eocene age, but far smaller than the Great Australian Bight mounds, are build-ups in the Pyrenees considered by Serra-Kiel and Reguant (1991) to be bryozoan mud mounds. They overlie a transgressive surface and contain abundant cheilostomes, including erect forms such as *Metrarabdotos* as well as large multilamellar encrusting colonies. Growth of the build-ups was attributed to the baffling effects of the bryozoans causing local retention of fine-grained sediment.

6.2.11 Neogene

The Miocene Sarmatian Stage (Serravalian–Tortonian) of the Parathethys is marked by an impoverished benthic biota that lived in 'abnormal' marine conditions, possibly resulting from fluctuating salinities and high alkalinity (Jasionowski 2006). Reefs developed in Sarmatian times commonly contain bryozoans (e.g. Hara and Jasionowski 2012). Cornée et al. (2009) described Hungarian Sarmatian lagoonal deposits containing microbial–nubeculariid–bryozoan–serpulid build-ups. Some dome-shaped build-ups 10 centimetres high and 1 metre wide are formed almost exclusively of multilayered colonies of the cheilostome *Schizoporella*. Others contain subordinate quantities of the cheilostomes *Conopeum* and *Cryptosula* and the cyclostome *Tubulipora*. A series of middle to late Miocene build-ups (Plate 2G) on the Kerch and Taman peninsulas, between the Black Sea and Sea of Azov, have bryozoans and algae as their major constituents (Goncharova and Rostovtseva 2009). These 'bryalgal' bioherms achieved their acme at the end of the Sarmatian when they reached up to 30 metres thick and were formed mainly by dense accumulations of bilaminate or unilaminate sheets of the cheilostome *Tamanicella lapidosa* (Viskova and Koromyslova 2012; Figure 6.2).

Figure 6.2 Sheet-like growth of the cheilostome *Tamanicella lapidosa* from a late Miocene bryozoan/algal reef (Miocene, Sarmatian; Kerch, Crimea). Scale bar = 1 mm.

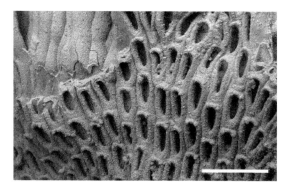

Muddy scleractinian reefs in the Miocene Kutai Basin of East Kalimantan, Borneo, contain a bryozoan fauna dominated by cheilostomes. The majority of the bryozoans encrust the undersides of thin, platy corals (Figure 6.1E). On average, 30% of the platy corals are bryozoan-encrusted, mostly by sheet-like colonies (Di Martino, Taylor, and Johnson 2015). A detailed study of one of the reefs inferred mesophotic conditions due to high terrigenous sediment input (Novak et al. 2013). A total of 62 bryozoan species were recorded from this late Burdigalian reef, exhibiting a variety of colony-forms: encrusting runners (3%), encrusting unilaminar sheets (67%), encrusting multilaminar (massive) (3%), erect fenestrate (5%), erect branching (7%), erect articulated (13%), and free-living (2%).

6.2.12 Quaternary

A series of subsurface bryozoan reef mounds occur in the Great Australian Bight, at depths greater than 7 metres below the modern seafloor (James et al. 2000, 2004). Individual mounds are as much as 65 metres thick and mounds can occur either singly or in complexes. The mounds are unlithified floatstones containing bryozoans set in a muddy matrix. Several colony-forms are represented, including branching, fenestrate, articulated, nodular, and encrusting. The youngest mounds developed during the last glacial lowstand, and it is believed that others also grew during glacial lowstands when upwelling introduced nutrients to this upper slope environment.

6.3 Colony-form and Palaeoenvironments

In a short but highly influential paper, Stach (1936) contended that there is a strong relationship between colony-form in cheilostome bryozoans and habitat. His assertion was based largely on functional morphological reasoning. For example, the rigidity and crosslinks in fenestrate ('reteporiform') colonies were inferred as adaptations to habitats with strong wave action and currents. Following on from Stach's work, bryozoan colony-form has been used frequently in palaeoenvironmental studies, particularly those focussing on cheilostome-dominated Cenozoic faunas. Attempts to apply the method to older deposits have been fewer, at least in part because they were developed for cheilostomes which are rare or lacking before the Late Cretaceous, and Kelly and Horowitz (1987) found that growth-form distributions in Mississippian palaeostomate bryozoans from the United States did not always match Stach's models.

Lagaaij and Gautier (1965) mapped the distributions of bryozoan species and colony-forms in the 150–800 μm sediment fraction collected by grab from the Mediterranean Sea off the Rhone delta. Although 105 species were recognized, the samples were greatly dominated by the articulated cheilostome *Cellaria fistulosa*. The virtual absence of bryozoans in the fluviomarine fan was attributed to high sedimentation, rate of deposition being identified as a major controlling factor in the distribution of bryozoans. In this respect, the articulated colonies of *C. fistulosa* seem better able to cope with sedimentation as they have the ability to shake off settling sediment passively. A problem with this study is that a high proportion of the small fragments of bryozoans used may have been transported or even remanié.

As pointed out by Hageman et al. (1997), colony-form analysis suffers from methodological problems that include whether to employ abundance or the number of species having particular colony-forms (see also Steinthorsdottir, Lidgard, and Håkansson 2006), and numerical versus volumetrical frequency of these colony-forms. Their study of recent bryozoan colony-forms over the Lacepede Shelf of southern Australia distinguished 17 colony-forms and recorded their distributions across two inner shelf environments (40–80 m depth), the outer shelf (80–180 m) and the slope (160–450 m). Almost all of the 17 colony-forms were found to be present in each of the four environments though in somewhat different proportions (Figure 6.3).

An environmental analysis (Hageman et al. 2003) of modern and fossil colonies of one cheilostome genus – *Celleporaria* – from southern Australia focussed on large multilayered colonies with hollow branches resulting from growth around sponges. These colonies inhabit low-energy or sub-swell wave-base settings on mud-silt substrates, where sedimentation rate is moderate and the nutrient level is mesotrophic. However, such conditions are not unique to a single setting, being found in: (i) deep water (>200 m) below the shelf break with nutrients supplied by upwelling during extreme sea-level low stands; and (ii) shallow-water (<50 m) nearshore embayments with a terrestrial nutrient supply during sea-level high stands.

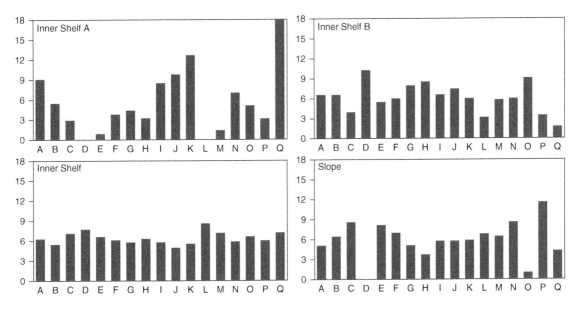

Figure 6.3 Relative importance of bryozoan colony-forms, based on combined abundance and species richness, in four 'physiographic provinces' on the Lacepede Shelf of southern Australia (based on Hageman et al. 1997, fig. 8). Letters indicate the 17 colony-forms that were recognized: A = cemented, unilaminar, solid substrate; B = cemented, unilaminar, flexible substrate; C = cemented, unilaminar, hollow cylinder; D = cemented, multilaminar, encrusting massive; E = cemented, erect, cylindrical branches; F = cemented, erect, unilaminar branches; G = cemented, erect, bilaminar branches; H = cemented, erect, fenestrate sheet; I = rooted, articulated, cylindrical branches; J = rooted, articulated, unilaminar branches; K = rooted, articulated, articulated zooids; L = rooted, rigid, encrusting sheet; M = rooted, rigid, branches; N = rooted, rigid, bilaminar sheet; O = rooted, rigid, fenestrate sheet; P = rooted, rigid, cone-disc; Q = free-living, motile, disc.

The ratio of erect to encrusting species (and colonies) has been used as a palaeoenvironmental indicator following Schopf (1969) who showed that this proportion increased with depth on the continental shelf off New England, USA. A similar pattern was found at Signy Island in the Antarctic by Barnes (1995): the ratio of encrusting to erect bryozoan species changed rapidly between 0- and 50-metre depth, from exclusively encrusting at 0–5 metres to almost equal numbers at 50 metres (Figure 6.4). With regard to erect colony-forms, foliaceous colonies were abundant in the shallowest water, giving way to erect flexible forms and then erect rigid forms with increasing depth.

Guha (2013) calculated the ratio of erect to encrusting bryozoan species for several Palaeogene formations in Kachchh, India. He used the stratigraphical changes in this ratio to infer two episodes of deepening (Rupelian–Chattian and Burdigalian) between two times of shallowing (Lutetian and Aquitanian).

A facies interpretation of Late Oligocene carbonates in Malta found the mid-ramp between 15- and 40-metre depth to contain encrusting multilamellar bryozoan colonies, the mid- to outer ramp (40–50 m) foliose colonies, and the outer ramp (>50 m) delicate branching colonies (Knoerich and Mutti 2003).

In the Permian subtropical to cool water sediments of Spitsbergen, Nakrem (1994) found the expected correlation between bryozoan colony-forms and sedimentary environments: narrow-branched ramose trepostomes and cryptostomes and delicate fenestrate colonies characterized quiet conditions, whereas thick-branched ramose trepostomes and cystoporates and robust fenestrate were found in high-energy deposits. Reid (2010) studied the palaeoenvironmental distributions of bryozoan colony-forms in the Permian of Tasmania, remarking on differences compared with equivalent colony-forms in modern environments. She found the primary control on distributional patterns to be water energy, with fenestrate and delicate branching colonies occurring in quiet-water, often offshore environments, whereas foliose and thick-branching colonies inhabited environments of moderate water energy.

There is a clear need for a much more nuanced approach to the application of colony-form as a tool for palaeoenvironmental inference. As Hageman, McKinney, and Jaklin (2012, p. 119) remarked "Our understanding of broader environmental and phylogenetic controls on the distribution of bryozoan colonial growth habits remains incomplete." The use of bryozoan colony-form as a palaeoenvironmental proxy must also take into consideration taphonomic factors influencing the preservation of particular colony-forms in the sedimentary record (e.g. Smith and Nelson 1994), and how evolution within the phylum has affected the temporal distributions of particular colony-forms through geological time (see Chapter 9.16).

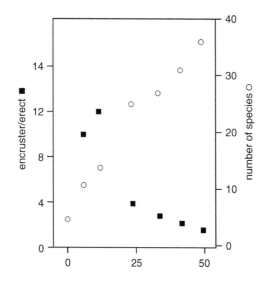

Figure 6.4 Depth gradients in ratio of encrusting to erect species (black squares) and numbers of cheilostome species (open circles) at Signy Island, Antarctica (redrawn from Barnes 1995).

6.4 Depth Distributions and Palaeobathymetry

Almost 40 years ago, Hayward (1981, p. 64) commented that our knowledge of bryozoan depth ranges was 'hopelessly outdated, still hampered by a shortage of reliable quantitative data'. Regrettably, this statement remains largely valid today.

A survey by Grant and Hayward (1985) of bryozoan diversity in shell gravels from 102 dredge stations in the English Channel ranging in depth from 22 to 106 metres revealed a clear increase in diversity with increasing depth (Figure 6.5). Gordon (1999, fig. 1) plotted the number of bryozoan species living today around New Zealand in 100-metre depth bins. The peak value of 493 species occurred in the 0–100-metre depth bin, a subdivision of which revealed substantially more species at 26–100 metres (461 spp.) than 0–25 metres (267 spp.). A rapid decline to fewer than 100 species at 700 metres was followed by a minor increase before a gentle rate of decrease into the deeper parts of the sea. A subsequent study of bryozoan diversity around New Zealand (Rowden, Warwick, and Gordon 2004) used a biodiversity index – average taxonomic distinctiveness (AvTD) – to seek correlations between diversity and depth and other environmental and geographical parameters. This index was found to be correlated with depth, while the high primary productivity and strong current flow that characterized two areas of particularly high bryozoan biodiversity – the Three Kings Plateau in the north and Foveaux Strait in the south – were also important factors.

Denisenko et al. (2016, fig. 3) showed a strong negative correlation between species richness and depth in bryozoans from the Faroe Islands: maximum species richness (120 spp.) occurred between 0 and 100 metres, as in New Zealand, declining to about 40 species at depths greater than 800 metres. At Signy Island in the Antarctic, Barnes (1995) found the total number of cheilostome bryozoan species to increase with depth between 0 and 50 metres, paralleling the increase in the proportion of erect vs. encrusting species discussed above (Figure 6.4).

A study using settlement panels deployed at 0, 5, 15, 60, 150, and 500 metres depth in the Azores recorded bathymetric variations in the sclerobionts present, including bryozoans, after one and two years of exposure (Wisshak et al. 2015). Bryozoan diversity peaked in the 150-metre panels (30 taxa) compared to the 60-metre (26 taxa) and 15-metre (22 taxa) panels, whereas the abundance of colonies was greatest in the 60-metre panels and the largest areal coverage occurred in the 15-metre panels. The proportion of cheilostome to cyclostome taxa decreased with depth as follows: 15 metres (4.5), 60 metres (4), 150 metres (2.8), 500 metres (0.6). Spot-like encrusters dominated only in the 500-metre panels.

The decrease in both bryozoan diversity and abundance with depth in the Pacific off Panama was explained by Schäfer, Herrera Cubilla, and Bader (2012) as a consequence of a reduction in the substrates available for colonization. Suitable substrates become rarer here with depth and are more frequently covered by fine sediment films that may inhibit larval settlement, while sediment entrained in suspension potentially interferes with feeding.

Regarding the application of particular bryozoan species as depth indicators, Hayward's (1981) review of depth distributions among cheilostomes described from the deep sea showed bathymetrical ranges of individual species commonly to exceed

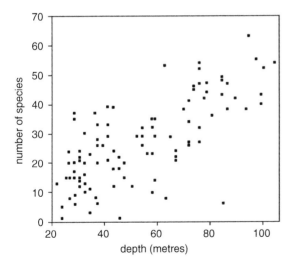

Figure 6.5 Positive relationship between bryozoan species diversity and depth in shell gravel communities from the English Channel (after Grant and Hayward 1985).

3000 metres. Kobluk et al. (1988) recorded the depth distributions of 75 bryozoan species inhabiting the fringing reef of Bonaire, finding the majority of species to have extensive depth ranges. Only a few, rarer species were limited to either shallow or deep settings. Elsewhere, many bryozoan species are also clearly stenobathic (see Moissette 2000a, table 2), with the Austral cheilostome *Rhabdozoum wilsoni* ranging remarkably from 8 to 800 metres (Cook and Bock 1994).

Transportation is capable of distorting bathymetric ranges. The depth distributions of three cyclostomes and one cheilostome species occurring commonly around New Zealand were found to be wide (Taylor, Gordon, and Batson 2004). For instance, the ramose cyclostome *Cinctipora elegans*, which occurred in 179 samples, had a depth range of 17–914 metres, although pristine examples occurred over a narrower range of 17–346 metres. It is likely that the poorly preserved specimens from depths greater than 346 metres were transported into deeper water.

Deep-water (c. 1 km) pelagic-hemipelagic muds of Quaternary age on the Labrador Slope of the Canadian Arctic contain layers rich in the shallow-water cyclostome *Idmidronea atlantica* (Thomas, Hardy, and Rashid 2003). The perfectly preserved branches of this bryozoan are interpreted to have been dislodged and/or transported into deep water by floating ice.

At the genus level, bathymetric distributions are often very large at the present day. For instance, *Poricella* (as *Tremogasterina*) ranges between 0 and 506 metres (Cook 1977c), *Steginoporella* from a few metres to 615 metres (Gordon, Voje, and Taylor 2017), and *Monoporella* from 3 to 355 metres (Dick 2008).

6.4.1 Fossil bryozoans as palaeodepth indicators

Can bryozoans be used as reliable indicators of palaeodepth? Although bryozoans are not strictly limited to the photic zone, as they feed mostly on phytoplankton it might be expected that they would predominantly inhabit well-illuminated parts of the ocean. Indeed, the diversity and abundance of bryozoans does tend to diminish with depth. The known depth ranges of some species are limited, but this may reflect sedimentary facies or biological associations, or simply represent an artefact of inadequate sampling. Moissette (2000b) ambitiously attempted to use depth ranges inferred from fossil associations of extant species to augment the likely depth distributions of some rare species in modern seas.

Monospecific or oligospecific assemblages containing conescharelliniform bryozoans such as *Batopora* in the Neogene of the Mediterranean seem to be indicative of deepening water based on independent evidence from foraminifera (Moissette 1996). The same author (Moissette 2000a) extended the use of bryozoan colony-forms in order to distinguish four depth zones in late Miocene (Messinian) sediments in Algeria, supporting his inferences using data on the bathymetric distributions of extant cheilostome species (Figure 6.6). Overlaps in the present-day bathymetric ranges of 17 extant species found in the Plio-Pleistocene Kolymbia limestone of Rhodes were used to infer a probable depositional depth of between 30 and 80 metres for this deposit (Steinthorsdottir, Lidgard, and Håkansson 2006).

Erect ramose cyclostome bryozoan species often exhibit considerable variability in the diameters of their branches. Schopf, Collier, and Bach (1980) studied *Heteropora pacifica* close to Friday Harbor in Washington State, USA. In each of four transects, branch diameter was found to decrease significantly from 9 to 15 to 30 metres depth, colonies becoming less robust and more gracile with increasing depth (Figure 6.7). However, this change seemed not to be related to ambient current strength as differences in current velocity at the same depth were not mirrored by branch thickness. Whatever the controlling factor/s, an empirical inverse relationship exists between branch diameter and depth in *Heteropora pacifica*. Brood (1976b) found the same relationship in another cyclostome identified as *Pustulopora delicatula* from 11 sampling stations ranging from 50 to 700 metres off the coast of East Africa. However, the branch diameter/depth relationship is not necessarily universal. A test of the correlation

Figure 6.6 Distribution of seven bryozoan colony-forms according to inferred palaeodepth in samples from the late Miocene (Messinian) of Algeria (redrawn from Moissette 2000a).

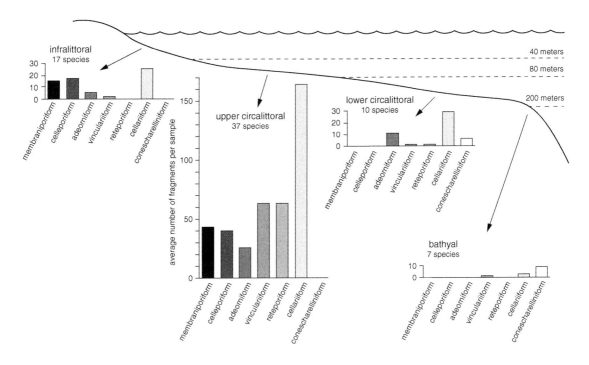

Figure 6.7 Contrasting morphologies of shallow- and deep-water colonies of the cyclostome *Heteropora pacifica* at Friday Harbor, Washington State (redrawn from Schopf, Collier, and Bach 1980).

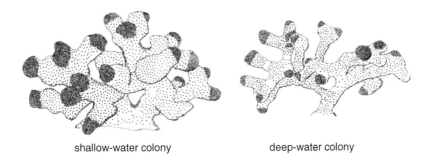

shallow-water colony　　　　deep-water colony

between depth and branch diameter was undertaken using three cyclostome species (*Cinctipora elegans*, *Diaperoecia purpurascens*, and *Erksonea* sp.) dredged from 15 benthic stations ranging from 38 to 668 metres depth on the New Zealand shelf (Taylor, Kuklinski, and Gordon 2007). *Cinctipora elegans* alone showed a clear correlation and then only at depths less than 200 metres. Subsequently, Figuerola et al. (2017c) found a significant, albeit minor, diminution in branch diameter with depth in the Antarctic cyclostome *Fasciculipora ramosa*. Branching colonies of the sponge *Haliclona*, the hydrozoan *Millepora*, and the sclaractinian coral *Pocillopora* show a similar pattern to that seen in bryozoans, with branches becoming narrower and colonies less compact in habitats where water movement is reduced (Kaandorp 1999). For the ramose cheilostome *Myriapora trunctata*, Berning (2007) believed food supply to be the most likely determinant of branch thickness.

Given the lack of contrary trends showing branch diameter increasing with depth, any geographical or stratigraphical clines of decreasing branch diameter in fossil populations of single bryozoan species can reasonably be taken to indicate deepening, quite probably reflecting lower current speeds and concomitant diminished plankton concentrations. It should, however, be noted that there are other factors that can control colony-form plasticity in species of erect bryozoans: for example, the cheilostome *Bugula neritina* develops longer internodes between bifurcations when colonies grow in high densities with conspecifics (Gooley, Marshall, and Monro 2010).

The potential for branch morphology to be used in palaeobathymetry was demonstrated by the research of Flor (1972) on the cyclostome *Spiropora verticillata* which is common in Late Cretaceous sites across northern Europe. Flor found branch fragments from nearshore sites – as indicated by facies analysis and palaeobiogeography – to be thicker and more variable than those from offshore sites.

6.5 Bryozoans as Sediment Producers

Bryozoans are locally the dominant source of carbonate sediments accumulating on the seafloor in temperate and polar seas at the present day (e.g. Henrich et al. 1995; Hauck et al. 2011). Correspondingly, in some sedimentary rocks – termed 'bryozoarites' by Lamarti-Sefian (1997) – bryozoans are the dominant grain type.

A particularly spectacular example of bryozoans forming carbonate rocks has been described from southern Florida (Hoffmeister, Stockmann, and Multer 1967). Here, the Pleistocene Miami Limestone consists of two main facies: an oolitic facies covering the south-eastern part of Florida, which is underlain to the north-east by a bryozoan facies that is mostly hidden beneath present-day groundwater level in the Everglades. The majority of the bryozoans consist of the ascophoran cheilostome *Schizoporella floridana* forming tubular, multilayered colonies 30 centimetres or more in diameter. These colonies are believed to have grown around rhizomes of the seagrass *Thalassia*. In places, at least 70% of the rock is estimated to comprise bryozoans, and the bryozoan facies of the Miami Limestone covers a total area of about 2000 square miles. As the thickness of this facies is often 2–3 metres, the volume of bryozoans in the limestone may exceed five million cubic metres.

6.5.1 Heterozoan vs. photozoan carbonates

The terminology of **photozoan** and **heterozoan** benthic carbonate 'factories' was introduced by James (1997) to encapsulate the differences between warm-water, tropical carbonate associations dominated by photoautotrophs such as zooxanthellate corals and coralline algae, and cool-water carbonate associations in which heterotrophs like bryozoans, molluscs, barnacles, and foraminifera are the main carbonate producers. For example, the cool-water carbonates occupying about 50 000 km² of the continental shelf around New Zealand consist predominantly of the skeletal remains of bryozoans and molluscs (Nelson et al. 1988), as do those off southern Australia (Wass, Conolly, and MacIntyre 1970). According to Smith (2014), erect, multibranched bryozoan colonies like those typical of these heterozoan carbonates can produce as much as 24 grams of $CaCO_3$ annually.

Temperature, light, and nutrient levels are the main environmental factors controlling photozoan vs. heterozoan associations (Mutti and Hallock 2003). Thus, in shallow, well-lit waters of the tropics today, oligotrophic nutrient levels favour photozoan associations of scleractinian corals and algae over heterozoan associations containing a significant bryozoan biomass. Heterozoan carbonate 'factories' in cool waters produce an estimated 20–30% of the sediment formed by photozoan 'factories' in the tropics but extend into deeper waters as they are not light dependent (Schlager 2003). Not only are carbonate accumulation rates lower in cool waters – typically less than 10 cm/kyr – but low-Mg calcite dominates over aragonite and high-Mg calcite, seafloor cementation is less common, and grain abrasion and dissolution result in a more destructive regime (Smith and Nelson 2003).

Halfar et al. (2006) mapped the distribution of carbonate factories in the warm-temperate Gulf of California. They found a transition from coral-dominated carbonates in the south where water temperatures were high (mean 25°C) but nutrient levels low, through red algal dominated carbonates, to mollusc-bryozoan carbonates in the north where water temperatures were low (mean 20°C) but nutrient levels high. The combination of low temperatures and high nutrients related to upwelling and tidal mixing appears to control the transition from photozoan to heterozoan assemblages here.

Several examples in the stratigraphical record of carbonates containing abundant bryozoans have been identified as heterozoan carbonates. A rich Late Carboniferous bryozoan fauna from Sonora, Mexico, was interpreted by Ernst and Vachard (2017b) as a heterozoan assemblage, probably deposited following regional climatic cooling. Comparison between bryozoan faunas from two localities in the Lower–Middle Permian of Oman allowed Ernst, Weidlich, and Schäfer (2008) to assign the older fauna to a heterozoan assemblage whereas the younger fauna was photozoan, indicating a transition from cool water to tropical carbonate production.

Heterozoan carbonates rich in bryozoans are developed at several stratigraphical levels through the Paleocene to Pliocene succession of the Chatham Islands in the Pacific Ocean east of mainland New Zealand (James et al. 2011). Despite their dominance, the bryozoan faunas from most of the neritic limestones and fossiliferous tuffs have yet to be described taxonomically, excepting the Late Paleocene–Early Eocene Red Bluff Tuff (Gordon and Taylor 1999) and the Early Eocene Tumaio Limestone (Gordon and Taylor 2015) in which 42 and 77 bryozoan species were found, respectively. A combination of upwelling and run-off from the volcanic massif of the islands is believed to have created mesotrophic environments suitable for bryozoans (James et al. 2011).

Transitions from photozoan- to bryozoan-rich heterozoan associations have been recorded from several parts of the stratigraphical column and used to infer climatic cooling. In the Svedrup Basin of the Canadian Arctic Archipelago, cooling and higher nutrient levels were interpreted as causing a transition from Late Carboniferous and earliest Permian photozoan inner ramp carbonates, to heterozoan assemblages in the post-Sakmarian, the latter containing abundant fenestrate, branching, and foliaceous bryozoans (Reid et al. 2007).

Shallow marine carbonates of the Calcare di Nago Formation in northern Italy span the Eocene–Oligocene boundary and the transition from greenhouse to icehouse conditions (Jaramillo-Vogel et al. 2013, 2016). This is marked by a change from a photozoan association dominated by larger benthic foraminifera, corals, and red algae, to a heterozoan association dominated by bryozoans. While this may have been driven by a temperature drop, an increase in nutrient levels could have played a role. Early Oligocene 'bryozoan beds' reach 12 metres in thickness and contain up to 86% bryozoans. Bryozoan-rich sediments of the same age from the Transylvania Basin of Romania were believed by Braga and Crihan (2006) to have been deposited close to the shelf edge at depths of c. 50–200 metres, with a combination of palmate and ramose cheilostomes occupying a shallower zone than ramose cyclostomes, which lived below swell-wave base at more than about 100 metres depth. According to Soták (2010), widespread Late Eocene bryozoan marls and other bryozoan-rich sediments in the Central Carpathian Palaeogene Basin were deposited during an interval of cooling when upwelling or increased run-off caused a rise in nutrient levels (see also Zágoršek 1996).

The northward drift of the Australian plate towards the tropics brought about a change from heterozoan to photozoan carbonate deposition in north-eastern Australia (Brachert et al. 1993). The Upper Oligocene in the Great Barrier Reef Province comprises weakly cemented, bryozoan carbonates representing temperate water deposition. Miocene warming saw a change to subtropical well-cemented, bryozoan-foraminifer carbonates with calcareous algae, and eventually to the tropical chlorophycean-scleractinian carbonates that characterize the region today.

6.5.2 Latitudinal changes in bryozoan contribution to carbonates through time

Bryozoans have been significant contributors to shallow-water carbonate sediments since the Ordovician. Carbonate sediments containing at least 10% bryozoans were distributed pan-latitudinally in the Palaeozoic. For instance, the Ordovician contains both low and high latitude examples of bryozoan-rich sediments (Taylor and Sendino 2010). Low latitude and high latitude (e.g. Haig et al. 2014) examples can also be found in the Permian. However, bryozoan-rich sediments have become almost entirely restricted to non-tropical latitudes since the Jurassic (Taylor and Allison 1998). Most Palaeozoic occurrences occur at palaeolatitudes of less than 30° whereas post-Palaeozoic examples formed predominantly between 30° and 45° (Figure 6.8). Unfortunately, the extreme rarity of bryozoan-rich sediments in the Triassic (cf. Baud et al. 2008) makes it impossible to establish a latitudinal pattern for this geological period and hence to test whether the Palaeozoic vs. post-Palaeozoic contrast is clade dependent (Triassic bryozoan faunas resemble Palaeozoic faunas more than post-Palaeozoic faunas in being dominated by palaeostomates, see Chapter 9.7) or clade independent.

Examples of bryozoans contributing significant amounts of carbonate sediments in the tropics today are rare. One exception is a mid-depth (40–100 m) bank in Hawaii where the presence of cold waters may be responsible (Agegian and Mackenzie 1989). Klicpera, Michel, and Westphal (2015) described carbonate sediments accumulating today at a latitude of 20°N off Mauritania. The facies here is heterozoan, with up to 10% of the sediment comprising bryozoan skeletal remains,

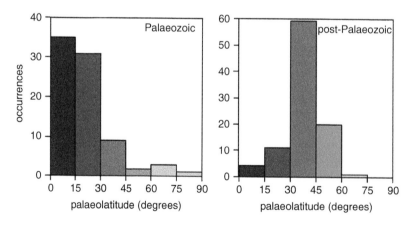

Figure 6.8 Contrast between the palaeolatitudinal distributions of bryozoan-rich deposits during the Palaeozoic and post-Palaeozoic, the former dominated by low latitude occurrences and the latter by mid latitude occurrences (redrawn form Taylor and Allison 1998. © The Trustees of the Natural History Museum, London).

which is attributed to eutrophic conditions caused by upwelling and low light levels on the seabed. Similarly, upwelling on the Pacific coast of Panama is responsible for the occurrence of a heterozoan association with up to 20% bryozoans at the low latitude of 7–9°N (Reijmer, Bauch, and Schäfer 2012), and in the Galápagos where upwelling also occurs and bryozoans can account for almost the same proportion of the sediment (Reymond et al. 2016).

A rare example of bryozoans in the Cretaceous tropics present in rock-forming abundance (Plate 3D) is the ramose malacostegine cheilostome *Heteroconopeum janieresiense* in the Late Cenomanian Pindiga Formation of north-eastern Nigeria (Taylor and Zaborski 2002). By comparison with present-day malacostegines, this bryozoan may have been tolerant of variable salinities, enabling it to colonize the upper Benue Trough which experienced intermittent hyposalinity.

The reason for the striking change in palaeo-latitudinal distribution through geological time is unclear. Bryozoans are known to be diverse and common in the Cenozoic to Recent tropics (information is sparse for the Jurassic and Cretaceous, see Chapter 7), but colonies tend to be small and often weakly mineralized. In a comparative study of bryozoans and other organisms colonizing settlement panels at different latitudes (Freestone, Osman, and Whitlach 2009), it was noted that compared to the common bushy bryozoan colonies at a site in Connecticut, USA, bryozoans from a tropical site in Belize were generally very small, delicate, and rare. Tropical bryozoans therefore contribute minor amounts of carbonate sediment relative to corals, molluscs, and calcareous algae. This is also true in Cenozoic sediments. For instance, Fournier, Montaggioni, and Borgomano (2004) mentioned the presence of bryozoans in several facies in the Oligo-Miocene build-ups offshore of Palawan in the Philippines, but always in small amounts of less than 5%.

Three factors should be considered with regard to the relative decline of tropical bryozoans: (i) nutrient levels, (ii) ecological displacement by calcareous algae and other faster growing taxa, and (iii) grazing pressure. The low nutrient levels typical of the tropics today may perhaps exclude large bryozoan colonies requiring a copious supply of phytoplanktonic food. If nutrient levels in the Palaeozoic tropics were higher than they are at the present day, then perhaps large bryozoans could have flourished at low latitudes. However, it has been argued that nutrient levels in general were low (oligotrophic to mesotrophic) during the Palaeozoic (Martin 1996).

Modern analogues have been sought for the common Palaeozoic shelf communities rich in ramose and other erect bryozoans. Gili et al. (2006) claimed that the epibenthic suspension feeding communities dominating some Antarctic seafloors today have archaic features and include 'Palaeozoic elements' as a result of long-term isolation, reduced terrestrial run-off, and favourable feeding conditions. McKinney and Hageman (2006) contrasted oligotrophy in the eastern part of the northern Adriatic Sea with the high nutrient conditions in the west where terrestrial run-off is greater. Epibenthic communities of suspension feeders including bryozoans reminiscent of some Palaeozoic communities occur in the east, whereas communities in the west are typically infaunal (see also McKinney, Hageman, and Jaklin 2007). Both of these studies are thus consistent with oligotrophic nutrient levels in the Palaeozoic. However, this view has been contested for the Adriatic by Zuchsin and Stachowitsch (2009) who argued that the presence of stable hard substrates and seasonally high productivity accounted for the prevalence of epifauna on the eastern side of the northern Adriatic.

6.6 Taphonomy

The taphonomy of bryozoans has received scant attention. In particular, very little is known about exactly how erect colonies break down into the branch fragments that are so common in the fossil record (e.g. Thomsen 1976). Breakage of branches is frequently coincident with the presence of the boring *Sanctum* in Late Ordovician trepostomes from the Cincinnati region (Key, Wyse Jackson, and Felton 2016), but it is clear that factors other than boring can also be responsible for fragmentation of colonies.

Smith and Nelson (1994) undertook tumbling experiments to estimate the differential resistance of different colony-forms to physical abrasion. They found free-living and fenestrate colonies to be the most resistant, encrusting the least resistant, and ramose and articulated colonies intermediate. A subsequent study identified multilayered encrusting colonies as being particularly prone to destruction through abrasion (Smith and Nelson 2006). The impact on the composition of fossil bryozoan assemblages of differential survival correlating with colony-form was highlighted, which has clear consequences for palaeoecological and palaeoenvironmental interpretations.

Colony-form is undoubtedly a major factor in the size and shape of sediment particles resulting from the breakdown of bryozoans (Taylor and James 2013). Encrusting colonies growing as epiphytes on algae and seagrasses generate silt- and sand-grade particles, dome-shaped colonies usually survive intact as pebble-grade particles, whereas erect colonies typically fragment to form gravels (Figure 6.9)

Comparisons were undertaken by McKinney (1996) of sclerobiont communities from North Carolina associated with two species of bivalves before and after bleaching to simulate taphonomic loss of organic material. Higher feeding tiers of these communities were lost following bleaching, which included a few erect bryozoans that had been attached to the bivalves, whereas most of the encrusting bryozoan survived.

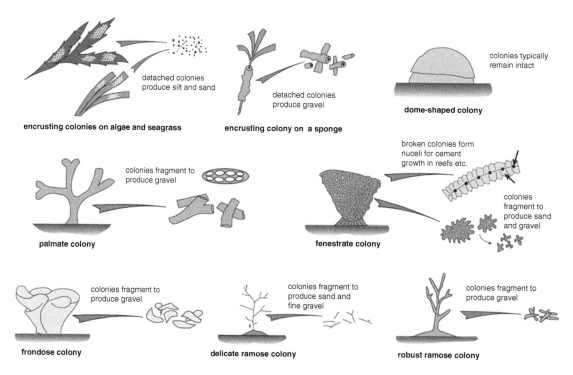

Figure 6.9 Taphonomy of bryozoans with different colony-forms and substrates (based on Taylor and James 2013, fig. 7. © The Trustees of the Natural History Museum, London).

Elongated fragments of erect bryozoans often become oriented according to ambient current flow after death. In a study of *Archimedes* from the Lower Carboniferous of West Virginia, Wulff (1990) measured the orientation of over 200 specimens on two bedding planes, allowing her to infer palaeocurrent directions.

Long distance transport of bryozoans – typically from shallow into deep waters – can occur in some depositional settings. While transportation may be suggested by the broken and abraded condition of the bryozoans concerned, similar damage can also occur locally without significant transportation if the skeletons are not quickly buried. Such may be the case when admixtures of damaged and pristine branches are found together in bryozoan-rich sediments. Conversely, it is possible for bryozoans that have been transported over long distances to remain in a relatively pristine condition. An example is the Pliocene Bowden Shell Bed of Jamaica (now Bowden Member, Layton Formation), which contains a shallow-water bryozoan fauna transported by sediment gravity flows and buried in a deep-water environment (Taylor and Foster 1998). Lagaaij (1968) described an interesting case of longshore transportation of exhumed Eocene and Pliocene fossil bryozoan fragments into Holocene and Recent sands off the Dutch coast. This research sounds a note of caution when using fragmentary bryozoans as indicators both of stratigraphical age and also of depositional environment.

An unusual mode of transportation has been recorded in the Antarctic, where perfectly preserved Holocene bryozoans (*c*. 3000 years BP) were discovered on the surface of the Ross Ice Shelf (Collen 1979; Hayward and Taylor 1984). The bryozoans, along with other benthic organisms, fine sediment, and pebbles, are believed to have become frozen to the base of the ice and plucked from the seabed. As a result of a combination of the processes of surface melting and basal ice accretion, the bryozoans migrated upwards through 70–105 metres of ice at a rate of a few centimetres per year, while lateral flow of the ice

transported them approximately 6 kilometres away from the site where they lived.

Bryozoans themselves can function as 'taphonomic engineers' through their ability to promote the preservation of organisms not normally fossilized by overgrowth, a process termed **bryoimmuration** (Wilson, Buttler, and Taylor 2019). The best examples come from the Upper Ordovician of the Cincinnati region of the US Midwest where aragonitic molluscs, including bivalves and monoplacophorans, are preserved as natural moulds on the undersides of calcitic trepostome bryozoan colonies that overgrew the molluscs before their aragonite shells were dissolved during early diagenesis (Plate 10I).

6.7 Palaeoclimatology and Zooid Size

Bryozoans are not generally thought of as fossils able to contribute much to palaeoclimatology. A few extant taxa with significant fossil records are, however, climatically constrained. Notable among these are several Cenozoic cheilostome genera that are characteristic of warmer, tropical or subtropical waters at the present day. These include *Steginoporella* (see Pouyet and David 1979), *Exechonella*, *Stylopoma*, *Metrarabdotos*, and *Drepanophora*.

Lagaaij's (1963a) classic study of the free-living cheilostome *Cupuladria canariensis* showed this widely distributed species – which is more likely to represent a species complex – to be confined between the 14°C surface isocrymes today. Its presence in the southern North Sea Basin during the Pliocene, notably in the Coralline Crag Formation (Taylor and Taylor 2012), implies temperatures appreciably higher than those pertaining here at the present day. Temperatures during Coralline Crag deposition are a matter of on-going debate as data from stable light isotopes (Vignols et al. 2018) and some faunal evidence (see Williams et al. 2009) suggests that they were not appreciably different from the present day or even cooler. This is not only at odds with the presence of the warm-water cheilostome genus *Metrarabdotos* (Cheetham 1967), but also with a taxon-independent line of bryozoan evidence – comparative zooid size – which supports warm water during Coralline Crag deposition (see below).

Morris (1976) noted the different temperatures required for the formation of female zooids with ovicells in three species of a cheilostome now regarded as belonging to *Celleporella* in the Pacific Ocean along the coast of California and Baja California. She applied this information to infer the summer palaeotemperatures pertaining during the deposition of some mid-Pliocene formations containing ovicell-bearing colonies of each of the three species.

An empirical relationship exists between zooid size and ambient temperature in a wide range of cheilostome bryozoans, with zooid size declining with increase in temperature at the time of budding. This was first shown by Menon (1972) who studied the effects of culturing the cheilostomes *Electra pilosa* and *Conopeum reticulum* at different temperatures. Zooid size was found to decline with increasing temperature in both species: in *Electra pilosa* mean zooid length was 686 µm at 6°C, 596 µm at 12°C, 586 µm at 18°C, and 577 µm at 22 °C; in *Conopeum reticulum* zooid length averaged 558 µm at 12°C, 519 µm at 18°C, and 500 µm at 22°C. Menon's findings were not a laboratory artefact as natural populations of various cheilostomes show similar relationships between zooid size and ambient temperature (e.g. Silén and Harmelin 1976).

There is some indication that the zooid size pattern also occurs among species inhabiting different climatic zones but belonging to the same genus: species from higher latitudes surveyed by Kuklinski and Taylor (2008) had larger zooids in six of the eight genera for which data was available (Figure 6.10). As the zooid size/temperature relationship has been found in not only autozooids but also in avicularia (Figure 6.11), it is not a direct consequence of temperature-related variations in the size of planktonic food. The pattern, which is found in other ectotherms (Atkinson 1994) as well as endotherms and is known as Bergmann's Rule, is probably a consequence of changes in surface area to volume relationships

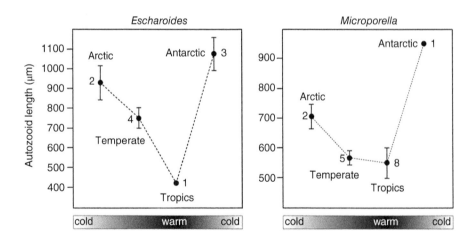

Figure 6.10 Latitudinal changes in autozooid size in species belonging to two genera of cheilostome bryozoans, showing decrease in size from the cold waters of the Antarctic and Arctic to the warm waters of the tropics. Mean values of average species size are indicated using solid circles (with standard errors) with the numbers of species in each climatic zone (redrawn from Kuklinski and Taylor 2008, fig. 3. © The Trustees of the Natural History Museum, London).

Figure 6.11 Decline in the size of avicularia with increasing summer temperature in the cheilostome *Schizoporella errata* along the southern shore of Cape Cod (based on the data of Schopf and Dutton 1976, table 1).

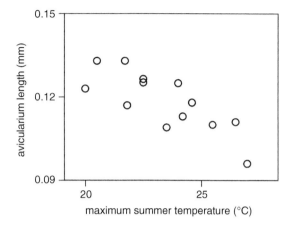

through metabolic scaling and the lower oxygen content of seawater at higher temperatures. In order for zooids to obtain the same amount of oxygen in warm waters, zooid size must decrease to maintain a favourable surface area to volume. For some cheilostome species, the relationship is mirrored by an increase in zooid size with depth as water temperature decreases (Stępień et al. 2017).

It is reasonable to assume that a zooid size/temperature relationship should also exist in stenolaemate bryozoans. This has yet to be established in living cyclostomes but a study of Late Ordovician palaeostomates (Jiménez-Sánchez, Taylor, and Gómez 2013) found that zooid size was greater in congeneric species from cold- than warm-water provinces in all 10 genera tested, with the exception of the trepostome *Trematopora*.

The size/temperature relationship occurs not only between colonies but also within them. O'Dea and Okamura (1999) monitored zooid size in *Conopeum seurati* colonies over a 15-month period, recording temperature, salinity and food availability at the same time. Zooid size within colonies showed a strong inverse correlation with

ambient temperature and also salinity but not with the level of chlorophyll *a* or colony growth rate. Variation in zooid size during growth in *Pentapora foliacea* has also been shown to track estimates of ambient temperature determined from $\partial^{18}O$ data (O'Dea 2005).

Okamura and Bishop (1988) used the inverse zooid size/temperature relationship when comparing zooid size in colonies of eight species from the Coralline Crag Fm. with conspecific colonies living in the adjacent North Sea today. They found the Pliocene colonies to have smaller zooids than their Recent counterparts, implying warmer conditions at the time of Coralline Crag deposition.

While temperature is generally thought to be the most important determinant of zooid size in cheilostome colonies, trophic factors may also be influential. In an experimental study of *Electra pilosa*, Hageman, Needham, and Todd (2009) found that zooids were stunted in size at very low food concentrations. As food concentration was increased, zooid size reached a maximum before a threshold was reached, after which size declined again. In addition, zooids of the cheilostome *Cribrilina annulata* differ in size according to the substrate encrusted – in the White and Barents seas, colonies growing on stones have larger zooids than those growing on the thalli of red algae (Yagunova and Ostrovsky 2008); the former also had larger ancestrulae but a smaller proportion of ovicells (Yagunova and Ostrovsky 2010). While molecular data is needed to ascertain whether colonies growing on stones and algae are truly conspecific or represent two cryptic species, their skeletal morphology would not allow a distinction to be made in fossil material.

At faunal level, an inverse relationship between zooid size and temperature does not always exist. Jackson and Hererra Cubilla's (2000) comparisons of cheilostome bryozoans on either side of the Isthmus of Panama found zooid size to be on average smaller in the eastern Pacific than the Caribbean, despite the lower temperatures in the former.

6.7.1 MART analysis

An ingenious technique for inferring temperature seasonality from the amount of within colony variation in zooid size was devised by O'Dea and Okamura (2000b). Their MART (mean annual range in temperature) method is based on the premise that size-related variability in colonies growing in constant temperatures will be less than those growing in highly seasonal environments. The premise was borne out using zooid size data from cheilostomes inhabiting environments experiencing varying levels of temperature seasonality (Figure 6.12), allowing the following equation to be formulated:

$$MART = -3 + 0.745(b)$$

where (b) is the mean intracolony CV of frontal area (length x width)

An early palaeontological application of the zooid MART method – subsequently referred to as zs-MART – was used by O'Dea and Okamura (2000a) to infer lower seasonal variation in temperature during deposition of the Pliocene Coralline Crag Fm. of eastern England than in the same region today (5.8°C vs 11.5°C). This was attributed to a stronger Gulf Stream and warmer winter temperatures during the Pliocene.

Testing of the zs-MART method using encrusting cheilostomes from Panama showed a good correspondence between predicted and real values (O'Dea 2003). The effect of temperature seasonality on within-colony variation in zooid size was also demonstrated by a study of free-living cupuladriid cheilostomes on either side of the Isthmus of Panama (O'Dea and Jackson 2002). A very low annual range in temperature characterizes the Caribbean side of the isthmus exhibits, whereas the Gulf of Panama on the Pacific side experiences seasonal upwelling resulting in high temperature variation through the year. Measurements of zooid size along transects within colonies, from the oldest at the centre to the youngest around the margins, showed no trends in colonies from the stable Caribbean but clear cyclicity in colonies from the thermally unstable Gulf of Panama. Mean CV

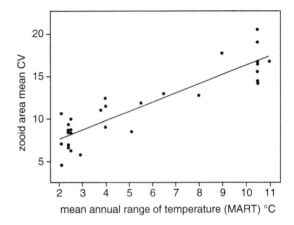

Figure 6.12 Correlation in 29 species of Recent cheilostomes between within-colony variability in zooid size (expressed as mean coefficient of variation of zooid area) and mean annual range of temperature (MART) at sites of collection (redrawn from O'Dea and Okamura 2000b).

of zooid size in Caribbean colonies was 8.2 compared with 15.3 in colonies from the Gulf of Panama. The contrast between cupuladriids from the two sides of the isthmus is also evident in profiles of skeletal $\partial^{18}O$ (Key et al. 2013).

A combined $\partial^{18}O$ isotopic and zs-MART study was undertaken using the cheilostome *Pentapora folicacea* from two localities on the coast of Wales (Knowles et al. 2010). Estimates of seasonal temperature variations obtained by the two methods were compared with temperatures recorded by a data logger deployed at one of the two localities. The annual range of temperature estimated using zooid size variability was close to that recorded by a data logger, whereas the skeletal oxygen isotope ratios gave a lower range, which was attributed to the continued thickening of the bryozoan skeleton during growth smoothing out the seasonal variations.

Okamura, O'Dea, and Knowles (2011, table 1) specified rules that should be applied for choosing zooids, colonies, and species appropriate for zs-MART analysis. These authors also reviewed limitations of the method and advantages. Among the latter are the low cost and simplicity compared to other methods for estimating temperature seasonality. A detailed exploration of various sources of errors in zs-MART analysis (McClelland et al. 2014), has enabled a revised zs-MART equation to be proposed:

$$MART = CV \text{ zooid area} - 6 \pm 4$$

Note that this includes an error bar of ±4°C, which limits the usefulness of the technique. The effect on size caused by the position of zooid relative to row bifurcations can further compromise zs-MART analysis and a more nuanced approach should be developed on a species-by-species basis.

6.7.2 MART applications

Several applications have been made of the bryozoan zs-MART technique. Dick et al. (2008) applied a slightly modified zs-MART technique to infer temperature seasonality through the Pleistocene of northern Japan. They found a decrease in average MART value from the lower (11.8°C) to upper Setana Formation (8.6°C) which could reflect the onset of colder conditions, although it is possible the younger samples came from a deeper water environment where the MART would be less.

A study of geographical patterns of seasonality in the early and mid-Pliocene of the North Atlantic (Knowles et al. 2009) found greater seasonality (c. 4.5 °C) in the southern Caribbean at about 4.25 Ma, consistent with upwelling prior to the closure of the Isthmus of Panama. Cheilostome bryozoans collected further to the north along the US Coastal Plain yielded MART estimates suggesting a slightly lower annual range in temperature than at the present day (Figure 6.13), possibly indicating that cold currents flowing along the east coast of North America did not extend as far south. On the other side of the Atlantic, seasonality was probably greater at c. 8°C during deposition of the Coralline Crag Formation than it is in the adjacent North Sea today (c. 4–5°C).

Clark et al. (2010) measured zooid sizes in early Pliocene cheilostomes from a diamictite on James Ross Island in the Weddell Sea. This allowed a MART of 4.3–10.3°C to be inferred, with a cluster of analyses of 6.6–7.7°C from colonies of *Dakariella* considered to be the most reliable estimates of seasonality. In contrast, the adjacent

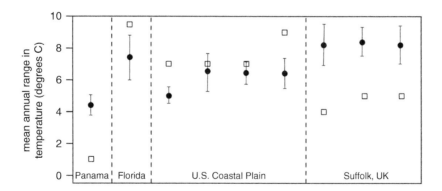

Figure 6.13 Mean annual range in temperature at the present day (square boxes) compared with Pliocene ranges estimated using MART analyses of zooid size (solid circles with standard deviation error bars (redrawn from Knowles et al. 2009. © The Trustees of the Natural History Museum, London). The four localities on the US Coastal Plain are (left to right): New Bern, North Carolina (Duplin Formation), Chuckatuck, Virginia (Yorktown Formation), Chipokes State Park, Virginia (Yorktown Formation), and Moor House, Virginia (Yorktown Formation). The three values in Suffolk represent (left to right) the Ramsholt, Sudbourne, and Aldeburgh members of the Coralline Crag Formation.

Weddell Sea today has a seasonal range of temperature of only about 2°C.

The potential of zs-MART to identify El Niño/La Niña–Southern Oscillation (ENSO) variability in ancient tropical environments of the eastern Pacific and south-western Caribbean was explored by Okamura et al. (2013). These authors reasoned that absence of ENSO variability would be reflected in a normal distribution of zs-MART estimates obtained from bryozoan assemblages. In contrast, the presence of ENSO variability should result in a broader range of zs-MART estimates deviating from normality due to the inclusion of years containing periods of warmer (El Niño events) or cooler (La Niña events) temperatures. Bryozoan zs-MART values suggested that ENSO variability characterized Caribbean environments in Panama during the Miocene and Pliocene, but not during the Pleistocene after closure of the Isthmus of Panama.

7 Biogeography

Determining geographical distributions of bryozoan species and higher taxa today – and more especially in the geological past – is hampered by inadequacies in taxonomy and our poor knowledge of faunas from some regions of the globe. For example, a recent study (Dick and Grischenko 2017) of the bryozoan fauna close to a marine station in Okinawa found that 86% of the 52 species detected were either new species or previously unrecorded in Japan. Remarkably, there has been no modern synthesis of the global biogeography of living marine bryozoans. The biogeography of present-day bryozoans undoubtedly bears the strong imprint of human activities, with many species having achieved pan-oceanic distributions through transportation on ships' hulls or in their ballast tanks. Less easy to explain are palaeogeographical distributions that imply transoceanic migrations of bryozoans that are known, or are assumed, to have had short-lived larvae. Rafting may have been a more important vector of dispersal than is generally acknowledged, with rafting events that are rare in ecological time being relatively common in the vastness of geological time. Bryozoans at the present day seem not to exhibit the strong latitudinal gradient of diversity increase towards the equator that characterizes many other marine and terrestrial phyla.

7.1 Cosmopolitan vs. Endemic Distributions

Uncritical reading of the older literature on bryozoans could easily lead to the erroneous impression that a large number of species living at the present day are distributed pan-latitudinally. Berning and Kuklinski (2008) referred to the paradox of bryozoan species which brood their larvae and yet have seemingly cosmopolitan distributions. For example, records of supposed *Microporella ciliata*, the type species of an ascophoran cheilostome described originally from the Mediterranean Sea, extend from the type area to the Red Sea, north-west Europe, the Arctic, the western North Atlantic, the Caribbean, Brazil, the coast of western America from Oregon to Chile, Galapagos, the Antarctic Peninsula, south-east Asia, Australia, and New Zealand (Taylor and Mawatari 2005). Even allowing for anthropogenic transportation (see below), the occurrence of this species across such a wide range of latitudes, and hence climatic zones, is implausible. Much of the problem lies in imprecise taxonomy: application of scanning electron microscopy has revealed small-scale differences in the morphology of the skeleton pointing to the presence of numerous species masquerading under the name *Microporella ciliata*.

Confidence in interpreting such slight differences in skeletal morphology as significant at the species level receives support from molecular studies, led by Jackson and Cheetham's (1990) classic paper on three genera of Panamanian cheilostomes. This research showed that morphologically distinct species belonging to *Steginoporella*, *Stylopoma*, and *Parasmittina* are also distinct in terms of allele frequencies determined using protein gel electrophoresis. Crucially, no cryptic species identifiable only on the basis of molecular information were found. Since this paper, the pendulum has swung from grouping to splitting in taxonomic studies of cheilostomes. However, subsequent studies have shown the need for some caution. Despite conspicuous morphological differences – especially in spinosity – that were formerly used to characterize several different species, an allozyme study (Schwaninger 1999) of the widely distributed cheilostome *Membranipora membranacea* revealed scarcely any genetic differences between morphologically distinct populations separated by long distances (e.g. on either side of the North Atlantic). Conversely, a combined morphological and molecular study of the anascan cheilostome *Cauloramphus* found that some genetically distinct clades were inseparable morphologically, and at least one putative cryptic species is present

(Dick, Hirose, and Mawatari 2012). Likewise, morphologically cryptic species have been identified using DNA barcoding in populations of *Celleporella hyalina* (Gómez et al. 2007a), a finding subsequently corroborated using mitochondrial genome evidence (Waeschenbach, Porter, and Hughes 2012).

A good example of endemism among recent bryozoans is provided by the taxonomic paper of Berning, Harmelin, and Bader (2017) on cheilostomes from north-east Atlantic deep waters (c. 200–1700 m). Of the 13 species described in this study, nine were restricted to a single seamount. The presence of so many endemic species was considered by Berning et al. to be related to the low dispersal abilities of their brooded larvae.

While some supposed cosmopolitan distributions of bryozoan species are based on incorrect taxonomy (e.g. Tompsett, Porter, and Taylor 2009), others remain even after careful study of skeletal morphology. For example, the ascophoran cheilostome *Scorpiodinipora costulata* has a scattered tropical and subtropical distribution in the Pacific (Philippines), Indian (Oman), south-east Atlantic (Ghana) and south-west Atlantic (Brazil) oceans, the Red Sea, and southeastern Mediterranean Sea (Harmelin et al. 2012). This is very hard to explain for a species having short-lived larvae, unless it was dispersed by shipping. Alternatively, specimens described as *Scorpiodinipora costulata* may comprise a cluster of morphologically cryptic species. Molecular studies are needed to discriminate between these options. Even more puzzling is the bipolar occurrence of the cheilostome *Callopora weslawski*. While yet to be confirmed genetically, this is the only known benthic animal species that broods its embryos to have been recorded from both the Arctic and Antarctic but nowhere between these two polar regions (Kuklinski and Barnes 2010).

7.2 Modes of Dispersal

Dispersal of bryozoans can occur in several different ways, of which larval dispersal followed by rafting are the most common in marine species. The statoblasts of freshwater phylactolaemate bryozoans can also be dispersed by migratory wildfowl, as already mentioned above (Chapter 1.3). Statoblasts – especially those with hooks – may become entangled in the feathers of birds which carry them from one isolated water body to another, with consequences for population genetic structures compared to species in which statoblasts are not dispersed in this way (Okamura, Hartikainen, and Trew 2019). The importance of this vector of dispersal is evident from matches between wildfowl migratory routes and the geographical distributions of some phylactolaemate species (see Wood 2002).

7.2.1 Larvae

Larvae are the primary means of dispersal in marine bryozoans. All living cyclostomes and the great majority of cheilostomes have short-lived larvae which are incapable of feeding and must settle and metamorphose soon after their release. The necessity of rapid settlement may be a consequence of the rapid consumption of stored energy resources in non-feeding bryozoan larvae (Jaeckle 1994), although there is evidence that the larvae of *Bugula neritina* are able to top up these resources with dissolved organic matter (Johnson and Wendt 2007). Coronate larvae of the cheilostome *Thalamoporella californica* can only metamorphose successfully if they settle within 10–12 hours of their release (Chaney, Soule, and Soule 1989). Cook (1968a) reported settlement of the larvae of *Steginoporella buskii* to occur 33–48 hours after release, whereas all larvae of another cheilostome, *Stylopoma duboisii*, had settled within two hours (Cook 1973). Grave (1930) reported that the larvae of the cheilostome *Bugulina flabellata* swam for 4–6 hours before settling. In *Bugula neritina*, colonies produced by larvae made to swim for extended periods were significantly reduced in size, implying a fitness cost (Wendt 1998).

These short larval durations do not facilitate large-scale dispersal. Accordingly, population genetic studies of cheilostomes with non-planktotrophic (lecithotrophic) larvae have shown the existence of divergent genetic lineages in allopatric populations, as in the epiphytic species *Celleporella hyalina* from the north-east Atlantic (Gómez et al. 2007a, 2007b). Similarly, significant degrees of genetic variation between colonies of the cyclostome *Crisia denticulata* over distances of just a few metres suggest small-scale dispersal

of larvae and/or sperm (Pemberton et al. 2007). At the other end of the spectrum are the planktotrophic cyphonautes larvae of malacostegine cheilostomes such as *Membranipora membranacea* in which there is the potential for long-distance dispersal as the feeding larvae can live for up to two months (Yoshioka 1982b). Comparison of the population genetic structure of two cheilostome species with planktotrophic larvae and two with non-planktotrophic larvae found the predicted higher levels of genetic heterogeneity between populations of the non-planktotrophic species (Watts and Thorpe 2006).

A phylogeographical study of *Membranipora membranacea* using mtDNA cytochrome oxidase 1 (COI) sequences identified several infraspecific clades on either side of the equator (Schwaninger 2008). Human-mediated dispersal was considered not to have been significant in this species complex. Instead, episodic long-distance dispersal, perhaps related to climatic change, seemed best to explain the distributional pattern seen today. A mid Miocene dispersal event from the North to the South Pacific was inferred, followed by dispersal into the South Atlantic and finally the North Atlantic during the late Miocene. Unfortunately, this species has no fossil record that might be used to evaluate pre-human patterns of distribution.

7.2.2 Rafting

Dispersal by rafting can be achieved on both anthropogenic and natural substrates. Plastic and other floating debris in the sea is routinely colonized by bryozoans (e.g. Thiel and Gutow 2005; Farrapeira 2011; Goldstein, Carson, and Eriksen 2014; Kiessling, Gutow, and Thiel 2015; Rech et al. 2018). Spectacular transoceanic rafting of coastal marine species was reported by Carlton et al. (2017) in the six years following the Japanese tsunami of 2011. A total of 289 species were carried eastwards across the Pacific to the shores of Hawaii and North America on objects ranging in size from small items of plastic to the hulls of boats. Bryozoans were reported on 317 objects of debris and comprised 49 species (McCuller and Carlton 2018). At least 13 of these species were not native to Hawaii or to North America. Runner-like colonies were the commonest colony-form recorded, followed by erect flexible or articulated colonies, with sheet- and mound-like colonies less frequent.

The most abundant organism fouling plastic debris at Rapa Nui (Easter Island) was found to be the cheilostome *Jellyella eburnea* (Rech et al. 2018), a species whose main natural substrate consists of floating shells of the squid *Spirula* (Taylor and Monks 1997) but which is now flourishing on plastic substrates in the ocean (e.g. Moyano 2005a).

Movements of obsolete ships furnish another means of dispersal for bryozoans and other species fouling their hulls (Davidson et al. 2008). In addition, cyphonautes larvae of cheilostomes have been found in the ballast tanks of ships (Carlton 1985), as have the statoblasts of freshwater phylactolaemates in ships crossing the Great Lakes of North America (Kipp et al. 2010), suggesting these vessels may act as significant vectors for long-distance dispersal. A study of 197 cheilostome species found in British waters correlating global geographical ranges with various traits (Watts, Thorpe, and Taylor 1998), found that the ability to raft and to foul anthropogenic substrates had a significant positive effect on the ranges of these species, underlining the importance of rafting for dispersal.

Anthropogenic rafting has undoubtedly been responsible for the spread of non-native bryozoan species. For example, among non-native species that are common in European waters but which may have arrived anthropogenetically are the anascan cheilostome *Tricellaria inopinata* (Dyrynda et al. 2000) and the ascophorans *Smittoidea prolifica* (De Blauwe and Faase 2004) and *Schizoporella japonica* (Ryland et al. 2014). Non-native species can make up a significant proportion of some faunas: Harmelin, Bitar, and Zibrowius (2016) surveyed the bryozoan fauna of the Mediterranean coast of Lebanon and recorded 93 species of which 27 (29%) were non-native. However, even for the Mediterranean, where the contemporary bryozoan fauna is relatively well known, possible native species that seem to have been overlooked are still being discovered (Rosso 2009). There is an as yet unfulfilled role for fossil evidence in testing hypotheses of anthropogenic introductions of bryozoans: human involvement could be falsified by finding pertinent species in sub-Recent sediments that antedate shipping.

Floating *Sargassum* weed is a natural substrate colonized by some species of bryozoans, such as the planktotrophic *Jellyella tuberculata* (see Thiel and Gutow 2005). Carapaces of marine turtles are very occasionally encrusted by bryozoans (Frazier, Winston, and Ruckdeschel 1992), providing another potential vector for long-distance dispersal.

A natural abiotic substrate capable of facilitating dispersal is pumice. In excess of 80 marine species, including numerous bryozoans, were carried more than 5000 kilometres across the Pacific to Australia in just 7–8 months on floating pumice erupted from a volcano in Tonga (Bryan et al. 2012). Rafting on pumice is therefore a very potent mode of long-distance dispersal, and one that has been possible throughout geological time. Unfortunately, proving the efficacy of rafting on pumice for the dispersal of bryozoans in the geological past would be challenging to say the least as there is little prospect of finding species of recognizably alien bryozoans on exotic pumice in the geological record.

While it is reasonable to propose rafting as a key factor explaining the wide dispersal of bryozoans with short-lived larvae, there are difficulties in applying it to some kinds of bryozoans. Cook and Chimonides (1994a) questioned the feasibility of rafting of the free-living cheilostome *Cupuladria*. Nevertheless, one species of this genus, *C. remota*, occurs on the Marquesas Islands 5250 kilometres distant from the Galapagos, the nearest shallow-water source area, and separated from it by deep water (2000–4000 m) environments totally hostile to this genus.

Vectors of dispersal remain obscure for many bryozoans, notably the cheilostome *Membraniporopsis tubigera*, which is considered to be a significant invasive species. Colonies of this weakly mineralized bifoliate species have been appearing sporadically in great numbers, recovered as drift debris along beaches and clogging fishing nets in South America (Gordon, Ramalho, and Taylor 2006; López Gappa et al. 2010), yet nothing is known about its larvae (cyphonautes or coronate?) or whether the species is dispersed through rafting.

7.3 Biogeography of Bryozoans at the Present Day

As noted above, there are no modern publications dealing with the global biogeography of present-day bryozoans and, unfortunately, regional studies are also surprisingly few. The table in the appendix of Schopf (1979) listed 23 bryozoan provinces and estimated the approximate number of species in each, mainly for the purpose of seeking a latitudinal diversity gradient.

Tilbrook and De Grave (2005) undertook a Detrended Correspondence Analysis on shallow-water, Indo-West Pacific cheilostomes. They found a split at family-, genus-, and species-level between Indian Ocean and Pacific Ocean cheilostome faunas. Among Pacific faunas, Hawaii stood out as being very different from the others. Species endemism was strongly correlated with species richness, i.e. the more species present in a fauna, the higher the proportion of endemics.

A study of bryozoan biogeography along the Pacific coast of Panama (Schäfer, Herrera Cubilla, and Bader 2012) compared the Gulf of Panama in the east with the Gulf of Chiriquí in the west. Species in the Gulf of Panama cluster had a much wider distribution in the Atlantic and Eastern Pacific than those in the Gulf of Chiriquí cluster which were more widely reported from the West Atlantic, Caribbean, and Indo-Pacific. This may reflect the more tropical 'Caribbean-type' environment of the Gulf of Chiriquí compared with the more open ocean, colder conditions of the Gulf of Panama.

Most other regional analyses have focused on southern South America and Antarctica. Moyano (1999) studied zoogeographical relationships between the Megallanic, Atlantic, Pacific, sub-Antarctic, and Antarctic regions, finding a significant number of species in common between the Megallanic and Antarctic regions. This he attributed to interchange via the Scotia Arc between the northernmost tip of the Antarctic Peninsula and the southernmost tip of South America (see also Moyano 2005b). A subsequent study by the same author (Moyano 2006) showed strong zoogeographical links between the bryozoans inhabiting Antarctica, South America, and Australasia, reflecting not only the Gondwanan origins of these regions but also the presence of guyots, submarine ridges, etc., that could facilitate dispersal. Data on the distributions of 1465 cheilostome and 167 cyclostome species from 29 regions across the Southern Hemisphere formed the foundation for a comprehensive analysis of the biogeography of the Southern Ocean benthos (Griffiths, Barnes, and Linse 2009). Echoing Moyano's findings, strong faunal links were found between South America and the Antarctic. The Antarctic Circumpolar Current was identified as having a major and long-term influence on distributions, the Southern Ocean forming a single unit for both cheilostomes and cyclostomes, with no evidence for an east–west biogeographical split. Newer research on the south-west Atlantic (Figuerola et al. 2014, 2017b) has revealed additional shared species in regions on either side of the Antarctic Circumpolar Current. The implication is that the Southern Ocean has not been quite as isolated through geological time as was previously believed.

7.4 Latitudinal Diversity Gradient

Latitudinal gradients in diversity have attracted considerable attention from biogeographers and other biologists (see reviews by Hillebrand 2004). More often than not, diversity increases towards the tropics in marine taxa (e.g. corals and molluscs), but there are some groups in which this is not the case (e.g. marine mammals). An early paper by Schopf (1970) based on a limited compilation of data from the taxonomic literature failed to find a clear latitudinal gradient for bryozoans in the Atlantic. Schopf's data from the Pacific showed higher species richness levels than the Atlantic, as well as an overall diversity decline from about 30°N to 40°S. Nonetheless, in a subsequent compilation, Schopf (1979) claimed that bryozoan species diversity was highest in the tropics and declined towards the poles.

The most detailed study of latitudinal gradients in bryozoans is that of Clarke and Lidgard (2000) who compiled data on 535 species from 304 sites in the North Atlantic. As mean assemblage species richness showed no significant variation with latitude, there was no compelling evidence for a latitudinal diversity gradient (Figure 7.1). However, after pooling their data into 10° latitudinal bins and applying correction techniques for undersampling, a peak of species richness was found at 10–30° with a steady decline north to 80°.

Kuklinski, Barnes, and Taylor (2006) found the diversity of encrusters, including bryozoans, on cobbles from north-west European sites (50–79°N) to show an increase into lower latitudes, mainly due to the impoverished species richnesses from Arctic sites. In the cosmopolitan, species-rich cheilostome genus *Microporella*, most species have been described from mid-latitudes, with diversity peaking at 40° (Taylor and Mawatari 2005, fig. 2). In their database of bryozoan species occurrences in the southern temperate and polar regions, Barnes and Griffiths (2008) found no latitudinal diversity gradient here. Instead, a longitudinal gradient of decreasing diversity from an Australasian centre was identified.

Further analyses are required to establish the existence and details of a latitudinal diversity gradient in bryozoans, and also how it might have arisen and been maintained. This should include the limited information currently available for the tropical Indo-West Pacific where preliminary indications (Di Martino et al. 2018) suggest the existence of a diversity hotspot similar to that seen in some other groups of marine animals (Renema et al. 2008). Differences in sampling effort may play a large part in estimates of bryozoan diversity (Dick and Grischenko 2017), a factor which should be considered during analyses of geographical patterns.

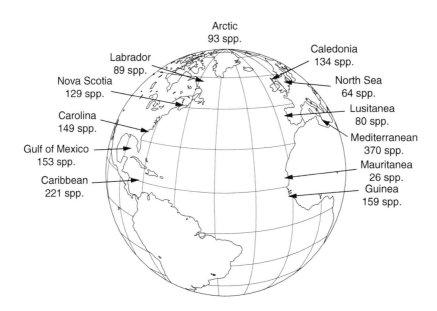

Figure 7.1 Beta diversity of North Atlantic bryozoan species (data from Clarke and Lidgard 2000).

7.5 Palaeobiogeography

Patchy knowledge of the geographical distributions of taxa through geological time, coupled with imprecise taxonomy, has hampered interpretations of bryozoan palaeobiogeography. Nevertheless, some biogeographical patterns have been identified, based mainly on the analysis of genus-level data. Palaeobiogeographers studying different geological periods have tended to address different issues, as reflected in the summaries given below. Notably, Palaeozoic palaeobiogeographical research has often focused on taxon distributions between ancient continental masses (e.g. Cocks and Fortey 1990); for research on the Jurassic and Cretaceous, Tethyan vs. Boreal provinciality has been a major theme (e.g. Hallam 1994); while the evolution of modern biogeography is often a dominant concern in Cenozoic palaeobiogeographical studies (e.g. Kocsis, Reddin, and Kiessling 2018).

7.5.1 Ordovician

The earliest bryozoans are known from the Tremadocian of China (Ma et al. 2015) and Arctic Russia (Ernst et al. 2014), but comprise only a few genera (Chapter 9.3). Bryozoan diversification during the Ordovician was a component of the Great Ordovician Biodiversification Event (GOBE), and by the Darriwilian 36 genera can be recognized, distributed between four provinces – Baltica, Belarus, Siberia, and North America – according to Buttler et al. (2013; see also Tuckey 1990a; Koromyslova 2011). In sharing 10 genera, Baltica and Belarus show the closest biogeographical relationship. The same four provinces can be recognized in the Sandbian but Baltica and North America exhibit the greatest affinity at this time, a relationship which continued into the Katian. Complicated palaeobiogeographical relationships exist between various regions containing Katian bryozoans. Some are surprising, such as the close affinity between China and North Africa. Bryozoan diversity was reduced in the Hirnantian, as apparently was provincialism.

Jiménez-Sánchez and Villas (2009) undertook a detailed analysis of late Katian bryozoan palaeobiogeography. They found that the late Katian marine transgression introduced bryozoans from tropical and subtropical regions onto the Mediterranean margin of Gondwana. A Mediterranean Province containing 22 endemic genera developed in the cooler waters at mid-latitudes estimated to have been located between 40°S and 55°S.

The three Late Ordovician extinction events distinguished by Tuckey and Anstey (1992) each affected these bryozoan provinces to varying degrees. The early Katian event mainly impacted bryozoans in Baltica and Siberia, whereas the late Katian event was felt most in North America, and the Hirnantian event was most severe in Baltica.

7.5.2 Silurian

Less information is available about Silurian bryozoan palaeobiogeography (see Tuckey 1990b), due to a paucity of bryozoan faunal studies. This is especially true for the Llandovery, reflecting both unfavourable clastic-rich facies and the relatively depauperate Silurian faunas following the Late Ordovician extinction. Six provinces – Kazakhstan, Mongolia, Siberia, China, North America, and Baltica – were distinguished in the Llandovery by Buttler et al. (2013), Siberia clustering with China, and North America with Baltica (as in the Ordovician). Wenlock bryozoan faunas are much better known, with North America and Baltica once again strongly linked. In the Ludlow, Australia was found to cluster with Kazakhstan, whereas Mongolia and North America grouped together, as did Baltica, Siberia, and China. Pridoli bryozoan faunas are sparse but those of Baltica and Kazakhstan clustered together.

Parsimony analysis of endemicity allowed McCoy and Anstey (2010) to recognize six bryozoan provinces in the Silurian of North America, Baltica, and Siberia, each containing several endemic genera. They found endemism to be high among bryozoans compared to other taxonomic groups that have been studied, and there was no significant change in the level of endemism through the Silurian.

7.5.3 Devonian

A review of Devonian bryozoan biogeography (Bigey 1985) noted the difficulty of defining bryozoan provinces because of poor knowledge of bryozoans from many parts of the world. Bigey therefore based her work on provinces that had been defined using other groups of marine animals. Early Devonian bryozoans were present in three realms – Malvinokaffric, Old World, and Eastern American – without any clear endemism. At the time of Bigey's publication, no bryozoans were known from the Malvinokaffric Realm in the Middle Devonian but some endemism existed within North America. Most bryozoans disappeared from sequences in Europe and North America at the end of the Frasnian.

A detailed analysis of bryozoan palaeobiogeography in the Famennian (Tolokonnikova and Ernst 2010) suggested migration from northern China to Kazakhstan and the western offshoots of Tian Shan, and from there to other regions.

7.5.4 Carboniferous

Ross (1981) documented the distribution of Carboniferous bryozoan genera among 11 regions across the globe. High levels of endemism typified some of these regions, such as the mid-continent and eastern USA. Various generic dispersals during the Carboniferous were recognized by Ross, which she believed were associated with transgressions, climatic changes, and tectonic events. An analysis of the palaeobiogeography of Eurasian Early Carboniferous bryozoans (Tolokonnikova et al. 2014) found a decrease in endemism from the Tournaisian into the Viséan. Two centres of radiation were identified: an eastern centre located at a palaeolatitude of 20–40°N, and a smaller western centre at 5–10°S.

7.5.5 Permian

Paralleling her work on Carboniferous bryozoan palaeobiogeography, Ross (1978) recorded the palaeogeographical distributions of Permian bryozoan genera across 10 regions. Tethyan bryozoans were found to be the most diverse, with many genera known only from this tropical region.

Other genera were apparently adapted to colder water conditions at higher latitudes. Sakagami (1985) noted the cosmopolitan distributions of several genera found in the Permian (e.g. *Fistulipora*, *Stenopora*, *Penniretepora*, and *Rhabdomeson*) and underscored the decrease in records of bryozoans into the Late Permian in which most known faunas are Tethyan. Ernst (2002) showed the bryozoan fauna of the Zechstein Sea, which covered parts of northern Europe, to be endemic at species level. Records of supposed Zechstein species elsewhere in the world are based on flawed taxonomy according to Ernst. A biogeographical analysis suggested that bryozoans in the Zechstein migrated into this warm sea from the colder waters of East Greenland to the north rather than via the tropical Tethys Sea (Sørensen, Håkansson, and Stemmerik 2007).

A more recent study has focussed on the high latitude Permian bryozoan biogeography of the Gondwanan and Boreal realms (Reid and James 2010). The expected lower diversity of these realms relative to the Tethyan Realm is apparent, with the particularly high latitude (>55°S) Gondwana Realm hosting fewer genera (61) than the Boreal Realm (>100) which was in part located at mid-latitudes. Three bipolar genera were recognized: *Diploporaria*, *Lyropora*, and *Metalipora*. Trepostomes and fenestrates, interpreted as more eurytopic, dominated at higher palaeolatitudes. The Lower Permian bryozoan fauna in the Zhongba area of south-western Tibet contains a mixture of Boreal and Gondwanan elements, which led Ernst (2016a) to suggest faunal migrations into this tropical Tethyan region from both the north and the south.

7.5.6 Triassic

Research in Triassic bryozoans and their palaeobiogeography has been fuelled by interest in the impact of the Late Permian extinctions that were severe enough to deplete bryozoan diversity throughout the Triassic. Most Triassic bryozoans are palaeostomate 'holdovers' from the Permian (Chapter 9.7). Several papers review or discuss Triassic bryozoan biogeography (Hu 1984; Sakagami 1985; Schäfer and Fois-Erickson 1986; Schäfer,

Cuffey, and Young 2003; Schäfer, Senowbari-Daryan, and Hamedani 2003; Powers and Pachut 2008). No Early Triassic bryozoans are known from the Tethyan Realm, the few examples described coming from non-tropical shelf environments elsewhere, including the Arctic. By contrast, Middle Triassic bryozoans occurred mostly in low palaeolatitudes on the margins of Tethys. Some genera, such as the trepostome *Pseudobatostomella*, were cosmopolitan, spanning all major climatic zones. Cosmopolitan trepostomes dominated in the Late Triassic, with *Dyscritella* joining *Pseudobatostomella* (Plate 4N) as one of the most abundant genera.

7.5.7 Jurassic

Jurassic bryozoans consist overwhelmingly of cyclostomes (Taylor and Ernst 2008). Although present, gymnolaemates are represented by only one cheilostome genus – *Pyriporopsis* – plus a few bioimmured and boring ctenostomes. Known occurrences of Jurassic cyclostomes are so strongly concentrated in Europe, particularly France, England, Germany, and Poland, that broader biogeographical patterns are difficult to discern. Only one small fauna is known from the Jurassic tropics: the Callovian Matmor Formation of Israel which contains six species in five genera (Wilson, Bosch, and Taylor 2014). There is some indication that a European diversity hot-spot migrated eastwards through time (Viskova 2009). At genus level, non-European Jurassic cyclostomes are a depauperate subset of the European fauna: there are no endemic genera, apart possibly from *Patulopora* which has been recognized so far only in the Western Interior Seaway of Utah (Taylor and Wilson 1999b).

7.5.8 Cretaceous

Continuing the pattern of the Jurassic, known occurrences of Early Cretaceous bryozoans are strongly centred on western Europe. One of the few palaeobiogeographical studies (Walter 1996) within this region documented the northwards spread into the Paris Basin and Germany during the Early Hauterivian transgression of a cyclostome fauna that had been present previously in the Late Valanginian of the Jura. An Argentinian fauna of bryozoans from the Agrio Formation (Valanginian–?Barremian) of the Neuquén Basin comprises six species, five of which belong to genera represented in coeval deposits in Europe (Taylor, Lazo, and Aguirre-Urreta 2009), implying no significant endemism at genus level.

The spectacular diversification of both cyclostome and cheilostome bryozoans that occurred in the Late Cretaceous (Chapter 9.9) was accompanied by a great increase in their global palaeogeographical distributions, especially during the Campanian and Maastrichtian (e.g. Guha and Nathan 1996; Taylor and Gordon 2007; Di Martino, Martha, and Taylor 2018; Dick, Sakamoto, and Komatsu 2018). Continuing the Jurassic pattern, tropical occurrences of Cretaceous bryozoans are sparse, although encrusting bryofaunas that inhabited the palaeotropics have been described from Campanian of the Middle East (Taylor 1995; Di Martino and Taylor 2013).

Voigt's (1983) study of European Late Cretaceous bryozoan faunas identified a range of taxa characteristic of northern (Boreal) and southern (Tethyan) provinces. The northern province coincides with the white chalk facies and associated nearshore deposits developed in the Anglo-Paris Basin and adjacent regions, whereas the southern province is best seen in the Aquitaine Basin where carbonate facies occur but not coccolith chalks. Among northern European genera are the cyclostome *Clinopora* and the cheilostome *Volviflustrellaria*. Southern European genera include the cheilostome *Heteroconopeum*, which also occurs as far south as Nigeria (Taylor and Zaborski 2002) and to the west in the south-eastern United States (Taylor and McKinney 2006).

Diverse bryozoan faunas in the Late Cretaceous of the Southern Hemisphere (Gondwana) are known from India, Madagascar, South Africa, Australia, and New Zealand. Genus-level composition generally resembles that of Northern Hemisphere faunas: fewer than 10% of the 98 named genera recognized in the Gondwanan Cretaceous are endemic (Taylor 2019). Preliminary data suggests that the proportions of major cheilostome groups may have been more similar between the Northern and Southern hemispheres in Cretaceous times than they are at the present day.

7.5.9 Cenozoic

Unfortunately, there is no global synthesis of Cenozoic bryozoan palaeobiogeography and regional studies are not numerous. Poor sampling for fossil bryozoans in the Indo-Pacific restricts our knowledge of their palaeobiogeography during the critical period of closure of the Tethyan Seaway in the Miocene (Harzhauser et al. 2007). One cheilostome genus (*Therenia*) appears to have survived west of the closure but is now extinct in the Indo-Pacific, while a few extant tropical genera (e.g. *Hiantopora, Emballotheca, Thalamoporella*) continued to live in the Mediterranean until they became extinct during the Messinian crisis. The effect of the closure of Tethys is evident in a molecular study of the common cheilostome *Electra pilosa* that identified separation of northeast Atlantic/Mediterranean clades from an Indo-West Pacific clade at this time (Nikulina, Hanel, and Schäfer 2007).

Paratethyan bryozoan palaeobiogeography has attracted some attention (e.g. Vávra 2000). Sedimentary basins of the Paratethys host a mixture of mostly endemic species and species that also occur in the Mediterranean and/or eastern Atlantic. Moissette, Dulai, and Muller (2006) identified 238 bryozoan species in the Hungarian Middle Miocene (Badenian), which were apportioned between the following palaeogeographical regions: Eastern Atlantic/Mediterranean (31%), Mediterranean (23%), Paratethyan (21%), cosmopolitan (13%), Indo-Pacific (4%), and unknown (8%). Vávra (1987, 2012) considered that many of the bryozoans found in the Miocene of the Vienna Basin migrated there from the west via the Rhone Basin and Alpine foredeep. Central Paratethyan bryozoan biotas share varying numbers of species with coeval biotas from the Mediterranean and Atlantic (Figure 7.2).

Cold-water bryozoans found by Rosso and Di Geronimo (1998) in deep water Early Pleistocene sediments of the Mediterranean, but not present in this sea today, represent Boreal Atlantic species that spread into the Mediterranean during thermal lows.

Turning to the Southern Hemisphere, the high levels of endemicity seen today in New Zealand bryozoan faunas can be traced back to the Palaeogene: 88% of the 190 species so far described from the Palaeogene are endemic, compared with 61% of species known today (Gordon and Taylor 2015). A late Paleocene–early Eocene bryofauna from the Chatham Islands contains taxa reminiscent of both the Masstrichtian–Danian of the Northern Hemisphere and the Neogene–Recent of the Southern Hemisphere (Gordon and Taylor 1999). Strong palaeogeographical links between bryozoans in the Cenozoic of Australasia and Patagonia are evident from the presence of some distinctive shared taxa, such as cinctiporid cyclostomes (Casadío et al. 2010) and the cheilostome *Melychocella* (Pérez, López-Gappa, and Griffin 2015). Biogeographical affinities are also apparent between South America and Antarctica through *Aspidostoma* and other cheilostome genera (Hara 2007). These match the biogeographical affinities based on recent taxa described above.

Biogeographical studies have also been published on a few individual cheilostome genera. These include *Adeonella* (Hayward 1983) and *Steginoporella* (Pouyet and David 1979). With one exception, the study of Kuklinski et al. (2013) on *Pseudoflustra*, these lack the phylogenetic underpinning that would greatly enhance their value in understanding the historical biogeography of the genera concerned. The ascophoran *Pseudoflustra* has no certain fossil record and consists of nine species living in the North Atlantic and Arctic oceans. These authors found the genus to consist of a paraphyletic clade of North Atlantic species and a monophyletic clade containing all of the Arctic species. Because the Arctic Ocean was fully glaciated until 18 Ka, it was hypothesized that the Arctic clade colonized this region very recently. However, this conclusion depends on the correctness of the morphological phylogeny, a supposition that needs to be tested with molecular data.

Figure 7.2 Paratethyan Middle Miocene biogeographical links showing numbers of bryozoan species recorded in each region with numbers along dashed lines indicating the numbers of species shared with the central Paratethys (after Moissette, Dulai, and Muller 2006).

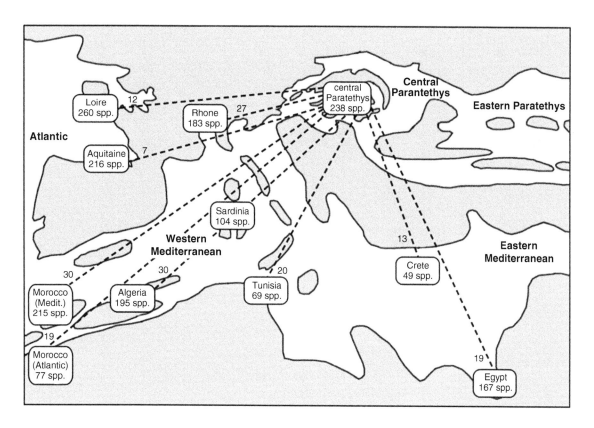

8 Phylogeny

Propelled by the molecular revolution, our understanding of bryozoan phylogeny has increased enormously during the last decade. Although the fossil record still furnishes the best information on the waxing and waning of bryozoans through geological time (Chapter 9), the phylogenetic framework necessary for a fuller understanding of evolutionary patterns is gradually emerging from molecular phylogenetic studies of living bryozoans. Furthermore, many of the skeletal morphological characters previously thought to be indicative of close genealogical affinities have been shown to be highly convergent, forcing a reconsideration of how we can reconstruct phylogeny from fossil evidence.

8.1 Relationships with Other Phyla

Problems placing bryozoans on the tree of life have led to their categorization as phylogenetic renegades (Jenner and Littlewood 2008; Giribet et al. 2009). This has not been helped by the use of contaminant sequences for bryozoans in some of the earlier analyses, as revealed by Waeschenbach, Taylor, and Littlewood (2012). Long branch attraction was a further issue, with some early molecular phylogenetic analyses recovering a sister-group relationship between bryozoans and another renegade phylum, Chaetognatha (e.g. Sun et al. 2009). Classical anatomy has generally favoured the grouping of bryozoans with brachiopods and phoronids in the 'superphylum' Lophophorata (Hyman 1959; Valentine 1973). While some molecular studies have cast doubt on the monophyly of Lophophorata (see references cited by Taylor and Waeschenbach 2015), others have supported this grouping (Nesnidal et al. 2013; Laumer et al. 2019).

Bryozoan Paleobiology, First Edition. Paul D. Taylor.
© 2020 Natural History Museum. Published 2020 by John Wiley & Sons Ltd.

Whether or not Lophophorata is a natural group, there is clear evidence that Bryozoa belongs in Lophotrochozoa together with molluscs and annelids as well as brachiopods, phoronids, and some minor phyla. Among the latter are the Entoprocta, a phylum of sessile suspension feeders resembling bryozoans in having a ciliated ring of tentacles for feeding but with the anus opening inside the ring. A sister-group relationship between Bryozoa and Entoprocta has long been suggested on morphological grounds, including larval anatomy (Nielsen 1987), and although supported by some molecular studies (e.g. Hausdorf et al. 2007), this relationship is not generally favoured.

After allowance for compositional biases in their phylogenomic data, Nesnidal et al. (2013, see also Dunn et al. 2014; Laumer et al. 2019) recovered the Lophophorata as a monophyletic clade, with Brachiopoda forming the sister-group to Bryozoa + Phoronida (Figure 8.1). This is the most likely topology based on current molecular evidence and has implications for the spate of palaeontological studies on the relationships between brachiopods and several extinct groups (e.g. hyoliths, tommotiids) believed by some to fall on the stem-group of the brachiopods + phoronids (e.g. Skovsted et al. 2009; Moysiuk, Smith, and Caron 2017). Recent work on patterns of lophophore

innervation (Temereva 2017) has found that the colonial phoronid *Phoronis ovalis* (Plate 5H) occupies an intermediate position between non-colonial phoronids and bryozoans, supporting a common ancestry between these two phyla.

The fossil record provides few major clues about the relationship between bryozoans and other phyla. Dzik (1991) proposed homology between the protoecium of stenolaemate bryozoans and the bulb-like initial part of the vermiform tubes of cornulitids, leading to the hypothesis that cornulitids are the solitary ancestors of colonial bryozoans. If correct, this would imply the presence of a biomineralized skeleton at the very origin of the phylum Bryozoa and its subsequent loss in the phylactolaemates which are generally deemed to be the most basal bryozoans (see below). More recent studies (Vinn and Mutvei 2009; Taylor, Vinn, and Wilson 2010; Vinn and Zaton 2012) have allied cornulitids with microconchids and tentaculitids as possible phoronid-like lophophorates that acquired calcareous tubes.

As all extant phoronids (and brachiopods) live in marine environments, it has been inferred that the earliest bryozoans were also marine animals, with freshwater environments colonized subsequently by phylactolaemates as well as by a few gymnolaemates (Koletic et al. 2014).

8.2 Inter-relationships of Bryozoan Classes

The exclusively freshwater Phylactolaemata has long been considered to be the most basal of the three bryozoan classes (e.g. Wood 1983). This is based on the more obviously trimeric body plan and the horseshoe-shaped lophophore resembling that of phoronids. An early morphological cladistic analysis recovered the relationship Phylactolaemata (Stenolaemata + Gymnolaemata) (Carle and Ruppert 1983), although another morphology-based cladistic analysis (Cuffey and Blake 1991) found a different topology: Stenolaemata (Phylactolaemata + Gymnolaemata). Apomorphies supporting the sister-group relationship between the predominantly marine Gymnolaemata and

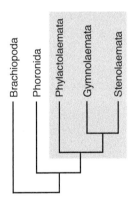

Figure 8.1 Phylogenetic relationships between the lophophorate phyla Brachiopoda and Phoronida and the three classes of the phylum Bryozoa (based on Taylor and Waeschenbach 2015, figs 1 and 2. © The Trustees of the Natural History Museum, London).

Stenolaemata include: (i) polypide cycling (see Chapter 1.1); (ii) funiculus lacking a basal lamina; (iii) parietal muscles; (iv) lack of an epistome; (v) anal budding direction (*fide* Jebram 1973); and (vi) complete body wall separating the zooids. Franzén's (1998) study of the comparative structure of sperm in bryozoans also supported the hypothesis that Stenolaemata and Gymnolaemata are more closely related to each other than either is to Phylactolaemata.

Fuchs, Obst, and Sundberg (2009) sequenced three genes (mitochondrial COI and nuclear 18S rDNA and 28S rDNA) in 32 bryozoan species to construct the first major molecular phylogeny of the phylum. They found the three classes (Phylactolaemata, Gymnolaemata, and Stenolaemata) to be monophyletic, some analyses recovering a sister-group relationship between Phylactolaemata and Stenolaemata, while others found Stenolaemata and Gymnolaemata to be sister-groups. However, more recent molecular analyses (Waeschenbach, Taylor, and Littlewood 2012; Nesnidal et al. 2013) have generally recovered the freshwater Phylactolaemata as the sister-group to a clade formed by Gymnolaemata + Stenolaemata (Figure 8.1).

8.3 Inter-relationships of Bryozoan Orders

The favoured relationship between the two gymnolaemate orders is that Cheilostomata represent a monophyletic clade nesting within the paraphyletic Ctenostomata. This topology has been suggested both on morphological (e.g. Todd 2000) and molecular (e.g. Waeschenbach, Taylor, and Littlewood 2012) grounds. It has long been suspected that cheilostomes evolved from a 'ctenostome-grade' ancestor which acquired a mineralized skeleton (see Banta 1975 and references therein). First known from the Triassic Muschelkalk of Germany (Todd and Hagdorn 1993; Figure 1.11D), bioimmured 'arachnidiid' ctenostomes resembling the earliest cheilostomes (e.g. *Pyriporopsis*; Figure 8.2C, D) occur abundantly in Middle and Late Jurassic clays of north-west Europe (Figure 1.11C, F), providing a suitable context for this

hypothesis of cheilostome origin (Taylor 1990a). In particular, bioimmured zooids of the ctenostome *Cardoarachnidium* possess an apparent operculum (Figure 8.2B), which along with the mineralized skeleton is one of the two key apomorphies of cheilostomes, allowing this genus to be placed in the stem-group of Cheilostomata (Todd 2000).

In an alternative interpretation, Dzik (1975) believed that cheilostomes evolved from a cyclostome resembling *Corynotrypa* (Figure 8.2E), entailing reduction in frontal calcification to convert the narrow cyclostome aperture to the broad large opesia of a primitive *Pyriporopsis*-like cheilostome. He also postulated the evolution of boring ctenostomes from cyclostomes in the Ordovician (Dzik 1981), a transition which would involve complete loss of the mineralized skeleton. These two interpretations become improbable when the differences between the soft tissue organizations of living cyclostomes and gymnolaemates (ctenostomes and cheilostomes) are considered. Nor are they favoured by molecular phylogenies. Equally unlikely is the notion, based on the shared box-like zooidal skeletons, that cheilostomes were descended from fenestrate stenolaemates, an idea that originated with Ulrich (1890) and which is still occasionally found in more modern literature (e.g. Xia 2002). Hillmer (1991, p. 123) believed the peculiar Late Ordovician genus *Schallreuterella*, which sometimes has apertures with straight proximal edges suggestive of a hinge line for an operculum, to be 'an early offshoot in cheilostome evolution', opening a 300-million-year gap in the fossil record of cheilostomes. It is more likely, however, that the similarity in aperture shape between *Schallreuterella* and some cheilostomes is due to evolutionary convergence.

There have also been suggestions that cheilostomes are polyphyletic (e.g. Jebram 1992). Requiring multiple origins of opercula and a mineralized skeleton, cheilostome polyphyly awaits testing following denser sampling of ctenostome and basal cheilostome taxa for molecular phylogenetic analysis. Until this has been done, it is most parsimonious to assume cheilostome monophyly.

Turning to the Stenolaemata, of the six orders (Table 1.2), only the Cyclostomata is generally considered to be extant (but see below). The phylogenetic

Figure 8.2 Basal encrusting bryozoans: (A) dried colony of the ctenostome *Arachnidium hippothoides*, the cuticle of the zooids appearing black against the grey of the encrusted shell in this backscattered electron image (Recent; UK); (B) mould bioimmuration of *Cardoarachnidium bantai*, a possible stem-group cheilostome, showing orifice closed by what appears to be an apparent operculum with a straight proximal edge (Jurassic, Oxfordian, Sandsfoot Clay; Weymouth, Dorset, UK); (C) crowded zooids of the oldest known cheilostome, *Pyriporopsis pohowskyi* (Jurassic, Oxfordian, or Kimmeridgian, Mabdi Fm.; Jebel Sakha, Yemen); (D) *Pyriporopsis portlandensis*, the first bona-fide Jurassic cheilostome to have been described (Jurassic, Tithonian, Portland Beds; ?Whitchurch, Buckinghamshire, UK); (E) the primitive cyclostome *Corynotrypa* (Ordovician, Sandbian, Bromide Fm.; Arbuckle Mountains, Oklahoma, USA). Scale bars: A = *c.* 200 μm; B = 50 μm; C = 200 μm; D, E = 500 μm.

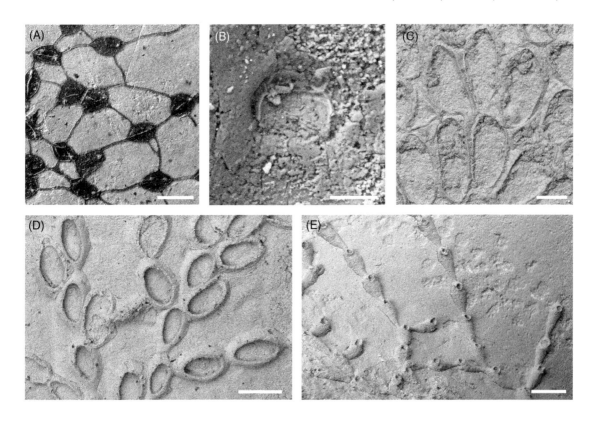

relationships between the other orders, all palaeostomates, have been inferred by McKinney (2000) and Ma, Buttler, and Taylor (2014) using cladistic methods (Figure 8.3). Cystoporata and Esthonioporata are interpreted to be sister-groups, as are Trepostomata and Cryptostomata + Fenestrata. A close relationship between fenestrates and ptilodictyine cryptostomes was earlier suggested by Tavener-Smith (1975) who argued that the longitudinal striations found on the reverse surfaces of fenestrate branches were vestiges of the interzooecial walls separating the autozooids of a bifoliate, ptilodictyine ancestor. The genus *Pseudohornera* appears to be transitional between ptilodictyines and fenestrates in having short zooids on the reverse side of the branches. The phylogenetic position of Timanodictyina is unclear but it is most likely to be close to the Cryptostomata + Fenestrata clade. However, it must be emphasized that inferring phylogenetic relationships between

Figure 8.3 (A) Phylogeny of stenolaemate orders (Timanodictyina excluded) (after Taylor and Waeschenbach 2015, fig. 5. © The Trustees of the Natural History Museum, London). Major apomorphies are: (1) loss of frontal exterior-walled skeleton; (2) maculae; (3) lunaria; (4) widely spaced diaphragms; (5) ramose colony-form; (6) polygonal cross section of zooids in endozone; (7) confined budding loci; (8) mesotheca; and (9) apertures on one side of branches only. (B) Tentative tree of stenolaemate orders (after Taylor and Waeschenbach 2015, fig. 6. © The Trustees of the Natural History Museum, London).

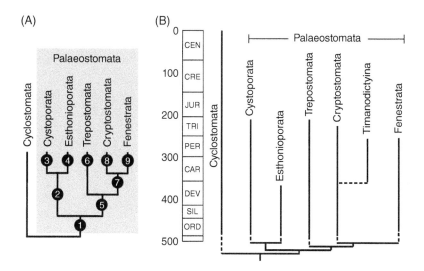

extinct stenolaemate orders is seriously hampered by a paucity of morphological characters coupled with high levels of convergence (see Chapter 9.11), and even the monophyly of the orders themselves is not always well founded.

A long-standing issue in bryozoan phylogeny has concerned the relationship between Palaeozoic stenolaemates – or more accurately the superorder Palaeostomata, which ranges into the Triassic – and the post-Palaeozoic stenolaemates traditionally classified in the order Cyclostomata. The two extreme views are that: (i) all palaeostomates became extinct in the Permian and Triassic, with all post-Palaeozoic cyclostomes descended from the small group of cyclostomes (suborder Paleotubuliporina) found in the Palaeozoic (e.g. Brood 1976c), specifically Crownoporidae (= Kukersellidae) which have porous interior walls and pseudoporous frontal walls (Figure 8.4); and (ii) post-Palaeozoic cyclostomes are polyphyletic, some (suborders Tubuliporina and Articulata) descended from

paleotubuliporine cyclostomes but others from palaeostomate orders (Borg 1965; Boardman 1984; Viskova 1992). The second view interprets trepostomes as ancestors of cerioporine cyclostomes (exemplified by *Heteropora*), cystoporates of rectangulate cyclostomes (exemplified by *Disporella*), and cryptostomes of cancellate cyclostomes (exemplified by *Hornera*).

Ernst and Schäfer (2006) proposed a third option: that post-Palaeozoic cyclostomes evolved independently from an *Arachnidium*-like ctenostome bryozoan (cf. Figure 8.2A) which acquired a biomineralized skeleton in the early Mesozoic. The main argument in favour of their hypothesis is the enormous gap in the fossil record between the paleotubuliporine family Crownoporidae with pores and pseudopores (see Taylor and Wilson 1996), and the appearance in the Late Triassic of the first post-Palaeozoic cyclostomes which also have these structures. Evaluating this proposal is currently difficult, although the finding of

Figure 8.4 Palaeozoic crownoporid cyclostomes that may belong in the stem-group of the post-Palaeozoic cyclostomes showing frontal walls containing prominent pseudopores: (A) *Cuffeyella arachnoidea*, an encrusting species with oligoserial branches (Ordovician, Katian, Lorraine Group; Cincinnati, Ohio, USA); (B) *Kukersella borealis*, an erect species with narrow cylindrical branches (Ordovician, DIII; Rakviere, Estonia). Scale bars = 200 μm.

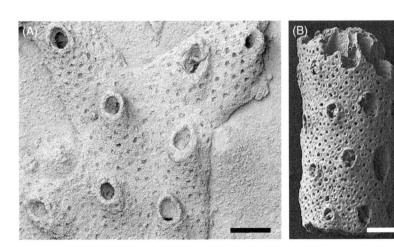

crownoporids later in the Palaeozoic would weaken the hypothesis, while molecular clock estimation of the timing of divergence between ctenostomes and cyclostomes (Early Palaeozoic vs. Mesozoic) could provide a more definitive test. Nevertheless, the strong resemblance of the skeletal microstructure between Palaeozoic and post-Palaeozoic cyclostomes – compared to differences with the other biomineralized order, the Cheilostomata – does not favour a separate post-Palaeozoic origin of cyclostomes from a soft-bodied ancestor.

8.4 Morphological Phylogenies

Several pre-cladistic phylogenies focusing on particular groups of bryozoans can be found in the literature. Ross (1964, text-fig. 5), for example, devised an evolutionary tree of 12 Ordovician ptilodictyine genera based on a combination of stratigraphical occurrence and morphological characters, interpreting hemisepta and cribrate colony-form as advanced traits for this group of genera. In a subsequent paper (Ross 1967), she used a similar

combination of morphology and stratigraphy to construct a tree of six Ordovician–Carboniferous trepostome genera belonging to the family Aisenvergiidae. Wass (1977, fig. 24) presented a tree, based on morphology and stratigraphical occurrences, for 18 Eocene–Recent genera of catenicellid cheilostomes, recognizing six main lineages.

The application of cladistic methods has led to the proposal of more explicitly testable hypotheses of relationships. An early attempt was made by Cuffey and Blake (1991) to use cladistics to investigate the relationships between 13 major groups of living and fossil bryozoans rooted on the Entoprocta as the out-group. This analysis recovered ctenostomes as polyphyletic, the stoloniferan and carnosan taxa separated by phylactolaemates. Stenolaemates were located crownward of the phylactolaemates and gymnolaemates, with cyclostomes positioned basally of the extinct orders now placed in Palaeostomata.

McKinney (2000) tested various hypotheses concerning the phylogenetic origin of phylloporonids, a group most often regarded as primitive fenestrates but by some considered to be trepostomes.

He assembled a matrix of 35 characters for seven phylloporinid species and 22 out-group taxa, and used the resulting cladogram to support the hypothesis that phylloporinids are basal fenestrates that arose from bifoliate cryptostomes (as was suggested by Tavener-Smith 1975) and subsequently gave rise to advanced fenestrates including fenestellids (see above).

Taylor and Weedon (2000) used a combination of skeletal ultrastructural, zooidal, and colonial morphological characters to infer the inter-relationships between 28 genera of post-Palaeozoic cyclostomes. Contrary to prior expectations, colonial characters performed slightly better than zooidal and ultrastructural characters in the sense of having the highest consistency indices in the single most parsimonious tree they recovered. The topology of the tree raised doubts over the phylogenetic status of several of the traditional suborders of cyclostomes: only two of the five suborders were found to be monophyletic.

Jiménez-Sánchez, Anstey, and Azanza (2010) undertook a cladistic analysis of ptilodictyine cryptostomes. Several cladistic analyses based on morphology of relationships between cheilostome species/genera within bryozoan genera/families have been published. These include Eurystomellidae (Gordon, Mawatari, and Kajihara 2002), Microporellidae (Taylor and Mawatari 2005), Doryporellidae (Grischenko, Mawatari, and Taylor 2000; Grischenko, Taylor, and Mawatari 2004), *Wilbertopora* (Cheetham et al. 2006; Figure 3.26), *Monoporella* (Dick 2008), *Macropora* (Gordon and Taylor 2008; Figure 8.5), *Pentapora* (Lombardi, Taylor, and Cocito 2010), *Pseudoflustra* (Kuklinski et al. 2013), and *Scrupocellaria* (Vieira et al. 2014). Support values for the trees have tended to be low, questioning the reliability of the topologies found.

8.5 Molecular Phylogenies

Molecular phylogenetics made a relatively late entry into bryozoology, in part because of the relatively few active bryozoan specialists but also a consequence of practical difficulties in obtaining sufficient amounts of bryozoan tissue not contaminated by associated organisms. Dick et al. (2000) made the first attempt to reconstruct bryozoan phylogeny using DNA sequence data, analyzing 16S rRNA sequences in 23 species. Hao et al. (2005) used the same gene to construct a phylogeny of cheilostomes and applied a molecular clock method to estimate clade divergence times. These authors calculated divergence of cheilostomes from the outgroup ctenostome at 245–282 Ma (Permo-Triassic) and ascophorans from anascans at 173–190 Ma (Middle Jurassic). Both dates are substantially older than indicated by fossil evidence and are difficult to accept. A molecular phylogeny of 27 bryozoan species using 18S rRNA, applying secondary structure to improve alignments, found cyclostomes to be monophyletic but both ctenostomes and cheilostomes polyphyletic, with the cheilostome *Scruparia* unexpectedly forming the sister taxon to the cyclostomes (Tsyganov-Bodounov et al. 2009).

The phylogenetic relationships among 23 species of phylactolaemates were inferred by Hirose, Dick, and Mawatari (2008) using 16S and 12S rDNA sequences. Contrary to expectations from previous morphology-based studies (e.g. Mukai 1999), but supporting two earlier molecular studies (Wood and Lore 2005; Okuyama, Wada, and Ishii 2006), phylactolaemates with compact gelatinous colonies were recovered as more basal than the branching colonies of plumatellids.

A good example of a molecular phylogenetic study focussing on a single genus is that of Fehlauer-Ale et al. (2015) on the common cheilostome genus *Bugula*. This used the mitochondrial gene cytochrome c oxidase subunit 1 (COI) and the large ribosomal RNA subunit (16S), combined with a morphological analysis. The molecular tree recovered comprised four major clades, and was mostly congruent with morphological characters.

In contrast, substantial discordance is evident between molecular and morphological phylogenies in cyclostome bryozoans (Figure 8.6). First highlighted by Waeschenbach et al. (2009), subsequent studies have found additional conflicts (Taylor, Waeschenbach, and Florence 2011; Taylor

Figure 8.5 Phylogenetic tree of the cheilostome *Macropora* based on morphological characters (after Gordon and Taylor 2008, fig. 17. © The Trustees of the Natural History Museum, London). Symbols indicate biogeographical provenances as indicated in the box.

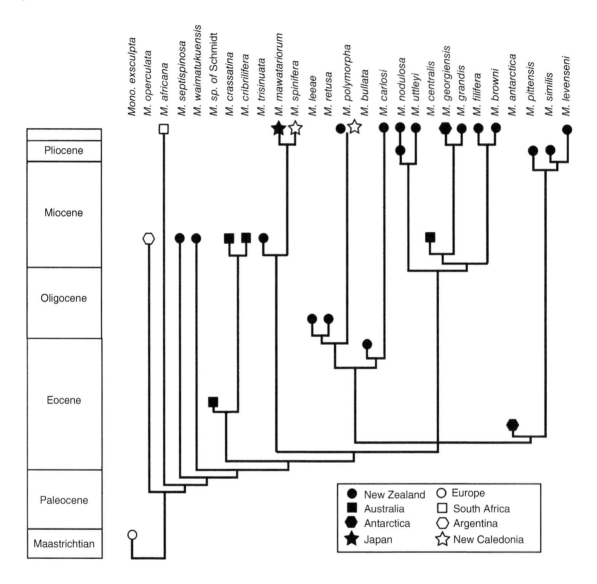

et al. 2015) that call into question the value of skeletal morphological characters in cyclostome phylogenetics and, by extension, classification. The conclusion to be drawn is that several aspects of the cyclostome skeleton are evolutionarily plastic. These include the presence of frontal exterior walls defining a fixed-walled organization (cf. free-walled organization in which these walls are lacking;

Figure 8.6 Molecular phylogeny of cyclostome bryozoans with traditional suborders mapped on the cladogram (after Taylor and Waeschenbach 2015, fig. 9. © The Trustees of the Natural History Museum, London). Note that while species belonging to the same genus cluster together, those belonging to the same suborder are not always grouped, e.g. the two species assigned to Cerioporina (*Favosipora rosea* and *Heteropora neozelanica*) are widely separated. The subordinal placement of *Cinctipora elegans* is contentious.

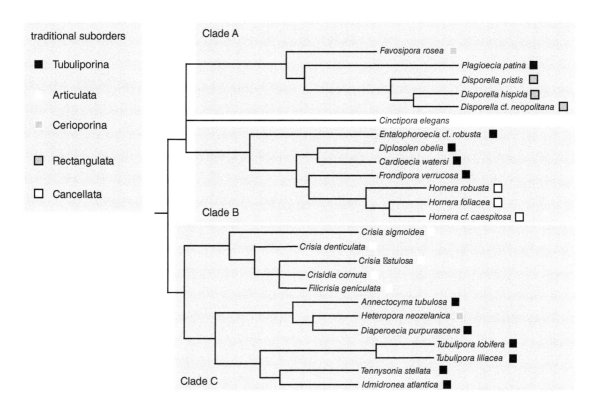

Figure 8.7), and the shapes of the gonozooids. With respect to the former, numerous examples were previously known of both fixed- and free-walled morphologies among cyclostomes that in other respects seemed to be closely related, and combinations of these two organizations have even been found within single colonies (reviewed by Taylor 2000). For example, in *Spiritopora* from northern New Zealand, zooids forming the extensive colony base have calcified exterior frontal walls (i.e. fixed-walled), whereas those forming the erect parts of the colony normally lack these walls and are free-walled (Taylor and Gordon 2003).

Gonozooid shape was particularly emphasized as a taxonomic character by Ferdinand Canu and Raymond S. Bassler in their numerous papers published during the first half of the twentieth century (e.g. Canu and Bassler 1926) and formed the basis for their family-level classification of cyclostomes that largely persists today. The striking contrast in the gonozooids of two New Zealand species that are very closely related based on molecular sequence data epitomizes the variability in this character: *Diaperoecia purpurascens* has a longitudinally elongated gonozooid that is not perforated by autozooidal peristomes

Figure 8.7 Fixed and free-walled organizations in stenolaemates as exemplified by the Recent cyclostomes *Diaperoecia purpurascens* and *Heteropora neozelanica*, respectively (after Taylor et al. 2015. © The Trustees of the Natural History Museum, London). In both cases, the ancestrula and first-budded generation of zooids forming the encrusting part of the colony are fixed-walled, a skeletal organization retained by the fixed-walled bryozoan during erect growth away from the substrate but not the free-walled bryozoan in which the erect zooids are free-walled, with autozooids separated by kenozooids (k). Interior walls are shown as thin solid lines, frontal exterior walls as dashed thick lines, and unmineralized, cuticular exterior walls as dotted lines. Note that many free-walled stenolaemates develop in different ways (e.g. Boardman and McKinney 1976, text-fig. 2).

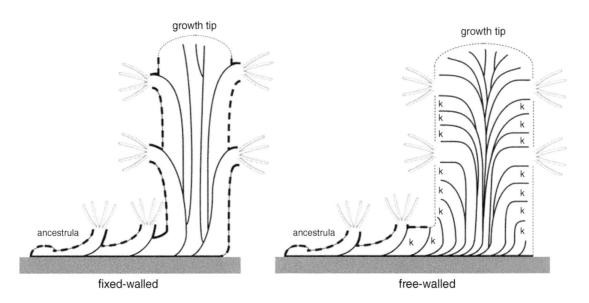

and has a terminal ooeciopore, whereas in its close relative *Heteropora neozelanica* the gonozooid is subcircular, perforated by autozooidal peristomes and has a more or less centrally placed ooeciopore (Taylor et al. 2015). In fact, just about the only skeletal character pointing to a close affinity between these two species is the crescentic pattern of pseudopores on the protoecium. Prompted by the discovery of this unexpected phylogenetically informative skeletal character, Jenkins and Taylor (2017) surveyed pseudopore patterns across a wide range of cyclostomes and mapped the data onto a molecular tree. However, their study found congruence between this promising character and the molecular tree to be incomplete.

The greatest progress has been made on the molecular phylogeny of cheilostomes, which are the dominant bryozoans in modern oceans. As much of this work is on-going (e.g. Orr et al. 2018), it would be premature to draw too many conclusions at this early stage but a preliminary tree (Figure 8.8) shows elements of cheilostome tree topology found by these continuing analyses.

Figure 8.8 Molecular phylogeny of selected cheilostome bryozoans based on the published trees of Waeschenbach, Taylor, and Littlewood (2012) and Orr et al. (2018). Taxa designated with asterisks are non-brooders with cyphonautes larvae; note that brooding has evolved three times among the taxa shown in this tree. Grey boxes indicate ascophoran-grade taxa with calcified frontal shields overlying an ascus, showing that this grade has evolved twice in this tree; the larger clade comprises taxa with umbunuloid or lepralioid frontal shields, the smaller clade contains taxa with gymnocystal frontal shields. Taxa in bold (species of *Microporella* and *Fenestrulina*) have an ascopore and were previously united in the family Microporellidae, which is clearly diphyletic in this tree. Branch lengths are arbitrary.

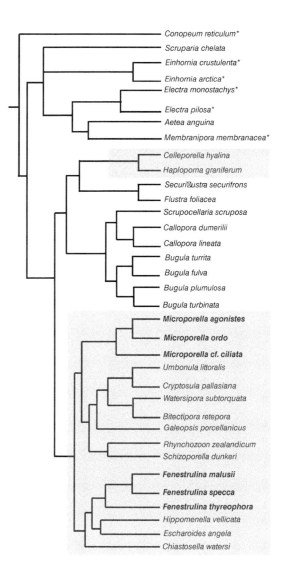

9

Evolution and Fossil History

The rich fossil record of bryozoans allows us to make several generalizations about the pattern of bryozoan evolution through geological time even in the absence of a robust phylogeny. Estimates of diversity at different taxonomic levels for geological periods and stages give a broad impression of the waxing and waning of bryozoans in time. Phases of major radiation and, conversely, of mass extinction are clearly evident. It is also possible to evaluate the diversity dynamics of co-existing clades, notably cyclostomes and cheilostomes that have co-existed for the last 150 million years. Convergence is a major theme of bryozoan evolution, both at colony and zooid levels, and the fossil record contains some striking examples of homeomorphy.

9.1 Phanerozoic Bryozoan Diversity

The work of Jack Sepkoski during the 1980s fuelled interest among palaeobiologists in the changing diversities of taxa through geological time, the extent to which such temporal diversity patterns are 'artefactual' (e.g. functions of taxonomic effort, or the availability of fossiliferous rocks) or represent true variations in global diversity, and the identity of the principal drivers of diversity change (e.g. biotic vs. abiotic factors).

Apart from the absence of bryozoans in the 'Cambrian Explosion' (see below), bryozoan family diversity through the Phanerozoic exhibits a pattern paralleling that of marine invertebrate families as a whole (Taylor and Larwood 1988, 1990), as to some extent does genus-level diversity (Figure 9.1). A steep rise in diversity during the Ordovician is followed by a plateau for most of the rest of the Palaeozoic except for a dip in diversity after the end of the Ordovician. Like most other marine phyla, bryozoan diversity plummeted at the end of the

Figure 9.1 Generic diversity of fossil bryozoans through time.

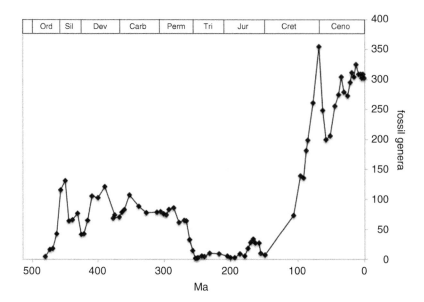

Palaeozoic, recovering very slowly during the early Mesozoic. It took until the mid-Cretaceous (*c.* 100 Ma) for diversity to regain the typical level pertaining during the Palaeozoic. Diversity rose steeply in the Late Cretaceous. A moderate dip in diversity in the early Palaeogene was followed by a rise towards the present day. Standing generic diversity of fossilizable bryozoans in modern seas is estimated to be at least three times greater than it was in the Palaeozoic. The extent to which sampling artefacts (e.g. outcrop area) have sculpted this pattern is as yet unclear. However, as species diversity in diverse bryozoan assemblages shows a generally similar trajectory through time (Figure 9.2), it seems likely that the global pattern at higher taxonomic levels has some biological reality.

Horowitz and Pachut (2000) used a database of bryozoan species names compiled from the literature to plot stage-by-stage diversity from the Ordovician to the present day (Figure 9.3). The pattern shows large fluctuations between stages as damping down is diminished when using species-level data. The seemingly equivalent levels of diversity seen in the Palaeozoic as in the Cenozoic are artefactual and reflect the longer durations of Palaeozoic stratigraphical stages. On average, Cenozoic bryozoan assemblages are twice as rich in species than are Palaeozoic assemblages (Figure 9.2), and there is as yet no evidence for greater provinciality leading to higher global species diversity of bryozoans in the Cenozoic.

9.2 Cambrian Bryozoans?

Bryozoans stand out among animal phyla that are routinely fossilized in having no unequivocal representatives of Cambrian age. The lack of an established Cambrian fossil record for the phylum conflicts with molecular clock estimates implying the origination of bryozoans well before their first fossil appearance in the Early Tremadocian (Ma et al. 2015), at about 485 Ma. For instance, in Erwin's (2015, fig. 1) time-calibrated phylogeny of

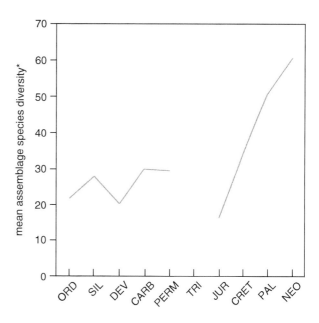

Figure 9.2 Mean species diversity in bryozoan assemblages through geological time. Note that only assemblages containing 10 or more species are included, which is why no data is available for the Triassic where no known assemblage contains this many species. The temporal pattern including a Palaeozoic diversity plateau, decline across the Palaeozoic–Mesozoic boundary, followed by a rise in diversity to levels roughly twice those of the Palaeozoic, parallels global diversity changes at genus and family levels (based on Taylor and James 2013, fig. 1. © The Trustees of the Natural History Museum, London).

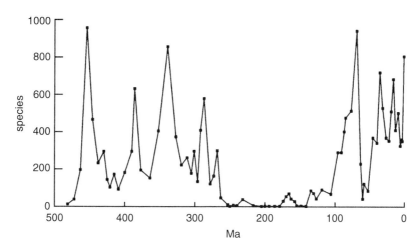

Figure 9.3 Phanerozoic species-level diversity of bryozoans based on the data published by Horowitz and Pachut (2000, table 2). The high yet variable diversity in the Palaeozoic is followed by low diversity in the Mesozoic until the Late Cretaceous when levels returned to those typical of the Palaeozoic.

the Metazoa, the bryozoan lineage extends back into the early part of the Ediacaran (*c.* 630 Ma), almost 150 million years before their advent in the fossil record.

Numerous Cambrian fossils have been claimed to be bryozoans. Long thought to be the oldest fossil bryozoan, Bassler (1911) described the supposed ctenostome *Heteronema priscum*, from the

Tremadocian of Estonia, which was then regarded as Late Cambrian in age. Later made the type species of *Marcusodictyon* (Bassler 1952), this fossil comprises a complex polygonal network on the outer surface of brachiopod shells. A restudy of *Marcusodictyon* (Taylor 1984b) revealed no diagnostic bryozoan features and the fossil was considered to be a phosphatic problematicum. Supposed bryozoans from the Lower Cambrian of Shropshire, England (Cobbold 1931; Cobbold and Pocock 1934) appear to be fragments of arthropod exoskeletons. Fritz (1947) erected *Archaeotrypa* containing two new species for two specimens of putative bryozoans from the Late Cambrian of Alberta, Canada. One (*A. secunda*) is represented by a tangential section, and the other (*A. prima*) by a longitudinal section. The thick crystalline walls of *A. secunda* contrast with the thin zig-zag walls of *A. prima*. The bryozoan identity of both of these species is questionable. Similar doubts have been cast on the affinities of several other putative Cambrian bryozoans (Elias 1954; Litherland 1975; Hampton 1979).

The most recent claim for a bryozoan of Cambrian age concerns a fossil named *Pywackia baileyi* from the Late Cambrian of Oaxaca in southern Mexico (Landing et al. 2010). Represented by a large suite of specimens, *Pywackia* comprises small, proximally tapering, spindle-shaped rods covered by shallow depressions of an irregularly polygonal shape. All examples are phosphatic, probably due to phosphatization of an originally calcareous skeleton. The depressions seem too shallow to represent the zooidal chambers of a bryozoan, and the original claims of zooidal polymorphism with distinct axial zooids could not be corroborated during a subsequent study of *Pywackia* (Taylor, Berning, and Wilson 2013). Furthermore, *Pywackia* shows a striking resemblance to the living pennatulacean octocoral *Lituaria*. Hageman (2018) provided further evidence from microstructure and taphonomy that *Pywackia* is not a bryozoan: a bryozoan model would require all specimens to have been abraded down to the same level of the endozone, which is very unlikely. Although it is far older than the first pennatulaceans, *Pywackia* is best interpreted as an octocoral with a pennatulacean-like functional morphology. It is not 'Earth's oldest bryozoan' (cf. Landing et al. 2015).

9.3 Great Ordovician Biodiversification Event

While bryozoans were at best cryptic players in the Cambrian Explosion, they were among the most prominent phyla involved in the 'Great Ordovician Biodiversification Event' (GOBE). The radiation of taxa constituting Sepkoski's 'Paleozoic Fauna' (Sepkoski 1984) began in the Ordovician and included stenolaemate bryozoans, along with articulate brachiopods, stromatoporoids, rugose and tabulate corals, pelmatozoan echinoderms, cephalopod molluscs, and ostracods (Harper 2006). Benthic biotas became dominated for the first time by suspension feeders, while both tiering and bioturbation increased. Servais et al. (2009) pointed to several factors that may have driven the GOBE, including the greatest continental dispersal and largest tropical shelf area of the Phanerozoic, warm climates, high sea levels, and intense volcanic activity providing a high nutrient input to the oceans.

The oldest undisputed bryozoan is the ptilodictyine cryptostome *Prophyllodictya simplex* (Plate 12G), which was found in the Early Tremadocian Nantzinkuan Formation of Hubei Province, South China (Ma et al. 2015). This region was part of the extensive Yangtze carbonate platform that was established during the Cambrian. The limestones are often dolomitized and very hard, factors not generally conducive to the preservation, detection, and extraction of bryozoans. Discovery of *P. simplex* was facilitated by silicification, causing branches to stand out on weathered surfaces. It seems likely that older bryozoans are present but as yet remain undetected in underlying carbonates on the Yangtze Platform. Overlying the Nantzinkuan Formation is the Late Tremadocian Fenhsiang Formation in which at least six bryozoan species are known (Xia, Zhang, and Wang 2007). Bedding planes in shale partings in the Fenhsiang Fm. reveal high abundances of the ramose trepostome *Orbiramus*.

Ernst (2017) compiled data on the ranges of 200 bryozoan genera through the Ordovician. Most of the recognized orders of stenolaemate bryozoans first appeared during stages of the Ordovician: Cyclostomata (Dapingian), Esthonioporata (Tremadocian), Trepostomata (Tremadocian),

Cystoporata (Tremadocian), Cryptostomata (Tremadocian), and Fenestrata (Dapingian). Only Timanodictyina appeared later. In addition, the oldest gymnolaemates – endolithic ctenostomes – date from the Ordovician (Dapingian). Bryozoan diversity through the Early and Middle Ordovician increased exponentially (Taylor and Ernst 2004), peaking in the early Katian at about 140 genera before declining to about 90 genera in the Hirnantian (Ernst 2017).

The evolution of a biomineralized skeleton in stenolaemates was a key event, not only because of its importance in generating the rich fossil record of bryozoans from the Ordovician onwards, but also in significantly expanding the adaptive repertoire of the phylum. A functional model explaining how biomineralization in stenolaemates might have originated was offered by Larwood and Taylor (1979). Beginning from an encrusting soft-bodied ctenostome-grade bryozoan resembling the extant *Arachnidium* (Figure 8.2A), the model hypothesizes separation from the outer body of a peritoneal layer comprising parietal muscles and associated membranes involved in lophophore eversion. The peritoneum is proposed to have formed the membranous sac, complete with annular musculature, that is an apomorphy of all living cyclostomes and is believed to have been present also in the primitive genus *Corynotrypa* (Figure 8.2E). This genus shows striking similarities in zooid shape, colony-form, and budding pattern (see Taylor and Wilson 1994) to *Arachnidium* and related extant ctenostomes (Figure 8.2A). Because the outer body wall no longer has to be deformed during lophophore eversion, it could be made rigid through the development of a calcified layer. Indeed, rigid calcification of the outer body wall is advantageous to lophophore eversion mechanism employed by cyclostomes (Chapter 3.1). More complete contiguity of zooids becomes possible if the walls between them are rigid and not deformed during lophophore eversion. Such contiguity is essential for the myriad of colony-forms that evolved in early stenolaemates. As pointed out by Schwaha, Wood, and Wanninger (2011), however, additional research using state-of-the-art techniques such as laser confocal microscopy are required to test models for the evolution of the membranous sac.

Palaeozoic cyclostomes, including *Corynotrypa* and the recently described *Zigzagopora* (Wilson and Taylor 2016; Plate 6B), all have 'weedy' encrusting or delicate branching colonies. They failed to diversify significantly in the Ordovician and, indeed, at any subsequent time in the Palaeozoic. Bryozoan radiation in the Ordovician was due almost entirely to diversification of the palaeostomate orders (Taylor and Ernst 2004). The free-walled organization of these orders allowed continued skeletal growth of the zooids and the development of a range of more robust colony-forms than is found among the exclusively fixed-walled cyclostomes of the Palaeozoic (Taylor and Larwood 1990).

9.4 End-Ordovician Extinction

The end-Ordovician extinction event is the first of the big-5 mass extinctions of the Phanerozoic (Raup and Sepkoski 1982). It is usually attributed to global cooling with glaciation, marine regression, and concomitant reduction in shallow shelf area (Sheehan 2001). Ernst (2017) distinguished three bryozoan extinctions in the Ordovician, of which the end-Ordovician event (Hirnantian) with a 16.5% loss in generic diversity was surpassed in severity by extinctions in the early Katian (22.7% loss) and late Katian (22.4% loss).

A first pulse of extinction removed a large number of endemic North American bryozoans (Tuckey and Anstey 1992). This was followed by a second pulse at the end of the Hirnantian when the low diversity bryozoan fauna of the Edgewood Province suffered only a small extinction. In contrast, bryozoans in Baltica showed a different pattern, with the second extinction pulse removing far more taxa than the first pulse.

9.5 Devonian Extinctions

Ernst's (2013) study of Devonian bryozoan dynamics concluded that the Givetian Taghanic Event had a more profound impact than either of the two younger extinctions, the Frasnian/Famennian

Event and the Hangenberg Event at the Devonian/Carboniferous boundary. The Frasnian/Famennian Event is marked by faunal turnover but no appreciable drop in bryozoan diversity or abundance. Data from Southern Siberia show good survival of bryozoans across both the Frasnian–Famennian and Devonian–Carboniferous boundaries but a mid-Famennian regional diversity decline linked to a rapid marine regression (Gutak, Tolokonnikova, and Ruban 2008).

Zaton and Borszcz (2013) noted the impoverished nature of early Famennian sclerobiont communities which are dominated by microconchids, an opportunistic group often present immediately after mass extinctions. Suspension-feeding encrusting animals in general, and bryozoans in particular, appear to have been far less common in the Carboniferous and Permian than they were in the Ordovician–Devonian. For example, the diverse communities of bryozoans, hederelloids, cornulitids, and microconchids often found encrusting Devonian brachiopod shells are rare in the Carboniferous and Permian (cf. Lescinsky 1997). The reason for the long-term scarcity of these sclerobionts is unknown but its coincidence with the claimed 'phytoplankton blackout' that lasted from the Devonian/Carboniferous boundary until the Late Triassic (Riegel 2008) deserves to be investigated.

9.6 Permian Mass Extinctions

Bryozoans were profoundly affected by the Late Permian extinctions (Powers and Bottjer 2009), making the end-Permian the most pivotal time in the entire evolutionary history of the phylum. Diversity at all taxonomic levels and geographical scales was severely reduced, and the ecological roles played by bryozoans in benthic communities were both diminished and altered. Unfortunately, the detailed pattern of bryozoan extinction at this time has been poorly studied. Song et al. (2012) mentioned the presence of bryozoans in their study that showed the existence of two extinction events during the Permian–Triassic biotic crisis, but bryozoans were not among the taxonomic groups for which species ranges were available for the seven Chinese stratigraphical sections they analyzed.

A study of bryozoan generic diversity in relation to location along an onshore–offshore gradient from the Early Permian to the Early Jurassic (Powers and Bottjer 2007) found that bryozoan diversity declined significantly well before both the end-Permian mass extinction, and also the end-Triassic mass extinction. In addition, offshore settings were the first to be affected, suggesting that both of these extinctions were related to the gradual encroachment of a factor, such as euxinia, that originated in deep water and spread across shallower waters.

Clapham and Payne (2011) identified the end-Permian extinction as a physiological crisis that selected against taxa, including bryozoans, with calcareous skeletons and weakly buffered respiratory physiologies. These taxa were less able to control intracellular pH and maintain respiratory efficiency under the high CO_2 levels that may have pertained at the end of the Permian, increasing their susceptibility to extinction from hypercapnia or ocean acidification (Knoll et al. 2007; Kiessling and Simpson 2011). Following Knoll (2003), bryozoans were classified as having poorly buffered (passive) physiologies in the latter two studies, while Clapham and Payne (2011) categorized them as 'moderately buffered' in a figure that showed no clear relationship between buffering and extinction rate.

Along with hypercapnia, ocean acidification (OA) has been hypothesized as among the most important kill mechanisms for marine life during the end-Permian mass extinction. Clarkson et al. (2015) used data from boron isotopes to suggest that OA drove the second phase of this mass extinction and was due to the rapid injection of large amounts of carbon which overwhelmed the buffering capacity of the oceans. At this time the shells of calcareous brachiopods appear to have increased their organic relative to carbonate components, which is consistent with a response to OA (Garbelli, Angiolini, and Shen 2017).

Interestingly, if Todd's (2000, fig. 6) time-calibrated morphological phylogeny of ctenostomes is correct, then a large number of lineages of these non-calcified bryozoans must have survived the end-Permian mass extinction, quite possibly proportionally more than among the calcified stenolaemates. There is evidence from experimental studies of the cheilostome *Myriapora* cultured in low pH conditions designed to simulate contemporary OA that the cuticle becomes thickened as a protective measure (Lombardi et al. 2011). Similar experiments have yet to be undertaken using living stenolaemates (cyclostomes). However, as these bryozoans seem to have uniformly thin cuticles (e.g. Nielsen and Pedersen 1979; see also Chapter 9.14) compared to gymnolaemates (i.e. ctenostomes + cheilostomes), it is conceivable that Permian stenolaemates were more vulnerable to OA being incapable of responding in the manner of gymnolaemates such as *Myriapora*.

9.7 Triassic Diversity and Mass Extinction

Few bryozoans have been described from the Triassic. Indeed, Powers and Pachut (2008) reported a mere 73 species. It can be argued that this figure is closer to the true preserved diversity than for other geological periods as the notorious scarcity of Triassic bryozoans has prompted any new finds to be recorded. A significant number of Triassic fossils originally identified as bryozoans have since been reinterpreted as belonging to other phyla, including sponges (Engeser and Taylor 1989; Schäfer and Grant-Mackie 1998), cnidarians (Schäfer and Grant-Mackie 1998), and serpulid polychaetes (Taylor 2014).

The great majority of bryozoans known from the Triassic belong to palaeostomate orders and have been categorized as 'Palaeozoic holdovers' (e.g. Schäfer, Senowbari-Daryan, and Hamedani 2003). Most are trepostomes (e.g. Plate 4N), an order which underwent a minor radiation in the Late Triassic, peaking at more than 20 species

in the Carnian (Powers and Pachut 2008). Trepostomes seem not to have survived into the Jurassic – the youngest known examples are recorded in the Rhaetian (e.g. Ernst, Schäfer, and Grant-Mackie 2015) – and it is conceivable that the order were victims of the Late Triassic mass extinction. As with the Permian mass extinction, there is some evidence that OA had a significant role in the Late Triassic mass extinction (Hautmann, Benton, and Tomasovych 2008; Greene et al. 2012), in which case the heavily calcified trepostomes of the Triassic may have been particularly hard hit. Cyclostome bryozoans, which were to become more important in the Jurassic, have no known fossil record in the Early and Middle Triassic and are represented in the Late Triassic by only three or four species.

9.8 Jurassic Cyclostome Radiation

Cyclostomes were the first bryozoans to radiate significantly in the Mesozoic, contributing the bulk of species to all bryozoan faunas of Jurassic and Early Cretaceous age (Taylor and Ernst 2008). Although only a handful of cyclostomes have been recorded in the Early Jurassic, they diversified appreciably in the Middle Jurassic, peaking at approximately 30 genera and 80 species in the Bathonian (Figure 9.4). Relatively few cyclostomes are known from the Upper Jurassic, the time of appearance of the first cheilostomes (Pohowsky 1973; Taylor 1994a). The occurrence of some Lazarus genera in the Middle Jurassic that reappeared in the Early Cretaceous implies a failure in the fossil record. However, it is possible that Late Jurassic cyclostomes were genuinely sparse compared to Middle Jurassic or Early Cretaceous cyclostomes: the maximum recorded diversity for a Late Jurassic cyclostome assemblage is just nine species whereas several Middle Jurassic faunas are known that contain more than 30 cyclostome species (Taylor and Ernst 2008, fig. 4).

Figure 9.4 Diversity of bryozoan genera and species through the stages of the Jurassic (after Taylor and Ernst 2008, fig. 3. © The Trustees of the Natural History Museum, London). Species curve incorporates extrapolated ranges. Genus data exclude form-genera such as '*Berenicea*'.

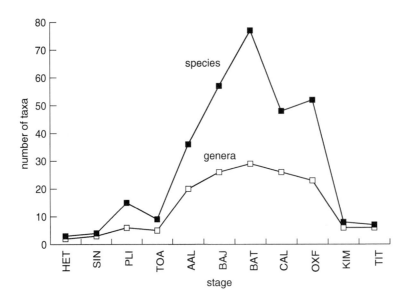

9.9 Cretaceous–Palaeogene Radiations

Cyclostomes diversified once again in the Cretaceous. The richest known Early Cretaceous fauna – the Late Aptian Faringdon Sponge Gravel of Oxfordshire (Pitt and Taylor 1990) – contains 49 cyclostome species but only one cheilostome. Higher cyclostome diversities were attained in the Upper Cretaceous, with global generic diversity in the order peaking in the Maastrichtian at more than 350 genera (McKinney and Taylor 2001).

Following the Late Jurassic origin of the order Cheilostomata (Taylor 1994a), the spectacular evolutionary radiation of cheilostomes was delayed until the mid-Cretaceous. Early cheilostome diversification occurred at a rate deviating from expectations of a simple model of exponential diversification (Patzkowsky 1995). It has been hypothesized that this radiation was triggered by

the evolution of brooded, non-planktotrophic larvae as the onset coincides with the first appearance in the fossil record of neocheilostomes possessing ovicells (Taylor 1988b). All known cheilostomes older than Late Albian are of malacostegine grade in terms of their overall skeletal morphology and lack ovicells.

By comparison with similar living cheilostomes, the earliest, malacostegine-grade cheilostomes, which lack ovicells, can be inferred to have possessed non-brooded cyphonautes larvae. These larvae are believed to be primitive for cheilostomes (e.g. Nielsen and Worsaae 2010); however, it is important to note that the absence of ovicells in itself is not unequivocal evidence that a cheilostome possessed cyphonautes larvae as some cheilostomes, such as *Gontarella*, have evolved internal brooding of non-planktotrophic larvae (e.g. Ostrovsky et al. 2006). Malacostegines did not contribute significantly to the explosive radiation of cheilostomes, which was almost entirely due to ovicellate neocheilostomes that can be inferred to have brooded their embryos and released non-planktotrophic, coronate larvae. The change from malacostegine- to neocheilostome-dominated cheilostome faunas is best seen in Texas. Here, six species of cheilostomes occur in the Early Albian Glen Rose Formation (Martha, Taylor, and Rader 2019), all of which are of malacostegine-grade (e.g. Figures 4.1C, 4.1G). In contrast, of the nine cheilostome species recorded from the Late Albian–early Cenomanian Washita Group, eight belong to the ovicellate neocheilostome genus *Wilbertopora* (Cheetham et al. 2006; Figures 3.25, 3.26) and only one is a malacostegine.

The ability to disperse is expected to be very different in cheilostomes with short-lived non-planktotrophic larvae than those with long-lived planktotrophic larvae in view of the positive correlation between larval duration and dispersal distance in marine animals (Shanks 2009). Once isolated, for example by rafting, connections between populations of non-planktotrophic species may not be easily regained, and genetic population structure may become more heterogeneous. The probability of allopatric speciation is therefore likely to be greater in non-planktotrophic than planktotrophic cheilostomes. Therefore, the evolution of non-planktotrophy could have resulted in accelerated speciation rates, triggering the mid-Cretaceous radiation of cheilostomes following almost 60 million years of low speciation rates and diversity among planktotrophic cheilostomes. This macroevolutionary hypothesis for cheilostome radiation has received some support from studies of comparative population genetic structure in planktotrophic vs. non-planktotrophic cheilostomes (Watts and Thorpe 2006). On the other hand, molecular phylogenetic has recently emerged that the transition from planktotrophy to non-planktotrophy occurred subsequently several times in cheilostome history (Heather Grant et al. in preparation; Figure 8.8), yet the later switches did not correlate with appreciable radiations in the groups involved. These molecular findings imply that the *de-novo* origin of non-planktotrophy has not always triggered evolutionary radiation.

Whether or not larval ecology had a significant role in cheilostome radiation, the Late Cretaceous and Palaeogene witnessed the first appearances and elaborations of zooid-level traits seemingly of major adaptive importance. As these include avicularia and frontal shields best interpreted as adaptations against small predators, an evolutionary arms race between predators and prey may also have been an important factor in cheilostome radiation. This was first suggested by Larwood and Taylor (1981) who noted the almost simultaneous evolution of defensive polymorphs in eleid cyclostomes and cheilostomes (see Jablonski, Lidgard, and Taylor 1997), placing the adaptive evolution of bryozoans in the context of the 'Mesozoic marine revolution' proposed by Vermeij (1977).

All pre-Late Albian cheilostomes have encrusting colonies. With regard to colony-forms, the oldest known erect cheilostome, *Jablonskipora kidwellae*, comes from the Late Albian of Devon in south-west England (Martha and Taylor 2017) and has narrow-branched ramose colonies. This species is a malacostegine, whereas nearly all of the diverse erect cheilostome species found in the Early Cenomanian are neocheilostomes.

Various planktonic groups, including dinoflagellates, a major item in the diets of present-day bryozoans, underwent evolutionary radiations between the Late Jurassic and Late Cretaceous, which have been correlated with diversifications among benthic animals (Fraaije et al. 2018). However, the pattern of dinoflagellate radiation, with a dip in the early Late Cretaceous, does not match bryozoan diversification well enough to favour a causal correlation. On a related theme, Scholz and Krumbein (2006) noted the coincidence in the appearance of diverse diatoms in the fossil record and the 100 Ma onset of cheilostome radiation, arguing for a causal link because diatoms are among the most important components of the microbial mats found on bryozoan colonies.

The Mesozoic radiations of cyclostomes and cheilostomes encompassed the parallel acquisitions of several traits, e.g. larval brood chambers, articulated colonies and, as already noted, mandibulate polymorphs. The contrast in the rates at which

Figure 9.5 Rates of first appearance of major traits in cyclostome (pale grey) and cheilostome (dark grey) bryozoans through the post-Palaeozoic in 25 million-year bins (based on Jablonski, Lidgard, and Taylor 1997, fig. 6). Novelties in cheilostomes appeared more rapidly than in cyclostomes and were strongly concentrated during the major radiation of the order in the mid-Cretaceous.

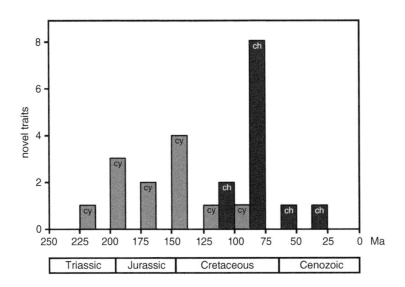

cyclostomes and cheilostomes acquired these novelties is revealing about their comparative evolutionary histories (Jablonski, Lidgard, and Taylor 1997). They appeared gradually through an interval of about 100 million years in cyclostomes but much more rapidly in cheilostomes, many early during the explosive radiation of cheilostomes (Figure 9.5).

9.10 End Cretaceous and Danian Extinctions

Of all mass extinctions, the K-Pg event at the end of the Cretaceous has attracted the most attention, not only because of the high profile of some of the victims – dinosaurs in particular – but also because of the continuing debate over the importance of the Chicxulub asteroid impact and Deccan Traps vulcanism as drivers. The small number of sections containing bryozoans on both sides of the K-Pg boundary limits analyses: bryozoans have not been recorded from the classic K–T boundary sections at the Brazos River in Texas, Gubbio in Italy, El Kef in Tunisia, or Zumaya in Spain (MacLeod et al. 1997).

Family-level data for bryozoans reveals only a minor K-Pg diversity drop, but the severity of the extinction at the end of the Maastrichtian for bryozoans becomes much more apparent at genus level for all three main metrics of extinction rate: extinctions per million years, extinctions per standing diversity, and extinctions per standing diversity per millions of years (McKinney and Taylor 2001). Interestingly, for the last of these two metrics, the extinction rate at the end of the Danian exceeds that for the K-Pg, which probably reflects the demise of the carbonate platform across northern Europe from which a high proportion of Danian bryozoan species have been described. Whatever the cause of the extinctions, recovery of bryozoans occurred very rapidly

(Stilwell and Håkansson 2012), in stark contrast to their delayed recovery after the Permian extinctions.

In the Danish Basin, where rich bryozoan faunas can be found in both the Late Maastrictian and Early Danian, extinction rate at species level has been estimated to be about 80% (Stilwell and Håkansson 2012). A study of bryozoan morphology immediately beneath the K-Pg boundary at Nye Kløv in Denmark (O'Dea et al. 2011), provided evidence of environmental instability. Cheilostomes belonging to four common species in the last 20 centimetres of Maastrichtian Chalk of this relatively complete section have smaller zooids, implying higher ambient temperatures. MART analysis suggests greater seasonality, while asymmetry of colony-form in free-living colonies and fewer avicularia point to unfavourable conditions prior to the impact event which is marked at Nye Kløv by an iridium-rich boundary clay.

Within Denmark there is a strong facies-related difference between Maastrichtian and Danian bryozoan faunas (Håkansson and Thomsen 1979), with pelagic Maastrichtian chalks containing high diversity faunas, whereas bryozoans are almost lacking in Danian chalks. In contrast, bryozoan limestones contain similar numbers of bryozoan species in the Danian as they did in the Maastrichtian.

Berthelsen (1962) weighed samples of Danian bryozoans from Danish localities and found that cyclostomes contributed more biomass than cheilostomes in his samples. McKinney et al. (1998) and McKinney, Lidgard, and Taylor (2001) confirmed the Early Danian spike in cyclostome skeletal dominance, though not diversity, by weighing the two groups in sorted bulk samples mostly from the Danish Basin and comprising predominantly erect colony-forms. The brief reversal of the temporal trend towards an increasing abundance of cheilostomes relative to cyclostomes was thought most likely to be due to the crash in phytoplankton levels at the K-Pg boundary, which may have favoured cyclostomes over cheilostomes with their greater energetic demands. A subsequent study (Sogot, Harper, and Taylor 2013), however, failed to find a similar cyclostome spike

in K-Pg samples collected from either the south-eastern United States or Maastricht in the Netherlands. Nor was there a Danian cyclostome spike among encrusting bryozoan colonies in the south-eastern United States and the Danish Basin. Furthermore, an increase in sheet-like encrusters compared to runners in the Danian is counter to the expectations of a phytoplankton crash as runners are expected to be better able to flourish in low nutrient conditions (see Okamura, Harmelin, and Jackson 2001).

Sogot, Harper, and Taylor (2014) tested the notion that there might be a diminution in bryozoan zooid size across the K-Pg boundary in accordance with the 'Lilliput effect' documented in various unitary organisms after mass extinction events. However, no significant change in zooid length was found in 59 bryozoan species from assemblages on either side of the K-Pg boundary. Furthermore, in only one of four genera were zooids smaller in species above than below the boundary.

Remarkably, among bryozoan species that survived the K-Pg extinction, one – the anascan cheilostome *Nellia tenella* – is apparently still alive today (Winston and Cheetham 1984). A common species in the tropical and subtropical Cenozoic, the oldest record of this 'living fossil' comes from the Upper Maastrichtian of Jamaica.

9.11 Convergence

Defined by Mahler et al. (2017, p. S14) as: 'the pattern of evolution in which species in two independently evolving lineages become phenotypically similar', convergence is a recurring theme of bryozoan evolution (e.g. Blake 1980). As long ago as 1919, W.D. Lang asserted that cheilostomes with spinocystal frontal shields (i.e. cribrimorph ascophorans) had arisen no fewer than 10 times from species with naked frontal surfaces (membraniporimorph anascans) during the Late Cretaceous alone. While this early claim of convergence at zooid-level requires confirmation, convergence in bryozoans is undoubtedly common at both colony and zooid level.

The kind of robust phylogenetic analysis necessary to confirm convergence (homoplasy) among taxa of bryozoans is usually lacking. However, convergence between bryozoans that are clearly only distantly related can be incontestable, especially when the taxa concerned are of very different geological age. For example, the unusual Carboniferous cryptostome *Worthenopora* bears a striking resemblance to some Cretaceous and younger onychocellid cheilostomes (Hageman 1991). Colonies of *Worthenopora* are palmate, like many onychocellids (see Taylor, Martha, and Gordon 2018), but more importantly have autozooids with depressed cryptocyst-like frontal walls of similar overall appearance to those of onychocellids (Figure 9.6). However, the presence of styles (Figure 2.8F) in *Worthenopora*, as well as its lamellar wall microstructure, are characteristics of palaeostomates not found in onychocellids, and *Worthenopora* is over 200 million years older than the earliest onychocellids which are known from the Cenomanian stage of the mid-Cretaceous.

While convergence in zooid morphology is moderately common among bryozoans, convergent evolution in colony-form has been rampant. Superficially identical colony-forms have evolved repeatedly in bryozoans, which can be understood in terms of the recurrent acquisition of adaptive colony morphologies within the constraints of the modular construction and suspension feeding ecology of the phylum. Some colony-forms have evolved many times (e.g. foliose colonies with 'boxwork' colonies: Plate 12A–C), while others have appeared just a few times.

9.11.1 Convergence in zooid morphology between eleid cyclostomes and cheilostomes

An especially striking instance of convergence at the zooid level (autozooids and mandibulate polymorphs: Figure 3.27) occurs between eleid cyclostomes and cheilostome bryozoans. Indeed, the cyclostome affinity of eleids was disputed by early bryozoologists (e.g. Waters 1891) until the seminal study of Levinsen (1912) emphasizing the tubular shape of the zooids and, crucially, the presence of typically cyclostome gonozooids (Figure 9.7). Morphological differences between eleids and other cyclostomes led Viskova (2016) to place them in a distinct order of stenolaemates, Melicerititida, despite the fact that skeletal microstructure and other traits point to a close relationship with some tubuliporine-grade cyclostomes (Taylor and Weedon 1996). Eleids date back to the Barremian (supposed Middle Jurassic examples

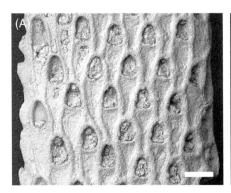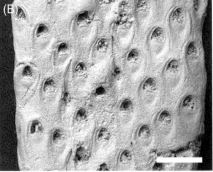

Figure 9.6 Homeomorphy is striking between the Palaeozoic cryptostome *Worthenopora* and some post-Palaeozoic onychocellid cheilostomes; both taxa have palmate bifoliate colonies and autozooids bounded by thick, raised walls and semi-elliptical openings: (A) *Worthenopora spinosa* (Lower Carboniferous, Warsaw Beds; Warsaw, Illinois, USA); (B) *Rhagasostoma aglaia* with avicularia, not seen in *Worthenopora*, distinguished by their long, curved rostra (Cretaceous, Campanian, Biron Fm.; Talmont, Charente Maritime, France). Scale bars = 500 μm.

Figure 9.7 Gonozooid of the eleid cyclostome *Reptomultelea oceani* with the ooeciopore arrowed; note the densely pseudoporous gonozooid roof and the numerous kenozooids laterally and distally of this structure (Cretaceous, Cenomanian; Cap de la Hève, Seine Maritime, France). Scale bar: 200 μm.

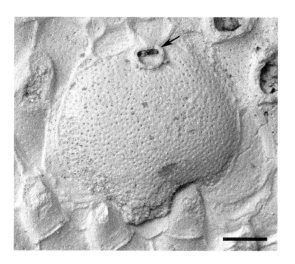

described by Viskova [2011] lack the diagnostic opercula) and underwent their main phase of diversification concurrently with cheilostomes in the Late Cretaceous (Taylor 1985). They are unique among stenolaemate bryozoans in having an operculum able to close and seal the skeletal aperture when the lophophore was retracted. The operculum was hinged along the proximal edge of the aperture, which was straight unlike that of other cyclostomes, and there are usually a pair of ridges on the interior surface of the operculum, perhaps for the attachment of the muscles used for closure (see reconstruction of Boardman 1983, fig. 38). Whereas calcification of the operculum seems to have been ubiquitous among eleids (Figure 3.1F), the opercula of cheilostomes normally comprise thickened cuticle and, as noted previously, calcification of the operculum in cheilostomes occurs in only a small minority of taxa widely scattered across the order (see Koromyslova 2014a; Figure 3.1B–E).

Convergence between eleid cyclostomes and cheilostomes extends beyond the shared presence of opercula: eleids also have mandibulate polymorphs (Figure 9.7A) resembling many cheilostome avicularia (e.g. Figure 3.23D). Like avicularia, these eleozooids originated through hypertrophy of the operculum relative to the zooid as a whole. The mandible, which is calcified in those species where it is known, ranges from large and spatulate ('rostrozooids': Figure 3.27A–E) to small and semi-circular ('demizooids': Figure 3.27F) (Taylor 1985), respectively mimicking the vicarious and inter-zooidal avicularia of cheilostomes.

9.11.2 Zooidal and colonial convergence between chiplonkarinid cheilostomes and cyclostomes

The long tubular zooids of stenolaemates are replicated by the unusual Cretaceous cheilostome *Chiplonkarina* (Taylor and Badve 1995; Plate 8N). When the type species of this genus was first described it was assigned without any hesitation to the cyclostome genus *Ceriopora*. This was hardly surprising in view of the ramose colonies with long tubular zooids that are thin-walled in the endozone and thick-walled in the surrounding exozone, and the overall similarity in zooidal morphology on the colony surface (Figure 9.8). The main clue to the cheilostome affinity of *Chiplonkarina* lies in the compound walls between the zooids, with a crenulated junction (Fig 2.3) between walls of adjacent zooids where the organic intercalary cuticle was originally located, a common feature in cheilostomes. The wall-perpendicular fibrous texture of the skeletal walls is another cheilostome trait present in *Chiplonkarina* that is not found among cyclostomes. A related genus – *Jablonskipora* – shows further evidence of cheilostome affinity in the presence of closure plates bearing impressions of the operculum (Martha and Taylor 2017). The continuously growing, cylindrical zooids of *Chiplonkarina* allowed branches to thicken, whereas thickening is generally achieved by multiple overgrowths in cheilostomes with box-like zooids (Figure 1.4B) that become fixed in size at the time of budding.

Figure 9.8 Homoemorphy between the cheilostome *Chiplonkarina* and cerioporine cyclostomes showing the resemblance between colony surfaces with autozooids and smaller kenozooids; note, however, wider zooecial walls with interzooidal boundary lines in the cheilostome compared with the sharper, undivided walls in the cyclostome: (A) *Chiplonkarina campbelli* (Cretaceous, Campanian or Maastrichtian, Kahuitara Tuff; Pitt Island, Chatham Islands, New Zealand); (B) *Ceriopora virginiana* (?Pliocene, James River or Waccamaw Fm.; Lee Creek Mine, Virginia, USA). Scale bars = 500 μm.

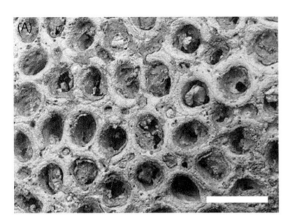
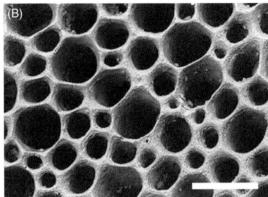

9.11.3 Parallel evolution of reticulate colonies

Erect colonies comprising meshworks of linked branches with zooids opening on one side only occur in three bryozoan suborders: Fenestrata, Cyclostomata, and Cheilostomata. Without close examination of zooidal morphology, it can be impossible to identify to which of these suborders a particular specimen belongs. As described above (Chapter 4.5.2), this 'fenestrate' colony-form is indicative of a unidirectional feeding current system, with water approaching the meshwork on the side bearing the autozooids, filtered of plankton by the lophophores, and passed through the gaps in the meshwork for expulsion from the side of the colony lacking lophophores. This advanced colony-form first evolved in Fenestrata during the Ordovician, disappeared with the extinction of this order at the end of the Permian, before re-appearing in Cretaceous cyclostomes and again in Palaeogene cheilostomes. Extant reticulate colonies mostly belong to the cheilostome family Phidoloporidae – sometimes referred to as 'lace corals' – but also occur in some other cheilostome genera such as *Petralia* and the cyclostome *Retihornera* (Smith, Taylor, and Spencer 2008).

A few post-Palaeozoic bryozoans bear a particularly strong resemblance to the Palaeozoic genus *Fenestella* and its close relatives (Figure 9.9A) in having sterile dissepiments forming cross-links between branches bearing two rows of autozooids. The Recent cyclostome *Fenestulipora* (Taylor and Gordon 1997; Figure 9.9B) and the Miocene cheilostome *Pirabasoporella* (Zágoršek et al. 2014) and its living relative *Jaculina* (Figure 9.9C) all have gracile colonies with this morphology.

Two of the most distinctive colony-forms typical of Palaeozoic Fenestrata both re-appeared in cyclostomes found in the Eocene Castle Hayne Limestone of North Carolina: *Hornera reteramae* with lyre-shaped colonies resembling *Lyropora* and *Lyroporidra* (McKinney, Taylor, and Zullo 1993), and *Crisidmonea archimediformis* which has colonies possessing an *Archimedes*-like screw axis (Taylor and McKinney 1996). The lyre-shaped colonies belonging to both Fenestrata and Cyclostomata have been interpreted as adapted to high-energy marine shoals. The low profile of their colonies and the ballast provided by the thickened marginal lyres endowed

Figure 9.9 Palaeozoic fenestellid bryozoan and two recent homeomorphs, all with biserial branches linked laterally by dissepiments to define fenestrules through which unidirectional currents flowed: (A) fenestellid *Spinofenestella* sp. (Carboniferous, Moscovian, Boggy Fm.; Sulphur, Oklahoma, USA); (B) cyclostome *Fenestulipora cassiformis* (Recent; New Zealand); (C) cheilostome *Jaculina parallelata* (Recent; Mediterranean). Scale bars: A = 200 μm; B = 1 mm; C = 500 μm.

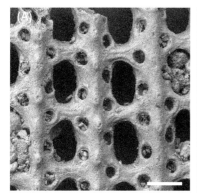 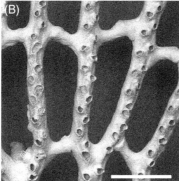 

stability in settings where bryozoans with delicate reticulate branches may have been excluded.

Far less common than the fenestrate colony-form but also planar and reticulate, are cribrate colonies having zooids opening on both sides of the branches that are characteristically broader and more flattened than in fenestrate colonies (Plate 8C–D). This colony-form occurs in Ordovician–Permian palaeostomates of the orders Cryptostomata and Cystoporata, the Jurassic cyclostome *Rorypora*, and at least two genera (*Nudicella* and *Adeona*) of Cenozoic–Recent cheilostomes (Taylor 2012). Cribrate colony-forms evidently evolved from bifoliate, palmate colonies by restriction of branch bifurcation to a single plane and regular anastomosis between branches to form a planar reticulate frond.

9.11.4 Parallel evolution of lunulitiform colonies among cheilostomes

Given their distinctive and unusual ecology, lunulitiform cheilostomes afford some of the most remarkable examples of evolutionary convergence among bryozoans (Figure 9.10). Particular traits of the autozooids and polymorphs have favoured the independent evolution of free-living lunulitiforms on multiple occasions from different groups of anascan-grade cheilostomes with encrusting colonies. Stages in the evolution of lunulitiform cheilostomes include: (i) growth beyond the edges of substrates of ever smaller size; (ii) development of cap-like, conical colonies of modest and more-or-less fixed size; (iii) prolongation of avicularian mandibles into hair-like vibracula setae; (iv) internalization of embryonic brooding and loss of prominent ovicells; and (v) acquisition of aragonitic biomineralization. Not all lunulitiform cheilostomes exhibit all of these traits and the sequence in which they appeared may have varied.

The first lunulitiform colonies appeared during the Coniacian stage of the Upper Cretaceous and belong to the family Lunulitidae (Voigt 1981a; Figure 9.10A). Like some later lunulitiforms, they probably evolved from an ancestor in the diverse family Onychocellidae (Cook and Chimonides 1983). Now extinct, Lunulitidae are locally abundant and diverse in the Chalk of north-west Europe (Figure 9.10B; Plate 6L–M), particularly in the Maastrichtian (Håkansson and Voigt 1996), and persisted into the Cenozoic (Figure 9.10C–D) where they were joined by three closely related families (Otionellidae, Lunulariidae, and Seleneriidae) recently united with Lunulitidae in the superfamily Lunulitoidea.

Figure 9.10 Homeomorphic lunulitiform cheilostomes belonging to Lunulitidae (A–D), Cupuladriidae (E–G), and Onychocellidae (H): (A) *Lunulites plana*, the oldest known lunulitiform species (Cretaceous, Coniacian, or Santonian, Craie de Villedieu; Villedieu, Loir-et-Cher, France); (B) *Lunulites tenax* (Cretaceous, Campanian; Norwich, Norfolk, UK); (C) *Discoradius tanzaniensis* (Oligocene, Rupelian; Pande Peninsula, Kilwa District, Tanzania); (D) underside of a colony of *Discoradius tanzaniensis* showing partly enveloped sand-grain substrate (Oligocene, Rupelian; Pande Peninsula, Kilwa District, Tanzania); (E) *Reussirella* sp. with older zooids at the centre having closure plates (Recent; Mauritania); (F) irregular colony of *Reussirella* sp. regenerated from a fragment (Pliocene, Coralline Crag Fm.; Sutton, Suffolk, UK); (G) tilted lateral view of a convex colony of *Cupuladria panamensis* (Recent; Nicaragua); (H) *Pseudolunulites irregularis* (Oligocene, Chattian; Bünde-Doberg, Nordrhein-Westfalen, Germany). Scale bars = 1 mm.

Lunulitiform colonies subsequently evolved independently in several other families during the Cenozoic. These include the calloporid *Bonellina* from the Oligocene Tortachilla Limestone of South Australia, which evidently enhanced its stability on the seafloor by attaching itself to dense grains of the mineral goethite (Schmidt 2007). Another Oligocene genus – *Pseudolunulites* (Figure 9.10H) from the Chattian of Germany, was considered to be an onychocellid by Cook and Voigt (1986) who noted the differences in its early astogeny from that of lunulitids.

The most speciose family of lunulitiform bryozoans living today is Cupuladriidae (Figure 9.10E–G), found in tropical and subtropical regions of the Atlantic, Pacific, and Indian oceans, with a few

records from the Mediterranean Sea. The oldest fossil cupuladriid is *Cupuladria ovalis* from the Paleocene of Senegal (Gorodiski and Balavoine 1962), but the family did not become diverse until the Neogene. Cupuladriids are characterized by the pairing of each autozooid with a small distal vibraculum. This recalls genera such as *Biselenaria* (Figure 4.7) from which cupuladriids possibly evolved.

The ability to grow freely beyond the edge of a substrate is a necessary precursory requirement for the evolution of the lunulitiform colony-form. This is facilitated by having strong basal calcification and zooids that are firmly bound by their shared lateral walls. Another important 'preadaptation' is the possession of regularly spaced avicularia with elongated mandibles capable of functioning as setae (Cook and Chimonides 1983). A common feature of lunulitiform cheilostomes is internalization of embryonic brooding, which may be because prominent ovicells would be disadvantageous to colonies moving through sediment when unburying themselves (Ostrovsky, O'Dea, and Rodríguez 2009).

All extant lunulitiform bryozoans have skeletons of aragonite; small amounts of calcite detected in some colonies of a few taxa (Steger and Smith 2005) are best explained by the inadvertent inclusion of calcitic substrates or epibionts in the analyses. However, the pristine preservation of nearly all Cretaceous examples of Lunulitidae in sediments leached of aragonitic fossils shows quite clearly that they were originally calcitic. Indeed, it is possible to chart the transition through time in this family from calcitic through bimineralic to aragonitic biomineralization (Taylor et al. 2009).

9.12 Palaeostomates and Post-Palaeozoic Cyclostomes Compared

It has become customary to interpret the palaeobiology of the dominant Palaeozoic bryozoan palaeostomate orders (i.e. Trepostomata, Cryptostomata, Cystoporata, Cryptostomata, Timanodictyina, and Fenestrata) by reference to the biology of living Cyclostomata (e.g. Borg 1965; Boardman 1971; Utgaard 1973; Tavener-Smith 1973b). This approach has led to major advances in our understanding of the growth and functional morphology of extinct palaeostomates.

Regardless of the exact phylogenetic relationships between palaeostomates and cyclostomes (see Chapter 8.3), there are, however, several key morphological differences between palaeostomates and cyclostomes. These differences have important implications for understanding the comparative biology of the two groups. With the sole exception of the enigmatic Cinctiporidae (Boardman, McKinney, and Taylor 1992; Schwaha et al. 2018), all adequately known species of living cyclostomes have gonozooids for brooding polyem-

bryonic larvae (Chapter 3.5). Gonozooids are unknown in some fossil cyclostome species, including all taxa (paleotubuliporines) from the Palaeozoic, but it is conceivable that at least some of these had gonozooids borne high on autozooidal peristomes, as in a few living cyclostome species (Harmelin 1974), which were broken off during fossilization. In contrast, nearly all palaeostomates lack polymorphs resembling the gonozooids of post-Palaeozoic cyclostomes (cf. the cystoporates *Lichenalia* [Figure 3.21B] and *Rhinopora*, see above). Weak evidence for polyembryony exists in the form of fused pairs of colonies of the Permian fenestrate *Septopora*, the argument being that fusion was the result of two polyembryonic larvae with full histocompatability settling next to one another on the same substrate (McKinney 1981a). However, fusion between genetically distinct colonies has been documented in cheilostome bryozoans such as *Hippodiplosia insculpta* (Craig and Wasson 2000), raising the possibility that the fused colonies of *Septopora* originated from nonclonal larvae, and therefore have no bearing on the issue of polyembryony in Palaeozoic bryozoans.

Using a different approach, Pachut and Fisherkeller (2010) compared the size of ancestrulae in some Palaeozoic trepostome stenolaemates from the Ordovician with those of polyembryonic cyclostomes as well as planktotrophic and non-planktotrophic cheilostomes. Their premise was that these three different reproductive modes would be characterized by larvae – and consequently ancestrulae – of different sizes. The similarity in size of the ancestrulae of cyclostomes and trepostomes, both of which are generally smaller than cheilostome ancestrulae, led them to infer that polyembryony probably occurred in these Palaeozoic stenolaemates. Making comparisons of ancestrula size between stenolaemates and cheilostomes is, however, compromised by differences in the patterns of larval metamorphosis between the two clades and the fact that cheilostomes typically have larger zooids than cyclostomes (McKinney 1993). Variations through time in the diameter of stenolaemate protoecia are greater than originally believed and have also changed through geological time. Four Ordovician trepostome genera have mean ancestrular widths (equivalent to protoecial diameters) of 97–130 μm (Pachut and Fisherkeller 2011, table 1), whereas four Ordovician cyclostome genera are smaller in size, with a range of 69–103 μm (unpublished data). Post-Palaeozoic cyclostome protoecia (the first skeletal part of the ancestrula produced directly from the settled larva) range from 83 μm to more than 700 μm in diameter (Taylor and Jenkins 2017), some of the largest occurring in hornerids that are known to have polyembryony. The modal size of protoecia in the dominant tubuliporine cyclostomes declined from the Jurassic (175–200 μm) to the Recent (125–150 μm) for unknown reasons. Thus, it is difficult to use the size of cyclostome ancestrulae as an indicator of reproductive mode.

Overall, therefore, evidence for polyembryony in palaeostomates is not compelling and the first reliable indication of polyembryony among stenolaemates comes with the oldest known cyclostome gonozooids in the Late Triassic cyclostome *Reptomultisparsa hybensis* (Taylor and Michalik 1991).

The other major difference between palaeostomates and cyclostomes is in the porosity of the skeletal walls. All living cyclostomes have interior walls penetrated by pores (Chapter 2.2) that link the two zooids on either side of the walls. In contrast, apart from a few ceramoporine cystoporates (Utgaard 1973; Figure 2.4C), interior walls in palaeostomates lack pores. As a result, any soft tissue connection between zooids has to occur via the membranes and hypostegal body cavity that drape over the distal ends of the interior walls. The formation of cuticle-covered frontal exterior walls in stenolaemates immediately severs such connections, which would isolate the zooids of palaeostomates in contrast to cyclostomes where pores in the interior walls maintain such linkages. Thus, palaeostomates tend not to have calcified exterior walls (**free-walled skeletal organization**, e.g. Figure 4.5B) whereas these may be present (**fixed-walled skeletal organization**, e.g. Figure 4.5A) or absent in cyclostomes (Figure 8.7). Structures constructed from calcified exterior

walls found in cyclostomes (e.g. long peristomes, terminal diaphragms, eleid opercula) are typically lacking in palaeostomates, suggesting a more limited adaptive repertoire, possibly as a consequence of the absence of porous walls.

Unlike terminal diaphragms, common in cyclostomes but uncommon in palaeostomates, calcified basal diaphragms are common in palaeostomates (Plate 4H) but are rare in cyclostomes (cf. Figure 3.9A). Basal diaphragms form the floors of living chambers. Given the lack of pores in these interior walls in palaeostomates, parts of the zooids proximal of basal diaphragms are isolated from the active tissues above. Therefore, thick palaeostomate colonies would have possessed a 'rind' of living tissues close to the surface that covered an inert skeletal framework (e.g. Madsen 1987). This is reminiscent of the soft tissue/hard skeleton relationships seen in corals for instance. In contrast, soft tissues extend deep into the interiors of cyclostome colonies in which the polypides can be extremely long (Chapter 3.1; Figure 3.6).

Despite differences in reproductive and skeletal organization, the few cyclostomes known from the Palaeozoic typically occur together with palaeostomates. While this suggests no major ecological differences between cyclostomes and palaeostomates, Smrecak and Brett (2014) in a study of sclerobionts on brachiopod shells in the Late Ordovician Cincinnatian found that cyclostomes were relatively more successful than other stenolaemates (i.e. palaeostomates) in the deep euphotic zone. In contrast to many palaeostomates, all Palaeozoic cyclostomes have small and gracile colonies, and may have been weeds in an ecological sense.

9.13 Frontal Shield Evolution in Ascophoran Cheilostomes

The overwhelmingly dominant bryozoans in modern seas are ascophoran-grade cheilostomes in which the frontal membrane forms the floor of an ascus and is protected by an overlying frontal shield. It has long been suspected that ascophorans are polyphyletic (e.g. Voigt 1991), suggesting that selective pressures, notably from micropredators (Chapter 5.2), may have favoured ascophoran-over anascan-grade morphologies where the frontal membrane used in lophophore extension is not protected by an overlying calcified layer. Initial molecular evidence provided by Knight, Gordon, and Lavery (2011) supported ascophoran polyphyly, a conclusion that has been corroborated by subsequent and on-going molecular analyses (Figure 8.8).

Morphological pathways leading to the evolution of the various kinds of ascophoran frontal shields (Plate 9D–H) were suggested by Gordon (2000) who proposed no fewer than nine different models. The most fundamental – and best established – is the evolution of the **spinocystal** frontal shield of 'cribrimorphs' by the bending of the erect spines surrounding the opesia in calloporid anascans so that they lie over the frontal membrane. The costal spines of cribrimorphs subsequently fused along the midline of the zooid and in many cases developed lateral fusions with neighbouring spines at intervals along their lengths. In primitive spinocystal frontal shields (Figure 3.5D), gaps between the spines allow seawater to enter into the space (ascus) between the frontal membrane and the spines to compensate for the displacement of the polypide when the lophophore is protruded. More complete fusion between spines often resulted in a frontal shield without such intercostal gaps, in which case seawater can only enter the ascus at the proximal end of the orifice where a sinus may be developed. Remarkably, the Cretaceous evolution of spinocystal frontal shields was reprised during the last 12 million years in the extant genus *Cauloramphus* (Dick et al. 2009). The morphologies of different species of *Cauloramphus* range from almost fully exposed frontal membranes, to spines overarching the frontal membrane but with gaps between them, to costate frontal shields that form a solid cover over a functional ascus floored by the frontal membrane.

In a second model of ascophoran origins, Gordon (2000) hypothesized that **gymnocystal** frontal shields like those found in Hippothoomorpha resulted from an increase in the extent of the gymnocyst at the expense of the spinocystal frontal shield. A Late Cretaceous genus, *Boreasina*, sometimes preserves a small spinocystal area proximal of the orifice, thus providing an intermediate state between spinocystal and fully developed gymnocystal frontal shields. Intermediate stages in the evolution of the porous gymnocyst characteristic of the extant ascophoran genus *Trypostega* (Figure 3.37B) can be seen in the Palaeogene genus *Trilophopora* (Figure 9.11C) which has a reduced spinocystal frontal shield located distally and centrally of a porous gymnocyst, and in some Palaeogene species of *Trypostega* with vestigial costae (Figure 9.11D).

The other two main types of frontal shield – **umbonuloid** and **lepralioid** – characterize the bulk of ascophorans and appear to be closely allied. Indeed, molecular trees imply multiple origins of one from the other. Umbonuloid frontal shields, however, considerably antedate lepralioid shields in the fossil record: the earliest cheilostome with an umbonuloid shield is the Santonian genus *Staurosteginopora*, whereas lepralioid shields are unknown until the Danian. From the Palaeogene onwards, lepralioid cheilostomes diversified rapidly (Gordon and Voigt 1996, fig. 9) and today constitute the bulk of ascophoran-grade cheilostomes. There is good evidence that umbonuloid shields have their origin in the kenozooids that can be observed surrounding the autozooids in many cribrimorphs (Gordon and Voigt 1996; Figure 9.11A). These kenozooids expanded outwards to overgrow the frontal surface of the autozooids. Early stages of this process can be seen in some Cretaceous cribrimorphs (Figure 9.11B). The underside of the resulting umbonuloid frontal shield is formed from the cuticle-covered surfaces of what were once the outer exterior walls of the kenozooids. Vestiges of the opesia of these kenozooids are represented by the areolar pores around the outer edge of the frontal shield.

Figure 9.11 Intermediate morphologies in ascophoran cheilostome showing the origin of umbonuloid (A–B) and trypostegid gymnocystal (C–D) frontal shields from cribrate (spinocystal) precursors: (A) in the cribrimorph *Ubaghsia langi* the costate frontal shield is surrounded by kenozooids with small opesia (Cretaceous, Coniacian or Santonian Chalk; Northfleet, Kent, UK); (B) similar kenozooids in a related pelmatoporid cribrimorph are beginning to overgrow the cribrate frontal shield, restricting its extent (Cretaceous, Campanian; Needs Camp, East London, South Africa); (C) *Trilophopora* sp. showing a frontal shield with a reduced costate component surrounded by a porous gymnocystal component; note the small polymorphs which can be compared to those present in the extant species *Trypostega venusta* (Figure 3.37B) (Eocene, Ypresian, London Clay; Lower Swanwick, Hampshire, UK); (D) in this un-named species of *Trypostega* the frontal shield consists almost entirely of porous gymnocyst apart from a raised region proximal of the orifice (arrowed) which is probably a vestige of the costate frontal shield (Eocene, Lutetian, Lower Bracklesham Beds; Selsey, Sussex, UK). Scale bars = 200 μm.

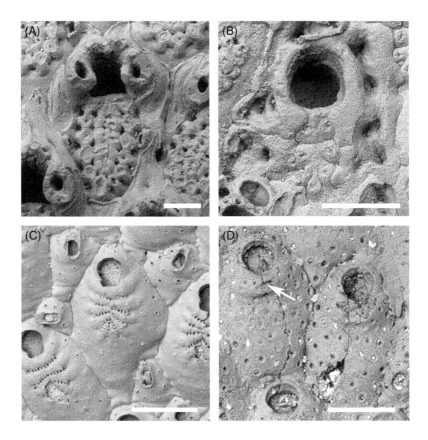

Lepralioid frontal shields are believed to have been derived from umbonuloid shields by the gradual replacement of the 'inside-out' umbonuloid frontal shield by a cryptocystal frontal shield often containing pseudopores which are lacking in umbonuloid shields.

9.14 Cyclostomes vs. Cheilostomes

Two orders of bryozoans with calcified skeletons are present in modern oceans, the dominant Cheilostomata and the subordinate Cyclostomata. They have co-existed since the Late Jurassic when cheilostomes first appeared in the fossil record. Species belonging to both orders can be found living together in many of the major habitat types containing bryozoans, often sharing the same substrate and possessing a similar range of colony-forms.

The number of species of cheilostomes living today is approximately nine times that of cyclostomes, and cheilostomes are correspondingly richer in genera and families too (Bock and Gordon 2013). The rise to dominance of cheilostomes is evident from several different metrics obtainable from the fossil record, including numbers of families and genera (Figure 9.12), and relative species richness and skeletal biomass of the two orders within communities (McKinney, Lidgard, and Taylor 2001).

In almost all respects, cheilostomes seem to be functionally superior to cyclostomes, as shown by McKinney (1993) who compared key functional traits in living cheilostomes and cyclostomes. Cheilostome zooids are typically larger, have more and longer tentacles, and mouths of greater diameter. They create feeding currents of higher velocity, giving them the potential to procure more food, which in turn may explain the typically higher colony growth rates of cheilostomes relative to cyclostomes.

Probably correlating with their higher growth rates and enhanced feeding capacities, encrusting cheilostomes overgrow cyclostomes in the majority of spatial competitive encounters between species belonging to the two groups. For example, McKinney (1992) studied overgrowth interactions between cyclostomes and cheilostomes in the northern Adriatic Sea off Rovinj, Croatia. Of 210 encounters, cheilostomes won 164, cyclostomes 16, and the remaining 30 were stand-offs. The dominance of cheilostomes was attributed to the feeding current system in these bryozoans that expelled filtered water centrifugally from the growing of the colony and towards other bryozoans competing for substrate space. In contrast, cyclostomes expel filtered water centripetally towards the centres of their own colonies.

In a seminal study of overgrowths between encrusting cyclostomes and cheilostomes through time, McKinney (1995a) scored overgrowths between multiserial species belonging to the two clades in 24 sclerobiont communities ranging from Albian to Holocene in age. He recorded between 50 and 320 interactions in each community. 'Mutual overgrowths' (reciprocal overgrowths and

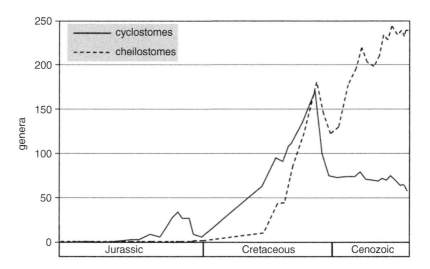

Figure 9.12 Generic diversity from the Jurassic to the Holocene in cyclostome and cheilostome bryozoans (based on Taylor and Waeschenbach 2015, figs 8 and 12. © The Trustees of the Natural History Museum, London).

Figure 9.13 Overgrowths between cheilostome and cyclostome bryozoans recorded from 24 fossil communities ranging from Albian to Holocene in age, showing the sustained dominance of cheilostomes through time (after McKinney 1995a, fig. 2).

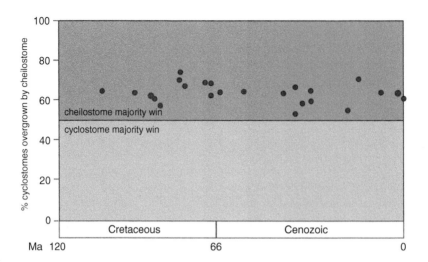

stand-offs) accounted for only 1–11% of interactions. Among the other 89% of overgrowths, cheilostomes won the majority in all 24 sclerobiont communities (Figure 9.13). They were dominant in 55–75% of overgrowth interactions, with an average of about two-thirds. No significant change was apparent through time in the ratio of cheilostome wins, the competitive asymmetry between the two clades remaining roughly constant for 100 million years.

Between the mid-Cretaceous and the present day, cheilostomes have increased massively in importance relative to cyclostomes. As mentioned above, this pattern is evident not only in global taxonomic diversity data (alpha diversity) but also in the compositions of bryozoan assemblages (Lidgard, McKinney, and Taylor 1993) which have become increasingly dominated by cheilostomes in terms of species numbers (Figure 9.14) and also abundance. Given the overall competitive superiority of cheilostomes over cyclostomes, the transition from cyclostome-to cheilostome-dominance is compatible with the competitive displacement of cyclostomes by cheilostomes. A coupled logistic model was developed by Sepkoski, McKinney, and Lidgard (2000) which showed how changes in generic diversity of the two clades was consistent with competitive displacement when a perturbation corresponding to the end-Cretaceous mass extinction was included. An analysis of the dynamics of generic diversity in the two clades showed that, whereas cyclostomes maintained low rates of origination through the Cenozoic, cheilostome origination rates were much more variable (McKinney and Taylor 2001). This dissimilarity in origination rates, rather than differences in extinction rates, appears to be the fundamental driver for the contrasting diversity histories of the two clades.

Compared to the fairly constant proportion of competitive encounters with cyclostomes won by cheilostomes through geological time (McKinney 1995a), Barnes and Dick (2000) found a significant latitudinal variability in outcomes among the present day bryozoan communities. Some high-latitude encrusting communities (e.g. Kodiak Island, Alaska) show a reversal of the usual competitive dominance, with cyclostomes winning more encounters than cheilostomes. Average growth rates of encrusters are slower (Bowden

Figure 9.14 Moving average trend of changes in the total number of bryozoan species (cyclostomes + cheilostomes) and of cheilostomes only from the Jurassic–Recent bryozoan assemblages described in the literature (based on Lidgard, McKinney, and Taylor 1993, figs 8 and 10).

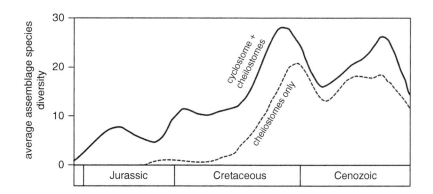

et al. 2006) at high latitudes, which may serve as geographical refuges for cyclostomes. As most of the overgrowths studied by McKinney (1995a) came from mid-palaeolatitudes, this may explain the apparent discrepancy between the consistent overgrowth dominance of cheilostomes through geological time and the latitudinal variability at the present day.

The fact that encrusting cyclostomes have persisted for 100 million years of geological time despite losing most competitive encounters for substrate space with co-occurring cheilostomes is remarkable and invites an explanation. Barnes (2002) considered two possibilities: (i) differential rates of recruitment, with cyclostomes colonizing newly available or disturbed hard substrates in advance of cheilostomes; and (ii) the ability of cyclostomes to maintain substrate space by blocking total colony overgrowth, resulting in stand-offs. Evidence for a decline through time in size among encrusting cyclostomes (McKinney 1995a, p. 475) may indicate that they have become more opportunistic, recruiting and reproducing earlier and developing 'spot' colonies *sensu* Bishop (1989). The majority of cheilostome species in most faunas have encrusting colony-forms but this is not always true for cyclostomes: in Neogene bryozoan faunas from Central America, over two-thirds

of the 179 cheilostome species are encrusters, whereas only about one-third of the 23 cyclostome species are encrusters, the remainder having erect colonies (Taylor 2001, fig. 3). Erect growth potentially serves as an escape mechanism for cyclostomes competing with cheilostomes that win most overgrowth encounters on substrate surfaces.

A potentially important, though hitherto overlooked, factor in the success of cheilostomes may be their ability to secrete thicker cuticles than cyclostomes. The cuticle can be as little as 0.2 μm in thickness in some crisiid cyclostomes (Nielsen and Pedersen 1979), compared with >1 μm in most cheilostomes (e.g. Carson 1978) and up to 10 μm thick in the cheilostome *Myriapora truncata* when exposed to low pH conditions (Lombardi et al. 2011). This thick cheilostome 'skin' is the basis for an assortment of traits that increase the adaptive repertoire of cheilostomes relative to cyclostomes. Cheilostomes have been able to evolve flexible flustriform colonies in which a thick cuticle rather than a mineralized skeleton gives the colony cohesion. No flustriform colonies are known among cyclostomes. Hollow spines in cheilostomes are prolongations of cuticle, beneath which a mineralized layer may develop, and are capable of serving various functions as described earlier (Chapter 3.8). The spine base is often basally articulated and the spine is open at the distal end, allowing lengthening and repair, the spine being an exterior wall. Cyclostomes do not have exterior-walled spines of this kind: instead, cyclostome spines are solid structures secreted as interior walls beneath a membrane and therefore more vulnerable. Stolons and rootlets (or rhizoids) formed by many cheilostomes are unmineralized cuticular tubes. Neither has an analogue among cyclostomes, and rooted colonies – an important ecological group of cheilostomes on particulate substrates at the present day – do not exist in cyclostomes. Lunulitiform colonies (Chapter 4.3.3) characterize several clades of cheilostomes inhabiting fine particulate substrates but have no cyclostome equivalents. Cuticular setae are essential for the functioning of lunulitiform colonies to support colonies above the sediment surface and allow them to regain the surface if buried. Even though mandibulate polymorphs (eleozooids) evolved independently in eleid cyclostomes, these are always calcified and would have been too brittle if prolonged into a hair-like setal morphology.

In summary, the body plan of cyclostomes seems inferior to that of cheilostomes in several key respects: cyclostomes are unable to extend their lophophores as high above the colony surface as can cheilostomes, limiting their access to faster flows and restricting their behavioural plasticity; cyclostome feeding currents do not cause filtered water to be expelled towards competitors; and the thin cyclostome cuticle cannot be employed to form hollow spines, setae, rootlets, stolons, and flexible flustriform colonies, thereby limiting the adaptive repertoire of cyclostomes compared to cheilostomes. These attributes may together explain the increasing dominance of cheilostomes over cyclostomes since the Cretaceous.

9.15 Colony-forms Through Geological Time

The proportions of distinct bryozoan colony-forms in bryozoan assemblages from the Ordovician to the Recent have not been constant: different colony-forms have dominated fossil bryozoan assemblages at different times, while during some geological periods, colony-forms common at other times were rare to lacking. All of the major colony-forms recognizable in bryozoans have evolved on multiple occasions, as discussed above (Chapter 9.11). Indeed, single genera often contain species with several differ colony-forms. For example, among the five species of the Australian Cenozoic genus *Nudicella*, colony-forms that are encrusting, palmate, foliose, cribrate, and narrow-branched ramose are all known to occur (Schmidt and Bone 2004). Plasticity of colony-form at low taxonomic levels means that changing proportions of bryozoan colony-forms through time cannot be explained simply by phylogenetic patterns leading to the dominance of particular higher taxa at specified times.

Taylor and James (2013) investigated temporal changes in nine major colony-forms from the Ordovician to Neogene by calculating the mean numbers of each of these colony-forms in post-1900 faunal studies that covered all groups of bryozoans with calcified skeletons (Figure 9.15). They found the following temporal trends: (i) a steady increase in the number and proportion of encrusting species through time that was interrupted by a drop in the Late Palaeozoic when encrusters seem to have been rare; (ii) a post-Triassic decrease in robust branching colonies; (iii) a rise in the proportion of fenestrate colonies through the Palaeozoic, followed by their absence in the Triassic and Jurassic, rarity in the Cretaceous and reappearance but in reduced numbers in the Cenozoic; (iv) a scarcity of articulated colonies until the Cretaceous; (v) the absence of free-living lunulitiform colonies until the Cretaceous. Trend (iii) was previously identified by McKinney (1986a) in general for erect bryozoans with autozooids opening on one side of the branches only (Figure 9.16).

Do these variations in major colony-forms signal changes in the environments colonized by bryozoans through geological time? It is reasonable to suppose that this might be the case for lunulitiform colonies, which are capable of inhabiting an adaptive zone of finely particulate sediments precluded for bryozoans without the ability to

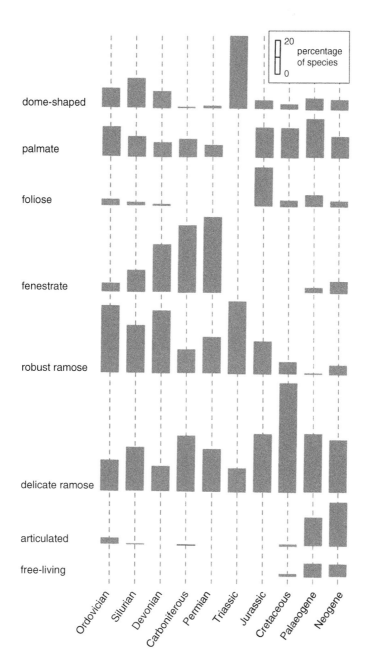

Figure 9.15 Average proportions of species in bryozoan assemblages through geological time having eight non-encrusting colony-forms (based on Taylor and James 2013, fig. 9. © The Trustees of the Natural History Museum, London).

Figure 9.16 Parallel increase through time in the proportion of unilaminate species of erect bryozoans in the Palaeozoic (solid line, lower scale bar) and post-Palaeozoic (dashed line, upper scale bar) (redrawn from McKinney 1986a, fig. 3). Note that both axes of the graph are logarithmic.

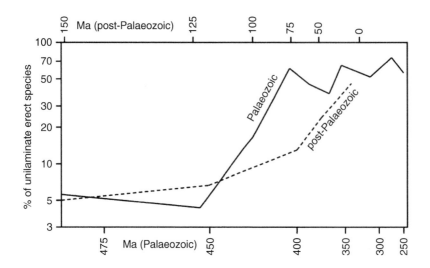

9.16 Evolutionary Tempo in Bryozoans

develop setae (see above). However, most of the other variations may instead indicate shifts through time in the repertoire of morphological strategies employed by bryozoans to inhabit the same general environments.

The landmark paper of Eldredge and Gould (1972) introducing the punctuated equilibrium concept ignited a debate about the tempo and mode of evolutionary change which still continues today. Their punctuated equilibrium model interpreted species as morphologically unchanged for long stretches of geological time (i.e. stasis), with rapid morphological change happening during speciation events. This view contrasted with the prevailing notion of phyletic gradualism in which morphological change is more or less continuous through time and not dependent on speciation.

Confidence in recognizing biological (genetic) species in some ascophoran-grade cheilostomes with complex morphologies and adequately sampled fossil records has allowed bryozoans to contribute to the debate on evolutionary tempo. In particular, research by Alan Cheetham and Jeremy Jackson on the cheilostomes *Metrarabdotos* (Figure 9.17) and *Stylopoma* from tropical America provides outstanding examples of punctuated equilibria (Cheetham 1986b; Jackson and Cheetham 1994). Gould (2002, p. 78) regarded the work on *Metrarabdotos* as 'perhaps the best documented and most impressive case of exclusive punctuated equilibrium ever developed'. Aside from the ability to recognize biological species from skeletal morphology alone, a major strength of this research is that the species concerned were studied from the standpoint of the morphometry of all of their preserved skeletal characters, not

Figure 9.17 Punctuated pattern of morphological evolution in the cheilostome *Metrarabdotos* in the Caribbean (based on Cheetham 1986b, fig. 5, with species names updated according to Cheetham, Sanner, and Jackson 2007). Each species shows stasis, remaining relatively unchanged in morphology for a considerable period of geological time.

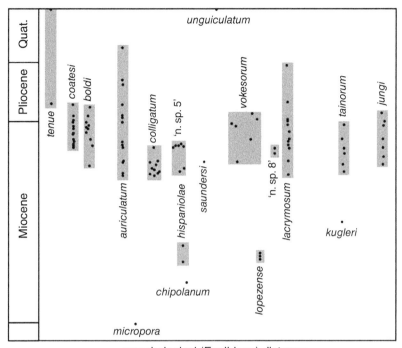

morphological (Euclidean) distance

just a selection of characters, unlike some other studies of evolutionary tempo in fossils.

In their study of the Ordovician trepostome *Perenopora* from eastern North America, Pachut and Anstey (2007) found a tempo consistent with what they referred to as 'punctuated speciation'. They calculated a speciation rate of 5.73 species per myr, with an average time required for speciation of 186 kyr compared with species durations that typically exceeded 5 myr. In a subsequent paper (Pachut and Anstey 2012), these authors found low rates of anagenesis in *Peronopora*, with only 8% of characters showing statistically significant anagenetic changes.

10 Prospective Future Research

This is not a book about bryozoan taxonomy. Nevertheless, it should be clear that taxonomy underpins all aspects of bryozoan palaeobiology, from tracking changes in diversity through time to interpreting the functional morphology of specific taxa. The availability of scanning electron microscopy for detailed studies of skeletal morphology, and more latterly molecular sequencing for taxon characterization, have allowed major advances in bryozoan taxonomy. Balanced against these, however, has been an alarming decline in the number of bryozoan taxonomists. Yet, it is clear that there remain many undescribed fossil species. For example, a 'collector curve' summarizing the cumulative addition of new Jurassic bryozoan species to the literature between 1800 and 2000 showed a sustained increase through time without any indication of levelling-off (Taylor and Ernst 2008, fig. 1), a clear sign that the number of described species is well below true species diversity. Even in regions of the world where palaeontological research has been actively undertaken for 200 years, knowledge of bryozoans has lagged well behind that of other groups of marine invertebrates: almost two-thirds of the 128 species monographed by Taylor and McKinney (2006) from the Campanian–Maastrichtian of the eastern United States were found to be previously undescribed, a remarkable proportion considering the number of active palaeontologists in that region.

Aside from taxonomy, there are numerous important gaps in our knowledge of fossil (and Recent) bryozoans, just a few of which are highlighted below.

Bryozoan Paleobiology, First Edition. Paul D. Taylor.
© 2020 Natural History Museum. Published 2020 by John Wiley & Sons Ltd.

10.1 Biomineralization

Knowledge of the processes involved in the growth of bryozoan skeletons is still in its infancy. The paucity of data on biomineralization in bryozoans themselves has led to an understandable tendency to assume that bryozoan skeletons grow in exactly the same way as those of the better-known molluscs and brachiopods. While this may indeed be true for certain aspects of biomineralization, the greater complexity of bryozoan skeletons, with numerous different types of skeletal wall even within an individual zooid, limits the strength of analogies with other metazoan phyla possessing calcareous skeletons. Likewise, while some ultrastructural fabrics can be superficially matched between bryozoans and, say, brachiopods, there are potentially important differences in the sizes of the crystallites, which are typically much smaller in bryozoans.

Studying biomineralizational processes *in-vivo* is notoriously difficult because of the necessity of working at extremely small scales while not perturbing sensitive processes during the collection of data. These problems are amplified in bryozoans by the challenges culturing living colonies in the laboratory of all but the most eurytopic species from nearshore habitats; put simply, bryozoans are not good 'lab rats'.

Technical issues aside, Taylor, Lombardi, and Cocito (2015) identified various key questions about bryozoan biomineralization, of which the following are worth highlighting: (i) what is the relationship between skeletal ultrastructural fabric and the organic and inorganic chemistry of the skeleton; (ii) how are transitions between different ultrastructural fabrics within bryozoan skeletal walls controlled; (iii) are specific ultrastructural fabrics indicative of slow or rapid skeletal growth; (iv) has bryozoan biomineralization been affected significantly by long-term (geological) changes in the environment (e.g. calcite versus aragonite seas); (v) what has been the effect on bryozoan biomineralization of episodes of ocean acidification in the geological past; and (vi) how resilient will bryozoans be to future environmental changes, especially global warming and ocean acidification? In addition, nothing is known about how different skeletal ultrastructures change during diagenesis and therefore how they may differ in modern and fossil skeletons.

10.2 Polymorphism

The ability of bryozoans to bud various kinds of polymorphic zooids within the constraint of a single genotype is intriguing. It is assumed that polymorphism entails the switching on and off of developmental genes, but the genes that control the often extreme differences between bryozoan polymorphs have yet to be identified. In addition, the extent to which the activity of these genes is prompted by environmental cues requires investigation. Better knowledge of triggers would not only provide insights into the significance of using polymorph proportions in palaeoenvironmental reconstructions, but also clarify the taxonomic value of polymorphs such as avicularia which are distributed patchily among some groups of cheilostomes.

With regard to fossil bryozoans, there is considerable scope for research on the temporal pattern of polymorph differentiation, comparable to the pioneering work of Cheetham (1973) which revealed progressive differentiation of avicularia from autozooids in species of *Poricellaria* and *Vincularia*, as well as one instance of avicularium reduction through time.

Finally, it is remarkable that the specific functions of cheilostome avicularia are still poorly understood in the great majority of species. This deficiency reflects the paucity of observations of avicularian behaviour in living colonies in their natural habitats.

10.3 Environmental Distributions of Colony-forms

For anyone collecting fossil bryozoans, the shape of the colony is the most striking feature. Different colony-forms represent alternative functional strategies for surviving in benthic communities, but what do individual colony-forms mean in terms of palaeoenvironments? Is it possible to use colony-form to reconstruct ancient seafloor environments, and to infer such parameters as current speed, depositional depth, rate of sedimentation, salinity, nutrient levels, etc.? Unfortunately, the environmental distributions of colony-forms in bryozoans living today are not well enough known to provide a satisfactory answer to this question. Some of the information needed to rectify this failing is undoubtedly available in the marine biological and oceanographical literature but awaits mining and analysis (cf. Achilleos, Smith, and Gordon 2019), while the application of benthic data-loggers and ROVs should lead to future advances.

10.4 Taphonomy

The taphonomy of bryozoans has been seriously neglected. Questions remain about, for example, how erect colonies are broken up prior to burial, the effects on bryozoans of various sizes, shapes, and mineralogies of passage through the taphonomically active zone where pH levels in the sediment decline and calcareous skeletons become vulnerable to dissolution, and the impact of bioturbating animals on the buried skeletal remains of bryozoans. Taphonomic filters are surely important in determining exactly which bryozoans survive into the fossil record and therefore have an impact on palaeoecological and evolutionary interpretations.

10.5 Small Palaeozoic Bryozoans

Research on Palaeozoic bryozoans has tended, not surprisingly, to have been focused on taxa with large and conspicuous colonies. Far less attention has been directed towards small species, including those with delicate erect colonies as well as thin encrusters. One issue is the difficulty in studying the internal morphology of small bryozoans using

thin sections, although silicified material can occasionally reveal external morphology in exquisite detail (e.g. Buttler 1994).

Aside from obtaining a better understanding of the palaeoecology of Palaeozoic benthic biotas, greater knowledge of small bryozoans is essential for phylogenetic reasons. Cyclostomes, in particular, have a paltry fossil record in the late Palaeozoic: 'advanced' forms with porous skeletons resembling post-Palaeozoic taxa are unknown in the Carboniferous and Permian where only primitive corynotrypids have been recorded to date. The absence in the late Palaeozoic fossil record of these porous cyclostomes has raised doubts over the close relationship between Palaeozoic cyclostomes and the diverse cyclostomes of the post-Palaeozoic (Ernst and Schäfer 2006).

10.6 Phylogeny and Classification

Phylogenetic studies have begun to reveal some alarming discordances between morphological trees and molecular trees based on the evidence of multiple genes, bringing into question the value of the morphological characters traditionally employed by systematists as indicators of phylogenetic affinity and which underlie taxonomic classifications. This is particularly clear in cyclostome bryozoans where both skeletal organization (fixed- vs. free-walled) and gonozooid morphology have been found wanting. With regard to freshwater phylactolaemates, characters of the statoblasts, which are routinely used for recognizing species, are discordant with the molecular phylogeny in Plumatellidae, the most diverse phylactolaemate family (Hirose, Dick, and Mawatari 2011). These results need not negate the use of skeletal organization and gonozooid morphology or statoblast characters for distinguishing between cyclostome and plumatellid species, respectively, but they do undermine the value of these characters in recognizing clades and therefore as a basis for supraspecific taxonomy.

Even for cheilostomes, in which the larger number and complexity of skeletal morphological characters has led to greater confidence in the phylogenetic value of morphology, an increasing number of discordances are being revealed between molecular trees and morphology that are best explained by previously unexpected morphological homoplasies.

Sequence data is currently available for only a small proportion of living bryozoan species. Much work needs to be done to construct a comprehensive, robust phylogenetic tree for the phylum. Integrating fossils into this tree will be a further challenge but one which is fundamental to our understanding of the evolutionary history of bryozoans in deep time.

References

Abbott, S.T. (1997) Mid-cycle condensed shell-beds from mid-Pleistocene cyclothems, New Zealand: implications for sequence architecture. *Sedimentology* **44**, 805–824.

Achilleos, K., Smith, A.M. & Gordon, D.P. (2019) The articulated bryozoan genus *Cellaria* in the southern Zealandian Region: distribution and associated fauna. *Marine Biodiversity* **49**, 2801–2812.

Adachi, N., Ezaki, Y. & Liu, J. (2011) Early Ordovician shift in reef construction from microbial to metazoan reefs. *Palaios* **26**, 106–114.

Adachi, N., Ezaki, Y. & Liu, J. (2012) The oldest bryozoan reefs: a unique Early Ordovician skeletal framework construction. *Lethaia* **45**, 14–23.

Adachi, N., Liu, J. & Ezaki, Y. (2013) Early Ordovician reefs in South China (Chenjiahe section, Hubei Province): deciphering the early evolution of skeletal-dominated reefs. *Facies* **59**, 451–466.

Addadi, L., Raz, S. & Weiner, S. (2003) Taking advantage of disorder: amorphous calcium carbonate and its role in biomineralization. *Advanced Materials* **15**, 959–970.

Agegian, C.R. & Mackenzie, F.T. (1989) Calcareous organisms and sediment mineralogy on a mid-depth bank in the Hawaiian Archipelago. *Pacific Science* **43**, 56–66.

Ager, D.V. (1961) The epifauna of a Devonian spiriferid. *Quarterly Journal of the Geological Society of London* **117**, 1–10.

Ahr, W.M. & Stanton, R.J., Jr (1994) Comparative sedimentology and paleontology of Waulsortian mounds and coeval level-bottom sediments, lower Lake Valley Formation (lower Mississippian), Sacramento Mountains, New Mexico. *Verhandlungen Geologischen Bundesanstalt, Wien* **50**, 11–24.

Alexander, R.R. & Scharf, C.D. (1990) Epizoans on Late Ordovician brachiopods from southeastern Indiana. *Historical Biology* **4**, 179–202.

Allen, G.R., Erdmann, M.V. & Dita Cahyani, N.K. (2016) *Sueviota bryozophila*, a new species of coral-reef goby from Indonesia (Teleostei: Gobiidae). *Journal of the Ocean Science Foundation* **20**, 76–82.

Alvarez, F. & Taylor, P.D. (1987) Epizoan ecology and interactions in the Devonian of Spain. *Palaeogeography, Palaeoclimatology, Palaeoecology* **61**, 17–31.

Alves, D.F.R., Barros-Alves, S.P., Lima, D.J.M., Cobo, V.J. and Negreiros-Fransozo, M.L. (2013) Brachyuran and anomuran crabs associated with *Schizoporella unicornis* (Ectoprocta, Cheilostomata) from southeastern Brazil. *Anais da Academia Brasileira de Ciências* **85**, 245–256.

Amado-Filho, G.M., Moura, R.L., Bastos, A.C., Salgado, L.T., Sumida, P.Y., Guth, A.Z. et al. (2012) Rhodolith beds are major $CaCO_3$ bio-factories in the tropical South West Atlantic. *PLoS ONE* **7**(4), e35171.

Anstey, R.L. (1987) Colony patterning and functional morphology of water flow in Paleozoic stenolaemate bryozoans. Pp. 1–8 in Ross, J.R.P. (ed.) *Bryozoa: present and past.* Western Washington University, Bellingham.

Anstey, R.L. (1989) Zooid orientation structures and water flow patterns in Paleozoic bryozoan colonies. *Lethaia* **14**, 287–302.

Anstey, R.L. (1990) Bryozoans. Pp. 232–252 in McNamara, K.J. (ed.) *Evolutionary trends.* Belhaven Press, London.

Anstey, R.L. & Perry, T.G. (1972) Eden Shale bryozoans: a numerical study (Ordovician, Ohio Valley). *Michigan State University, Publications of the Museum, Paleontological Series* **1**, 1–80.

Atkins, D. (1955) The cyphonautes larvae of the Plymouth area and the metamorphosis of *Membranipora membranacea* (L.). *Journal of the Marine Biological Association of the United Kingdom* **34**, 441–449.

Atkinson, D. (1994) Temperature and organism size – a biological law for ectotherms? *Advances in Ecology* **25**, 1–58.

Bader, B. (2000) Life cycle, growth rate, and carbonate production of *Cellaria sinuosa*. Pp. 136–144 in Herrera Cubilla, A. & Jackson, J.B.C (eds) *Proceedings of the 11th International Bryozoology Association Conference.* Smithsonian Tropical Research Institute, Balboa.

Bader, B. & Schäfer, P. (2004) Skeletal morphogenesis and growth checks in the Antarctic bryozoan *Melicerita obliqua. Journal of Natural History* **38**, 2901–2922.

Bader, B. & Schäfer, P. (2005) Bryozoans in polar latitudes: Arctic and Antarctic bryozoan communities and facies. *Denisia* **16**, 263–282.

Balduzzi, A., Barbieri, M. & Gristina, M. (1991) Morphology and life strategies of *Aetea* (Bryozoa: Cheilostomata) living on some western Mediterranean *Posidonia oceanica* meadows. *Bulletin de la Société des sciences naturelles de l'Ouest de la France, Mémoire* **H.S. 1**, 1–12.

Balson, P.S. & Taylor, P.D. (1982) Palaeobiology and systematics of large cyclostome bryozoans from the Pliocene Coralline Crag of Suffolk. *Palaeontology* **25**, 529–554.

Bancroft, A.J. (1986) Ovicells in the Palaeozoic bryozoan order Fenestrata. *Palaeontology* **29**, 155–164.

Bancroft, A.J. (1988) Palaeocorynid-type appendages in Upper Palaeozoic fenestellid Bryozoa. *Palaeontology* **31**, 665–675.

Banta, W.C. (1973) Evolution of avicularia in cheilostome Bryozoa. Pp. 295–303 in Boardman, R.S., Cheetham, A.H. & Oliver, W.A. (eds) *Animal colonies: development and function through time.* Dowden, Hutchinson & Ross, Stroudsburg.

Banta, W.C. (1975) Origin and early evolution of cheilostome Bryozoa. *Documents des Laboratoires de Géologie de la Faculté des Sciences de Lyon, Hors Série* **3**, 565–582.

Banta, W.C. & Crosby, M.M. (1994) *Electra venturaensis*, a new species (Bryozoa: Cheilostomata: Membraniporidae) from southern California. *Proceedings of the Biological Society of Washington* **107**, 544–547.

Banta, W.C. & Holden, P.M. (1974) Bud size alone does not control zoid [sic] bifurcation in *Schizoporella unicornis floridana* (Bryozoa, Cheilostomata). *Chesapeake Science* **15**, 104–109.

Banta, W.C., Gray, N. & Gordon, D.P. (1997) A cryptocystal operculum and a new method of lophophore protrusion in the cheilostome bryozoan *Macropora levinseni. Invertebrate Biology* **116**, 161–170.

Banta, W.C., McKinney, F.K. & Zimmer, R.L. (1974) Bryozoan monticules; excurrent water outlets? *Science* **185**, 783–784.

Barnes, D.K.A. (1995) Sublittoral epifaunal communities at Signy Island, Antarctica. 2. Below the ice foot zone. *Marine Biology* **121**, 565–572.

Barnes, D.K.A. (2000) Life patterns of encrusting and erect Antarctic Cheilostomatida. Pp. 145–153 in Herrera Cubilla, A. & Jackson, J.B.C. (eds) *Proceedings of the 11th International Bryozoology Association Conference.* Smithsonian Tropical Research Institute, Balboa.

Barnes, D.K.A. (2002) Polarization of competition increases with latitude. *Proceedings of the Royal Society, Series B* **269**, 2061–2069.

Barnes, D.K.A. & Clarke, A. (1998) The ecology of an assemblage dominant: the encrusting bryozoan *Fenestulina rugula*. *Invertebrate Biology* **117**, 331–340.

Barnes, D.K.A. & Griffiths, H.J. (2008) Biodiversity and biogeography of southern temperate and polar bryozoans. *Global Ecology and Biogeography* **17**, 84–99.

Barnes, D.K.A., Webb, K. & Linse, K. (2006a) Growth rate and its variability in erect Antarctic bryozoans. *Polar Biology* **30**, 1069–1081.

Barnes, D.K.A., Webb, K. & Linse, K. (2006b) Slow growth of Antarctic bryozoans increases over 20 years and is anomalously high in 2003. *Marine Ecology Progress Series* **314**, 187–195.

Bartley, J.W. & Anstey, R.L. (1987) Growth of monilae in the Permian trepostome *Tabulipora carbonaria*: evidence for periodicity and a new model of stenolaemate wall calcification. Pp. 9–16 in Ross, J.R.P. (ed.) *Bryozoa: present and past*. Western Washington University, Bellingham.

Bassler, R.S. (1911) The early Paleozoic Bryozoa of the Baltic Provinces. *United States National Museum Bulletin* **77**, 1–382.

Bassler, R.S. (1952) Taxonomic notes on genera of fossil and Recent Bryozoa. *Journal of the Washington Academy of Science* **42**, 381–385.

Bassler, R.S. (1953) Bryozoa. Pp. 1–253 in Moore, R.C. (ed.) *Treatise on invertebrate paleontology, part G*. Geological Society of America and University of Kansas Press, Lawrence.

Bastos, A.C., Moura, R.L., Moraes, F.C., Vieira, L.S., Braga, J.C., Ramalho, L.V., Amado-Filho, J.M., Magdalena, U.R. & Webster, J.M. (2018) Bryozoans are major modern builders of South Atlantic oddly shaped reefs. *Scientific Reports* **8**(9638), 1–11.

Baud, A., Nakrem, H.A., Beauchamp, B., Beatty, T.W., Embry, A.F. & Henderson, C.M. (2008) Lower Triassic bryozoan beds from Ellesmere Island, High Arctic, Canada. *Polar Research* **21**, 428–440.

Bayer, M.M. & Todd, C.D. (1997) Evidence for zooid senescence in the marine bryozoan *Electra pilosa*. *Invertebrate Biology* **116**, 331–340.

Bayer, M.M., Todd, C.D., Hoyle, J.E. & Wilson, J.F.B. (1997) Wave-related abrasion induces formation of extended spines in a marine bryozoan. *Proceedings of the Royal Society, Series B* **264**, 1605–1611.

Belcher, A.M., Wu, X.H., Christensen, R.J., Hansma, P.K., Stucky, G.D. & Morse, D.E. (1996) Control of crystal phase switching and orientation by soluble mollusc-shell proteins. *Nature* **381**, 56–58.

Benedix, G., Jacob, D.E. & Taylor, P.D. (2014) Bimineralic bryozoan skeletons: a comparison of three modern genera. *Facies* **60**, 389–403.

Berge, J.A., Leinaas, H.P. & Sandøy, K. (1985) The solitary bryozoan, *Monobryozoon limicola*, Franzén (Ctenostomata), a comparison of mesocosm and field samples from Oslofjorden, Norway. *Sarsia* **70**, 91–94.

Berner, R.A. (1975) The role of magnesium in the crystal growth of calcite and aragonite from sea water. *Geochimica et Cosmochimica Acta* **39**, 489–504.

Berning, B. (2007) The Mediterranean bryozoan *Myriapora truncata* (Pallas, 1766): a potential indicator of (palaeo-) environmental conditions. *Lethaia* **40**, 221–232.

Berning, B. (2008) Evidence for sublethal predation and regeneration among living and fossil ascophoran bryozoans. *Virginia Museum of Natural History, Special Publication* **15**, 1–7.

Berning, B., Harmelin, J.-G. & Bader, B. (2017) New Cheilostomata (Bryozoa) from NE Atlantic seamounts, islands, and the continental slope: evidence for deep-sea endemism. *European Journal of Taxonomy* **347**, 1–51.

Berning, B. & Kuklinski, P. (2008) North-east Atlantic and Mediterranean species of the genus *Buffonellaria* (Bryozoa, Cheilostomata): implications for biodiversity and biogeography. *Zoological Journal of the Linnean Society* **152**, 537–566.

Berning, B., Reuter, M., Piller, W.E., Harzhauser, M. & Kroh, A. (2009) Larger foraminifera as a substratum for bryozoans (Late Oligocene, Tethyan Seaway, Iran). *Facies* **55**, 227–241.

Berning, B., Tilbrook, K.J. & Ostrovsky, A.N. (2014) What, if anything, is a lyrula? *Studi Trentini di Scienze Naturali* **94**, 11–20.

Berry, M.S. & Hayward, P.J. (1984) Nervous and behavioral response to light in colonies of the free-living bryozoan, *Selenaria maculata* (Busk). *Experientia* **40**, 108–110.

Berthelsen, O. (1962) Cheilostome Bryozoa in the Danian deposits of east Denmark. *Danmarks Geologiske Undersøgelse* **83**, 1–290.

Best, B.A. & Winston, J.E. (1984) Skeletal strength of encrusting cheilostome bryozoans. *Biological Bulletin* **167**, 390–409.

Best, M.A. & Thorpe, J.P. (1985) Autoradiographic study of feeding and the colonial transport of metabolites in the marine bryozoan *Membranipora membranacea*. *Marine Biology* **84**, 295–300.

Best, M.A. & Thorpe, J.P. (1991) The uptake of ^{14}C radioisotope tracer from dissolved glycine and glucose by the marine bryozoan *Flustrellidra hispida* (Fabricius). *Bulletin de la Société des Sciences Naturelles de l'Ouest de la France, Mémoire* **H.S. 1**, 13–23.

Best, M.A. & Thorpe, J.P. (1996) The effect of suspended particulate matter (silt) on the feeding activity of the intertidal ctenostomate bryozoan *Flustrellidra hispida* (Fabricius). Pp. 39–45 in Gordon, D.P., Smith, A.M. & Grant-Mackie, J.A. (eds) *Bryozoans in space and time*. NIWA, Wellington.

Best, M.A. & Thorpe, J.P. (2002) Use of radioactively labelled food to assess the role of the funicular system in the transport of metabolites in the cheilostome bryozoan *Membranipora membranacea* (L.). Pp. 29–35 in Wyse Jackson, P.N., Buttler, C.J. & Spencer Jones, M.E. (eds) *Bryozoan studies 2001: Proceedings of the 12th International Bryozoology Association Symposium*. Balkema, Lisse.

Bigey, F. (1981) Overgrowths in Palaeozoic Bryozoa: examples from Devonian forms. Pp. 7–17 in Larwood, G.P. & Nielsen, C. (eds) *Recent and fossil Bryozoa*. Olsen & Olsen, Fredensborg.

Bigey, F. (1985) Biogeography of Devonian Bryozoa. Pp. 9–23 in Nielsen, C. & Larwood, G.P. (eds) *Bryozoa: Ordovician to Recent*. Olsen & Olsen, Fredensborg.

Bijma, J. & Boekschoten, G.J. (1985) Recent bryozoan reefs and stromatolite development in brackish inland lakes, SW Netherlands. *Senckenbergiana maritima* **17**, 163–185.

Bishop, J.D.D. (1988) Disarticulated bivalve shells as substrates for encrustation by the bryozoan *Cribrilina puncturata* in the Plio-Pleistocene Red Crag of eastern England. *Palaeontology* **31**, 237–253.

Bishop, J.D.D. (1989) Colony form and the exploitation of spatial refuges by encrusting Bryozoa. *Biological Reviews* **64**, 197–218.

Bishop, J.D.D. (1998) Fertilization in the sea: are the hazards of broadcast spawning avoided when free–spawned sperm fertilize retained eggs? *Proceedings of the Royal Society, Series B* **265**, 725–731.

Bishop, J.D.D., Manríquez, P.H. & Hughes, R.N. (1989) Water-borne sperm trigger vitellogenic egg growth in two sessile marine invertebrates. *Proceedings of the Royal Society, Series B* **267**, 1167–1169.

Blake, D.B. (1973) Acanthopore morphology and function in the bryozoan family Rhabdomesidae. *Journal of Paleontology* **47**, 421–435.

Blake, D.B. (1979) The Arthrostylidae and articulated growth habits in Paleozoic bryozoans. Pp. 337–344 in Larwood, G.P. & Abbott, M.B. (eds) *Advances in bryozoology*. Academic Press, London.

Blake, D.B. (1980) Homeomorphy in Paleozoic bryozoans: a search for explanations. *Paleobiology* **6**, 451–465.

Blake, D.B. (1983) Introduction to the Suborder Rhabdomesina. Pp. 530–549 in Boardman, R.S., Cheetham, A.H., Blake, D.B., Utgaard, J., Karklins, O.L., Cook, P.L., Sandberg, P.A., Lutaud, G. & Wood, T.S. (eds) *Bryozoa: treatise on invertebrate paleontology, part G, revised*. Geological Society of America and University of Kansas, Boulder and Lawrence.

Boardman, R.S. (1971) Mode of growth and functional morphology of autozooids in some Recent and Paleozoic tubular Bryozoa. *Smithsonian Contributions to Paleobiology* **8**, 1–51.

Boardman, R.S. (1983) General features of the class Stenolaemata. Pp. 49–137 in Boardman, R.S., Cheetham, A.H., Blake, D.B., Utgaard, J., Karklins, O.L., Cook, P.L., Sandberg, P.A., Lutaud, G. & Wood, T.S. (eds) *Bryozoa: treatise on invertebrate paleontology, part G, revised*. Geological Society of America and University of Kansas, Boulder and Lawrence.

Boardman, R.S. (1984) Origin of the post-Triassic Stenolaemata (Bryozoa): a taxonomic oversight. *Journal of Paleontology* **58**, 19–39.

Boardman, R.S. (1999) Indications of polypides in feeding zooids and polymorphs in Lower Paleozoic Trepostomata (Bryozoa). *Journal of Paleontology* **73**, 803–815.

Boardman, R.S. (2001) The growth and function of skeletal diaphragms in the colony life of Lower Paleozoic Trepostomata (Bryozoa). *Journal of Paleontology* **75**, 225–240.

Boardman, R.S. & Buttler, C. (2005) Zooids and extrazooidal skeleton in the Order Trepostomata (Bryozoa). *Journal of Paleontology* **79**, 1088–1104.

Boardman, R.S. & Cheetham, A.H. (1973) Degrees of colony dominance in stenolaemate and gymnolaemate Bryozoa. Pp. 121–220 in Boardman, R.S., Cheetham, A.H. and Oliver, W.J. (eds) *Animal colonies*. Dowden, Hutchinson and Ross, Stroudsburg.

Boardman, R.S. & McKinney, F.K. (1976) Skeletal architecture and preserved organs of four-sided zooids in convergent genera of Paleozoic Trepostomata (Bryozoa). *Journal of Paleontology* **50**, 25–78.

Boardman, R.S., McKinney, F.K. & Taylor, P.D. (1992) Morphology, anatomy, and taxonomy of the Cinctiporidae, new family (Bryozoa: Stenolaemata). *Smithsonian Contributions in Paleobiology* **70**, 1–81.

Bobin, G. (1968) Morphogenèse du termen et des épines dans les zoécies d'*Electra verticillata* (Ellis et Solander) (Bryozoaire Chilostome, Anasca). *Cahiers de Biologie Marine* **9**, 53–68.

Bobin, G. (1977) Interzooecial communications and the funicular system. Pp. 307–333 in Woollacott, R.M. & Zimmer, R.L. (eds) *Biology of bryozoans*. Academic Press, New York.

Bock, P.E. & Cook, P.L. (1994) Occurrence of three phases of growth with taxonomically distinct zooid morphologies. Pp. 33–36 in Hayward, P.J., Ryland, J.S. & Taylor, P.D. (eds) *Biology and palaeobiology of bryozoans*. Olsen & Olsen, Fredensborg.

Bock, P.E. & Cook, P.L. (2004a) Dimorphic brooding zooids in the genus *Adeona* Lamouroux from Australia (Bryozoa: Cheilostomata). *Memoirs of Museum Victoria* **61**, 129–133.

Bock, P.E. & Cook, P.L. (2004b) A review of Australian Conescharellinidae (Bryozoa: Cheilostomata). *Memoirs of Museum Victoria* **61**, 135–182.

Bock, P.E. & Gordon, D.P. (2013) Phylum Bryozoa Ehrenberg, 1831. *Zootaxa* **3703**, 67–74.

Boero, F., Bouillon, J. & Gravili, C. (2000) A survey of *Zanclea*, *Halocoryne* and *Zanclella* (Cnidaria, Hydrozoa, Anthomedusae, Zancleidae) with description of new species. *Italian Journal of Zoology* **67**, 93–124.

Bone, Y. (1991) Population explosion of the bryozoan *Membranipora aciculata* in the Coorong Lagoon in late 1989. *Australian Journal of Earth Sciences* **38**, 121–123.

Bone, Y. & James, N. P. (1997) Bryozoan stable isotope survey from the cool-water Lacepede Shelf, southern Australia. *SEPM Special Publication* **56**, 93–105.

Bone, E.K. & Keough, M.J. (2005) Responses to damage in an arborescent bryozoan: effects of injury location. *Journal of Experimental Marine Biology and Ecology* **324**, 127–140.

Bone, E.K. & Keough, M.J. (2010) Does polymorphism predict physiological connectedness? A test using two encrusting bryozoans. *Biological Bulletin* **219**, 220–230.

Borg, F. (1926a) On the body-wall in Bryozoa. *Quarterly Journal of Microscopical Science* **70**, 583–598.

Borg, F. (1926b) Studies on Recent cyclostomatous Bryozoa. *Zoologiska Bidrag från Uppsala* **10**, 181–507.

Borg, F. (1965) A comparative and phyletic study on fossil and recent Bryozoa of the suborders Cyclostomata and Trepostomata. *Arkiv för Zoologi* **17**, 1–91.

Borszcz, T., Kuklinski, P. & Zatoń, M. (2013) Encrustation patterns on Late Cretaceous (Turonian) echinoids from southern Poland. *Facies* **59**, 299–318.

Botquelen, A. & Mayoral, E. (2005) Early Devonian bioerosion in the Rade de Brest, Armorican Massif, France. *Palaeontology* **48**, 1057–1064.

Bowden, D.A. (2005) Seasonality of recruitment in Antarctic sessile marine benthos. *Marine Ecology Progress Series* **297**, 101–118.

Bowden, D.A., Clarke, A., Peck, L.S. & Barnes, D.K.A. (2006) Antarctic sessile marine benthos: colonisation and growth on artificial substrata over three years. *Marine Ecology Progress Series* **316**, 1–16.

Brachert, T.C., Betzler, C., Davies, P.J. & Feary, D.A. (1993) Climatic change: control of carbonate platform development (Eocene–Miocene, Leg 133, northeastern Australia). *Proceedings of the Ocean Drilling Program, Scientific Results* **133**, 291–300.

Bradstock, M. & Gordon, D.P. (1983) Coral-like bryozoan growths in Tasman Bay, and their protection to conserve commercial fish stocks. *New Zealand Journal of Marine and Freshwater Research* **17**, 159–163.

Braga, G. & Crihan, I.M. (2006) Up-dating of the taxonomy, stratigraphy and palaeoecology of Bryozoa rich sediments from Mera (NW Transylvania-Romania). *Courier Forschungsinstitut Senckenberg* **257**, 21–33.

Brancato, M.S. & Woollacott, R.M. (1982) Effect of microbial films on settlement of bryozoan larvae (*Bugula simplex*, *B. stolonifera* and *B. turrita*). *Marine Biology* **71**, 51–56.

Breton, G. (2017) Les sclérobiontes des huîtres du Cénomanien supérieur du Mans (Sarthe, France). *Annales de Paléontologie* **103**, 173–183.

Brett, C.E. & Liddell, W.D. (1978) Preservation and paleoecology of a Middle Ordovician hardground community. *Paleobiology* **4**, 329–348.

Brey, T., Gerdes, D., Gutt, J., Mackensen, A. & Starmans, A. (1998) Growth and productivity of the high Antarctic bryozoan *Melicerita obliqua*. *Marine Biology* **132**, 327–333.

Brey, T., Gerdes, D., Gutt, J., Mackensen, A. & Starmans, A. (1999) Growth and age of the Antarctic bryozoan *Cellaria incula* on the Weddell Sea shelf. *Antarctic Science* **11**, 408–414.

Brice, D. & Mistiaen, B. (1992) Épizoaires des brachiopodes Frasniens de Ferques (Boulonnais, Nord de la France). *Geobios Mémorie Special* **14**, 45–58.

Bromley, R.G. (2004) A stratigraphy of marine bioerosion. *Geological Society of London, Special Publication* **228**, 455–479.

Brood, K. (1972) Cyclostomatous Bryozoa from the Upper Cretaceous and Danian in Scandinavia. *Stockholm Contributions in Geology* **26**, 1–464.

Brood, K. (1976a) Bryozoan palaeoecology in the Late Silurian of Gotland. *Palaeogeography, Palaeoclimatology, Palaeoecology* **20**, 187–208.

Brood, K. (1976b) Cyclostomatous Bryozoa from the coastal waters of East Africa. *Zoologica Scripta* **5**, 277–300.

Brood, K. (1976c) Wall structure and evolution in cyclostomate Bryozoa. *Lethaia* **9**, 377–389.

Bryan, S.E., Cook, A.G., Evans, J.P., Hebden, K., Hurrey, L., Colls, P., Jell, J.S., Weatherley, D. & Firn, J. (2012) Rapid, long-distance dispersal by pumice rafting. *PLoS ONE* **7**, e40583.

Buge, E. (1979) Campagne de la Calypso au large des côtes Atlantiques de l'Amérique du Sud (1961–1962). 1. 34. Bryozoaires cyclostomes. *Annales de l'Institut océanographique, Paris* **55** (Suppl.), 207–261.

Buge, E & Fischer, J.C. (1970) *Atractosoecia incrustans* (d'Orbigny) (Bryozoa Cyclostomata) espèce bathonienne symbiotique d'un Pagure. *Bulletin de la Société Géologique de France, Paris, Série 7,* **12**, 126–133.

Buge, E. & Voigt, E. (1972) Les *Cellulipora* (Bryozoa, Cyclostomata) du Cénomanien française et la famille des Celluliporidae. *Geobios* **5**, 121–150.

Burgess, S.C., Ryan, W.H., Blackstone, N.W., Edmunds, P.J., Hoogenboom, M.O., Levitan, D.R. & Wulff, J.L. (2017) Metabolic scaling in modular animals. *Invertebrate Biology* **136**, 456–472.

Buss, L.W. (1979) Habitat selection, directional growth and spatial refuges: why colonial animals have more hiding places. Pp. 459–497 in Larwood, G. & Rosen, B.R. (eds) *Biology and systematics of colonial organisms*. Academic Press, London.

Buss, L.W. (1980) Bryozoan overgrowth interactions – the interdependence of competition of space and food. *Nature* **281**, 475–477.

Buss, L.W. (1981) Mechanisms of competition between *Onychocella alula* (Hastings) and *Antropora tincta* (Hastings) on an Eastern Pacific rocky shoreline. Pp. 39–49 in Larwood, G.P. & Nielsen, C. (eds) *Recent and Fossil Bryozoa*. Olsen & Olsen, Fredensborg.

Buss, L.W. & Jackson, J.B.C. (1979) Competitive networks: nontransitive competitive relationships in cryptic coral reef evironments. *American Naturalist* **113**, 223–234.

Buttler, C.J. (1989) New information on the ecology and skeletal ultrastructure of the Ordovician cyclostome bryozoan *Kukersella* Toots, 1952. *Paläontologishe Zeitschrift* **63**, 215–227.

Buttler, C.J. (1991) Possible brooding structure in rhinoporid cystoporate bryozoans. *Bulletin de la Societe des Sciences Naturelles de l'Ouest de la France Mémoire* **H.S. 1**, 61–70.

Buttler, C.J. (1994) Ordovician (Ashgill) micro-bryozoans from the Kildare Limestone, Ireland. Pp. 37–40 in Hayward, P.J., Ryland, J.S. & Taylor, P.D. (eds) *Biology and palaeobiology of bryozoans*. Olsen & Olsen, Fredensborg.

Buttler, C.J., Cherns, L. & Massa, D. (2007) Bryozoan mud-mounds from the Upper Ordovician Jifarah (Djeffara) Formation of Tripolitania, north-west Libya. *Palaeontology* **50**, 479–494.

Buttler, C.J. & Wilson, M.A. (2018) Paleoecology of an Upper Ordovician submarine cave-dwelling bryozoan fauna and its exposed equivalents in northern Kentucky, USA. *Journal of Paleontology* **92**, 568–576.

Buttler, C.J., Wyse Jackson, P.N., Ernst, A. & McKinney, F.K. (2013) A review of the Early Palaeozoic biogeography of bryozoans. *Geological Society, London, Memoirs* **38**, 145–155.

Cadée, G.C. & McKinney, F.K. (1994) A coral-bryozoan association from the Neogene of northwestern Europe. *Lethaia* **27**, 59–66.

Cahuzac, B. & d'Hondt, J.-L. (2017) D'abondantes populations de *Pectinatella magnifica* (Leidy, 1851) (Bryozoaires dulçaquicoles) dans les Landes, à Dax. Présentation illustrée de l'espèce. *Bulletin de la Société Linnéenne de Bordeaux* **45**, 265–287.

Cairns, S.D. & Barnard, J.L. (1984) Redescription of *Janaria mirabilis*, a calcified hydroid from the eastern Pacific. *Bulletin of the Southern California Academy of Sciences* **83**, 1–11.

Cancino, J.M., Castañeda, B. & Orellana, M.C. (1991) Reproductive strategies in bryozoans: experimental test of the effects of conspecific neighbours. *Bulletin de la Societe des Sciences Naturelles de l'Ouest de la France Mémoire* **H.S. 1**, 81–88.

Cancino, J.M. & Hughes, R.N. (1988) The zooidal polymorphism and astogeny of *Celleporella hyalina* (Bryozoa: Cheilostomata). *Journal of Zoology* **215**, 167–181.

Canu, F. & Bassler, R.S. (1926) Studies on the cyclostomatous Bryozoa. *Proceedings of the United States National Museum* **67**, 1–124.

Capel, K.C.C., Segal, B., Bertuol, P. & Lindner, A. (2012) Corallith beds at the edge of the tropical South Atlantic. *Coral Reefs* **31**, 75.

Carle, K.J. & Ruppert, E.E. (1983) Comparative ultrastructure of the bryozoan funiculus: a blood vessel homologue. *Zeitschrift für zoologische Systematik und Evolutionsforschung* **21**, 181–193.

Carlton J.T. (1985) Transoceanic and interoceanic dispersal of coastal marine organisms: the biology of ballast water. *Oceanography and Marine Biology: A Review* **23**, 313–371.

Carlton, J.T., Chapman, J.W.C., Geller, J.B., Miller, J.A., Carlton, D.A., McCuller, M.I., Treneman, N., Steves, B.P. & Ruiz, G.M. (2017) Tsunami-driven rafting: Transoceanic species dispersal and implications for marine biogeography. *Science* **357**, 1402–1406.

Carson, R.J.M. (1978) Body wall morphology of *Pentapora foliacea* (Ellis and Solander) (Bryozoa, Cheilostomata). *Journal of Morphology* **156**, 39–52.

Carter, J.W. (1982) Natural history observations on the gastropod shell-using amphipod *Photis conchicola* Alderman, 1936. *Journal of Crustacean Biology* **2**, 328–341.

Carter, J.W., Carpenter, A.L., Foster, M.S. & Jessee, W.N. (1985) Benthic succession on an artificial reef designed to support a kelp-reef community. *Bulletin of Marine Science* **37**, 86–113.

Carter, M.C., Bishop, J.D.D., Evans, N.J. & Wood, C.A. (2010) Environmental influences on the formation and germination of hibernacula in the brackish-water bryozoan *Victorella pavida* (Ctenostomata: Victorellidae). *Journal of Experimental Marine Biology and Ecology* **383**, 89–95.

Carter, M.C. & Gordon, D.P. (2007) Substratum and morphometric relationships in the bryozoan genus *Odontoporella*, with a description of a new paguridean-symbiont species from New Zealand. *Zoological Science* **24**, 47–56.

Carter, M.C., Gordon, D.P. & Gardner, J.P.A. (2010) Polymorphism and variation in modular animals: morphometric and density analyses of bryozoan avicularia. *Marine Ecology Progress Series* **399**, 117–130.

Carter, M.C., Lidgard, S., Gordon, D.P. & Gardner, J.P.A. (2011) Functional innovation through vestigialization in a modular marine invertebrate. *Biological Journal of the Linnean Society* **104**, 63–74.

Cartwright, P. (2003) Developmental insights into the origin of complex colonial hydrozoans. *Integrative Comparative Biology* **43**, 82–86.

Casadío S., Nelson, C., Taylor, P.D, Griffin, M. & Gordon, D.P. (2010) West Antarctic Rift system: a possible New Zealand–Patagonia Oligocene paleobiogeographic link. *Ameghiniana* **47**, 129–132.

Centurión, R. & López Gappa, J. (2001) Bryozoan assemblages on hard substrata: species abundance distribution and competition for space. *Hydrobiologia* **658**, 329–341.

Chaix, C. & Cahuzac, B. (2005) Le genre *Culicia* (Scléractiniaire): systématique, écologie et biogéographie au Cénozoïque. *Eclogae Geolicae Helvetiae* **98**, 169–187.

Chaney, H.W. (1983) Histocompatibility in the cheilostome bryozoan *Thalamoporella californica*. *Transactions of the American Microscopical Society* **102**, 319–332.

Chaney, H.W., Soule, D.F. & Soule, J.D. (1989) Systematics and zoogeography of *Thalamoporella gothica* and its allied species (Bryozoa, Cheilostomata). *Bulletin of Marine Science* **45**, 338–355.

Chave, K.E. (1954) Aspects of the biogeochemistry of magnesium. 1. Calcareous marine organisms. *Journal of Geology* **62**, 266–283.

Checa, A.G., Jiménez-López, C., Rodríguez-Navarro, A. & Machado, J.P. (2007) Precipitation of aragonite by calcitic bivalves in Mg-enriched marine waters. *Marine Biology* **150**, 819–827.

Cheetham, A.H. (1967) Paleoclimatic significance of the bryozoan *Metrarabdotos: Transactions of the Gulf Coast Association of Geological Societies* **17**, 400–407.

Cheetham, A.H. (1971) Functional morphology and biofacies distribution of cheilostome Bryozoa in the Danian stage (Paleocene) of southern Scandinavia. *Smithsonian Contributions to Palaeobiology* **6**, 1–87.

Cheetham, A.H. (1972) Cheilostome Bryozoa of Late Eocene age from Eua, Tonga. *US Geological Survey Professional Paper* **640-E**, E1–E26.

Cheetham, A.H. (1973) Study of cheilostome polymorphism using principal components analysis. Pp. 385–409 in Larwood, G.P. (ed.) *Living and fossil Bryozoa*. Academic Press, London.

Cheetham, A.H. (1975) Preliminary report on Early Eocene Cheilostome bryozoans from site 308–leg 32, Deep-Sea Drilling Project. *Initial Report of the Deep Sea Drilling Project* **32**, 835–851.

Cheetham, A.H. (1986a) Branching, biomechanics and bryozoan evolution. *Proceedings of the Royal Society, Series B* **228**, 151–171.

Cheetham, A.H. (1986b) Tempo of evolution in a Neogene bryozoan: rates of morphologic change within and across species boundaries. *Paleobiology* **12**, 190–202.

Cheetham, A.H. & Hayek, L.-C. (1983) Geometric consequences of branching growth in adeoniform Bryozoa. *Paleobiology* **9**, 240–260.

Cheetham, A.H., Hayek, L.-C. & Thomsen, E. (1981) Growth models in fossil arborescent bryozoans. *Paleobiology* **7**, 68–86.

Cheetham, A.H., Jackson, J.B.C. & Sanner, J. (2001) Evolutionary significance of sexual and asexual modes of propagation in Neogene species of the bryozoan *Metrarabdotos* in tropical America. *Journal of Paleontology* **75**, 564–577.

Cheetham, A.H., Sanner, J. & Jackson, J.B.C. (2007) *Metrarabdotos* and related genera (Bryozoa: Cheilostomata) in the Late Paleogene and Neogene of tropical America. *Journal of Paleontology* **81** (Supplement 1), 1–91.

Cheetham, A.H., Sanner, J., Taylor, P.D. & Ostrovsky, A.N. (2006) Morphological differentiation of avicularia and the proliferation of species in the mid-Cretaceous *Wilbertopora*

Cheetham, 1954 (Bryozoa: Cheilostomata). *Journal of Paleontology* **80**, 49–71.

Cheetham, A.H. & Thomsen, E. (1981) Functional morphology of arborescent animals: strength and design of cheilostome bryozoan skeletons. *Paleobiology* **7**, 355–383.

Chimonides, P.J. & Cook, P.L. (1981) Observations on living colonies of *Selenaria* (Bryozoa, Cheilostomata) II. *Cahiers de Biologie Marine* **22**, 207–219.

Christ, N., Immenhauser, A., Wood, R.A., Darwich, K. & Niedermayr, A. (2015) Petrography and environmental controls on the formation of Phanerozoic marine carbonate hardgrounds. *Earth-Science Reviews* **151**, 176–226.

Christophersen, C. (1985) Secondary metabolites from marine bryozoans: a review. *Acta Chemica Scandinavica B* **39**, 517–529.

Cigliano, M., Cocito, S. & Gambi, M.C. (2007) Epibiosis of *Calpensia nobilis* (Esper) (Bryozoa: Cheilostomida) on *Posidonia oceanica* (L.) Delile rhizomes: effects on borer colonization and morpho-chronological features of the plant. *Aquatic Botany* **86**, 30–36.

Clapham, M.E. & Payne, J.L. (2011) Acidification, anoxia, and extinction: a multiple logistic regression analysis of extinction selectivity during the Middle and Late Permian. *Geology* **39**, 1059–1062.

Clark, G.F., Stark, J.S. & Johnston, E.L. (2017) Tolerance rather than competition leads to spatial dominance of an Antarctic bryozoan. *Journal of Experimental Marine Biology and Ecology* **486**, 222–229.

Clark, N., Williams, M., Okamura, B., Smellie, J., Nelson, A., Knowles, T., Taylor, P., Leng, M., Zalasiewicz, J. & Haywood, A. (2010) Early Pliocene Weddell Sea seasonality determined from bryozoans. *Stratigraphy* **7**, 199–206.

Clarke, A. & Lidgard, S. (2000) Spatial patterns of diversity in the sea: bryozoan species richness in the North Atlantic. *Journal of Animal Ecology* **69**, 799–814.

Clarkson, M.O., Kasemann, S.A., Wood, R.A., Lenton, T.M., Daines, S.J., Richoz, S.,

Ohnemueller, F., Meixner, A., Poulton, S.W. & Tipper, E.T. (2015) Ocean acidification and the Permo-Triassic mass extinction. *Science* **348**, 229–232.

Cleary, D. & Wyse Jackson, P.N. (2007) *Stenophragmidium* Bassler, 1952 (Trepostomida: Bryozoa) from the Mississippian of Ireland and Britain. *Irish Journal of Earth Sciences* **25**, 1–25.

Cobbold, E.S. (1931) Additional fossils from the Cambrian rocks of Comley, Shropshire. *Quarterly Journal of the Geological Society of London* **87**, 459–512.

Cobbold, E.S. & Pocock, R.W. (1934) The Cambrian area of Rushton (Shropshire). *Philosophical Transactions of the Royal Society, Series B* **223**, 305–409.

Cocito, S. (2004) Bioconstruction and biodiversity: their mutual influence. *Scientia Marina* **68** (Supplement 1), 137–144.

Cocito, S., Ferdeghini, F., Morri, C. & Bianchi, C.N. (2000) Patterns of bioconstruction in the cheilostome bryozoan *Schizoporella errata*: the influence of hydrodynamics and associated biota. *Marine Ecology Progress Series* **192**, 153–161.

Cocito, S., Novosel, M., Pasari, Z. & Key, M.M., Jr (2006) Growth of the bryozoan *Pentapora fascialis* (Cheilostomata, Ascophora) around submarine freshwater springs in the Adriatic Sea. *Linzer Biologie Beiträge* **38**, 15–24.

Cocito, S., Sgorbini, S. & Bianchi, C.N. (1998) Aspects of the biology of the bryozoan *Pentapora fascialis* in the northwestern Mediterranean. *Marine Biology* **131**, 73–82.

Cocks, L.R.M. & Fortey, R.A. (1990) Biogeography of Ordovician and Silurian faunas. *Memoirs of the Geological Society of London* **12**, 97–104.

Collen, J.D. (1979) Marine invertebrates from the Ross Ice Shelf, Antarctica. *Search* **10**, 274–275.

Collins, J.S.H. (2011) *Venipagurus mariae* gen. et. sp. nov. (Decapoda, Anomura, Venipaguroidea superfam. nov.) from the Lower Lias of Dorset, England. *Bulletin of the Mizunami Fossil Museum* **37**, 17–21.

Condra, G.E. & Elias, M.K. (1944) Study and revision of *Archimedes* (Hall). *Special Papers of the Geological Society of America* **53**, 1–243.

Conlan, K.E., Lenihan, H.S., Kvitek, R.G. & Oliver, J.S. (1998) Ice scour disturbance to benthic communities in the Canadian high Arctic. *Marine Ecology Progress Series* **166**, 1–16.

Conti, S. & Serpagli, E. (1987) Functional morphology of the cap-like apparatus in autozooids of a Paleozoic trepostome bryozoan. *Lethaia* **20**, 1–20.

Cook, P.L. (1963) Observations on live lunulitiform zoaria of Polyzoa. *Cahiers de Biologie Marine* **4**, 407–413.

Cook, P.L. (1965) Notes on some Polyzoa with conical zoaria. *Cahiers de Biologie Marine* **6**, 435–454.

Cook, P.L. (1966) Some 'sand fauna' Polyzoa (Bryozoa) from eastern Africa and the northern Indian Ocean. *Cahiers de Biologie Marine* **7**, 207–223.

Cook, P.L. (1968a) Observations on living Bryozoa. *Atti della Società Italiana di Scienze Naturali, e del Museo Civico di Storia Naturale Milano* **108**, 155–160.

Cook, P.L. (1968b) Polyzoa from West Africa, the Malacostega, Pt. I. *Bulletin of the British Museum (Natural History), Zoology Series* **16**, 115–160.

Cook, P.L. (1973) Settlement and early colony development in some Cheilostomata. Pp. 65–71 in Larwood, G.P. (ed.) *Living and fossil Bryozoa*. Academic Press, London.

Cook, P.L. (1977a) Colony-wide water currents in living Bryozoa. *Cahiers de Biologie Marine* **16**, 31–47.

Cook, P.L. (1977b) Early colony development in *Aetea* (Bryozoa). *American Zoologist* **17**, 55–61.

Cook, P.L. (1977c) The genus *Tremogasterina* Canu (Bryozoa, Cheilostomata). *Bulletin of the British Museum (Natural History). Zoology Series* **35**, 103–165.

Cook, P.L. (1979) Some problems in interpretation of heteromorphy and colony integration in

Bryozoa. Pp. 193–210 in Larwood, G. & Rosen, B.R. (eds) *The biology and systematics of colonial organisms.* Academic Press, London.

Cook, P.L. (1985) Bryozoa from Ghana: A preliminary survey. *Annales Musee r. de l'Afrique centrale, Sciences zoologiques, Tervuren* **238**, 1–315.

Cook, P.L. (2001) Notes on the genera *Nordgaardia* and *Uschakovia* (Bryozoa: Bugulidae). *Memoirs of Museum Victoria* **58**, 215–222.

Cook, P.L. & Bock, P.E. (1994) The astogeny and morphology of *Rhabdozoum wilsoni* Hincks (Anasca, Buguloidea). Pp. 47–50 in Hayward, P.J., Ryland, J.S. & Taylor, P.D. (eds) *Biology and palaeobiology of bryozoans.* Olsen & Olsen, Fredensborg.

Cook, P.L., Bock, P.E. & Gordon, D.P. (2018) Class Gymnolaemata, order Ctenostomata. Pp. 33–60 in Cook, P.L., Bock, P.E., Gordon, D.P. & Weaver, H. (eds) *Australian Bryozoa, volume 2: taxonomy of Australian families.* CSIRO Publishing, Melbourne.

Cook, P.L. & Chiminodes, P.J. (1978) Observations on living colonies of *Selenaria* (Bryozoa, Cheilostomata), I. *Cahiers de Biologie Marine* **19**, 147–158.

Cook, P.L. & Chimonides, P.J. (1980) Further observations on water current patterns in living Bryozoa. *Cahiers de Biologie Marine* **21**, 393–402.

Cook, P.L. & Chimonides, P.J. (1983) A short history of the lunulite Bryozoa. *Bulletin of Marine Science* **33**, 566–581.

Cook, P.L. & Chimonides, P J. (1985) Larval settlement and early astogeny of *Parmularia* (Cheilostomata). Pp. 71–78 in Nielsen, C. & Larwood, G.P. (eds) *Bryozoa: Ordovician to Recent.* Olsen & Olsen, Fredensborg.

Cook, P.L. & Chimonides, P.J. (1994a) Notes on the family Cupuladriidae (Bryozoa), and on *Cupuladria remota* sp. n. from the Marquesas Islands. *Zoologica Scripta* **23**, 251–268.

Cook, P.L. & Chimonides, P.J. (1994b) Notes on the genus *Anoteropora* (Bryozoa, Chcilostomatida). *Zoologica Scripta* **23**, 51–59.

Cook, P.L. & Lagaaij, R. (1976) Some Tertiary and Recent conescharelliniform Bryozoa. *Bulletin of the British Museum (Natural History), Zoology Series* **29**, 317–376.

Cook, P.L. & Voigt, E. (1986) *Pseudolunulites* gen. nov., a new kind of lunulitiform cheilostome from the Upper Oligocene of northern Germany (Bryozoa). *Verhandlungen des Naturwissenschaftlichen Vereins in Hamburg* **28**, 107–127.

Cornée, J.J., Moissette, P., Saint Martin, J.P., Kázmér, M., Tóth, E., Görög, A., Dulai, A. & Müller, P. (2009) Marine carbonate systems in the Sarmatian (Middle Miocene) of the Central Paratethys: the Zsambek Basin of Hungary. *Sedimentology* **56**, 1728–1750.

Corneliussen, E.F. & Perry, T.G. (1970) The ectoproct *Batostoma*? *cornula* (Cumings & Galloway) and its enigmatic intrazooecial spines (Fort Atkinson Limestone (Cincinnatian), Wilmington, Illinois). *Journal of Paleontology* **44**, 997–1008.

Courtney Mustaphi, C.J., Githumbi, E.N., Shotter, L.R., Rucina, S.M., Marchant, R. (2016) Subfossil statoblasts of *Lophopodella capensis* (Sollas, 1908) (Bryozoa, Phylactolaemata, Lophopodidae) in the Upper Pleistocene and Holocene sediments of a montane wetland, Eastern Mau Forest, Kenya. *African Invertebrates* **57**, 39–52.

Cowen, R. & Rider, J. (1972) Functional analysis of fenestellid bryozoan colonies. *Lethaia* **5**, 145–164.

Craig, S.F. (1994) Intraspecific fusion in the encrusting bryozoan *Fenestrulina* sp. Pp. 51–54 in Hayward, P.J., Ryland, J.S. & Taylor, P.D. (eds) *Biology and palaeobiology of bryozoans.* Olsen & Olsen, Fredensborg.

Craig, S.F., Slobodkin, L.B., Wray, G.A. & Biermann, C.H. (1997) The 'paradox' of polyembryony: a review of cases and a hypothesis for its evolution. *Evolutionary Ecology* **11**, 127–143.

Craig, S.F. & Wasson, K. (2000) Self/non-self recognition and fusion in the bryozoan *Hippodiplosia insculpta.* Pp. 189–196 in Herrera Cubilla, A. & Jackson, J.B.C. (eds) *Proceedings of the 11th International Bryozoology*

Association Conference. Smithsonian Tropical Research Institute, Balboa.

Creary, M.M. (2003) A simplified field guide to the bryozoan species found on the roots of the red mangrove (*Rhizophora mangle*) in and around Kingston Harbour, Jamaica, WI. *Bulletin of Marine Science* **73**, 521–526.

Crowley, S.F. & Taylor, P.D. (2000) Stable isotope composition of modern bryozoan skeletal carbonate from the Otago Shelf, New Zealand. *New Zealand Journal of Marine and Freshwater Research* **34**, 331–351.

Cuffey, R.J. (2006) Bryozoan-built reef mounds – the overview from integrating recent studies with previous investigations. *Courier Forschungsinstitut Senckenberg* **257**, 35–47.

Cuffey, R.J. & Blake, D.B. (1991) Cladistic analysis of the phylum Bryozoa. *Bulletin de la Société des Sciences Naturelle de l'Ouest de la France, Mémoire* **H.S. 1**, 97–108.

Cuffey, R.J. & Fine, R.L. (2005) The largest known fossil bryozoan reassembled from near Cincinnati. *Ohio Geology* **2005**(1), 1–4.

Cuffey, R.J., Cawley, J.C., Lane, J.A., Bernarsky-Remington, S.M., Ansari, S.L., McClain, M.D., Ross-Phillips, T.L. & Savill, A.C. (2000) Bryozoan reefs and bryozoan-rich limestones in the Ordovician of Virginia. *Proceedings 9th International Coral Reef Symposium* **1**, 6 pp.

Cuffey, R.J., Dodge, C.H. & Skema, V.W. (2014) *Demafinga pennsylvanica* gen. et sp. nov., the first-known fossilised soft-bodied fleshy-branched ctenostome bryozoan (mid-Pennsylvanian, northwest Pennsylvania). *Studi Trentini di Scienze Naturali* **94**, 33–38.

Cuffey, R.J., Xiao, C., Zhu, Z., Spjeldnaes, N. & Hu, Z.-X. (2012) The world's oldest-known bryozoan reefs: Late Tremadocian, mid-Early Ordovician; Yichang, Central China. Pp. 13–27 in Ernst, A., Schäfer, P. & Scholz, J. (eds) *Bryozoan studies 2010.* Lecture Notes in Earth System Sciences **143**. Springer, Berlin.

Cummings, S.G. (1975) Zoid [sic] regression in *Schizoporella unicornis floridana* (Bryozoa, Cheilostomata). *Chesapeake Science* **16**, 93–103.

Darrell, J.G. & Taylor, P.D. (1989) Scleractinian symbionts of hermit crabs in the Pliocene of Florida. *Memoir of the Association of Australasian Palaeontologists* **8**, 115–123.

Datillo, B.F., Freeman, R.L., Peters, W.S., Heimbrock, W.P., Deline, B., Martin, A.J., Kallmeyer, J.W., Reeder, J. & Argast, A. (2016) Giants among micromorphs: were Cincinnatian (Ordovician, Katian) small shelly phosphatic faunas dwarfed? *Palaios* **31**, 55–70.

David, L. & Pouyet, S. (1984) Stratégie adaptative des Bryozoaires Chilostomes abyssaux. *Geobios, Mémoire Spécial* **8**, 373–378.

Davidson, I.C., McCann, L.D., Fofonoff, P.W., Sytsma, M.D. & Ruiz, G.M. (2008) The potential for hull-mediated species transfers by obsolete ships on their final voyages. *Diversity and Distributions* **14**, 518–529.

Day, R.W. & Osman, R.W. (1981) Predation by *Patiria miniata* (Asteroidea) on bryozoans: prey diversity may depend on the mechanism of succession. *Oecologia* **51**, 300–309.

De Blauwe, H. & Faase, M. (2004) *Smittoidea prolifica* Osburn, 1952 (Bryozoa, Cheilostomatida), a Pacific bryozoan introduced to The Netherlands (Northeast Atlantic). *Bulletin de l'Institut Royal des Sciences naturelles de Belgique, Biologie* **74**, 33–39.

De Burgh, M.E. & Fankboner, P.V. (1978) A nutritional association between the bull kelp *Nereocystis luetkeana* and its epizootic bryozoan *Membranipora membranacea*. *Oikos* **31**, 69–72.

De Yoreo, J.J. & 14 others (2015) Crystallization by particle attachment in synthetic, biogenic, and geologic environments. *Science* **349**, aaa6760.

Denayer, J. (2018) From rolling stones to rolling reefs: a Devonian example of highly diverse macroids. *Lethaia* **51**, 564–580.

Denayer, J. & Aretz, M. (2012) Discovery of a Mississippian reef in Turkey: the Upper Viséan Microbial-Sponge-Bryozoan-Coral Bioherm from Kongul Yayla (Taurides, S. Turkey). *Turkish Journal of Earth Sciences* **21**, 375–389.

Dendy, A. (1889) On the anatomy of an arena-ceous Polyzoon. *Proceedings of the Royal Society of Victoria* n.s. **1**, 1–11.

Dendy, J.S. (1963) Observations on bryozoan ecology in farm ponds. *Limnology and Oceanography* **8**, 478–482.

Denisenko, N.V., Hayward, P.J., Tendal, O.S. & Sørensen, J. (2016) Diversity and biogeographic patterns of the bryozoan fauna of the Faroe Islands. *Marine Biology Research* **12**, 360–378.

Desrochers, A., Bourque, P.-A. & Neuweiler, F. (2007) Diagenetic versus biotic accretionary mechanisms of bryozoan–sponge buildups (Lower Silurian, Anticosti Island, Canada). *Journal of Sedimentary Research* **77**, 564–571.

Di Geronimo, I., Rosso, A. & Sanfilippo, R. (1992) Bryozoans as sedimentary instability indicators. *Rivista Italiana di Paleontologia e Stratigrafia* **98**, 229–242.

Di Geronimo, I., Rosso, A. & Sanfilippo, R. (1995) Circalittoral to infralittoral communities encrusting the Pleistocene gravels of Motta S. Giovanni (Reggio Calabria, Italy). *Geobios, Mémoire Speciale* **18**, 119–130.

Di Martino, E., Jackson, J.B.C., Taylor, P.D. & Johnson, K.G. (2018) Differences in extinction rates drove biogeographic patterns of tropical marine biodiversity. *Science Advances* **4**(4), eaaq1508.

Di Martino, E., Martha, S.O. & Taylor, P.D. (2018) The Madagascan Maastrichtian bryozo-ans of Ferdinand Canu – systematic revision and scanning electron microscopic study. *Annales de Paléontologie* **104**, 101–128.

Di Martino, E. & Taylor, P.D. (2013) First bryo-zoan fauna from a tropical Cretaceous carbon-ate: Simsima Formation, United Arab Emirates–Oman border region. Cretaceous. *Cretaceous Research* **43**, 80–96.

Di Martino, E. & Taylor, P.D. (2014) A brief review of seagrass-associated bryozoans, Recent and fossil. *Studi Trentini di Scienze Naturali* **94**, 79–94.

Di Martino, E., Taylor, P.D., Cotton, L.J. & Pearson, P.N. (2017) First bryozoan fauna from the Eocene–Oligocene transition in Tanzania.

Journal of Systematic Palaeontology **16**, 225–243.

Di Martino, E., Taylor, P.D. & Johnson, K.G. (2015) Bryozoan diversity in the Miocene of the Kutai Basin, East Kalimantan, Indonesia. *Palaios* **30**, 109–115.

Di Martino, E., Taylor, P.D., Kudryavstev, A.B. & Schopf, J.W. (2016) Calcitization of aragonitic bryozoans in Cenozoic tropical carbonates from East Kalimantan, Indonesia. *Facies* **62**, 11, 15 pp.

Dick, M.H. (2008) Unexpectedly high diversity of *Monoporella* (Bryozoa: Cheilostomata) in the Aleutian Islands, Alaska: taxonomy and distribution of six new species. *Zoological Science* **25**, 36–52.

Dick, M.H., Freeland, J.R., Williams, L.P. & Coggeshall-Burr, M. (2000) Use of 16S mito-chondrial ribosomal DNA sequences to investi-gate sister-group relationships among gymnolaemate bryozoans. Pp. 197–210 in Herrera Cubilla, A. & Jackson, J.B.C. (eds) *Proceedings of the 11th International Bryozoology Association Conference.* Smithsonian Tropical Research Institute, Balboa.

Dick, M.H. & Grischenko, A.V. (2017) Rocky-intertidal cheilostome bryozoans from the vicinity of the Sesoko Biological Station, west-central Okinawa, Japan. *Journal of Natural History* **51**, 141–266.

Dick, M.H., Grischenko, A.V. & Mawatari, S.F. (2005) Intertidal Bryozoa (Cheilostomata) of Ketchikan, Alaska. *Journal of Natural History* **39**, 3687–3784.

Dick, M.H., Hirose, M. & Mawatari, S.F. (2012) Molecular distance and morphological diver-gence in *Cauloramphus* (Cheilostomata: Calloporidae). Pp. 29–44 in Ernst, A., Schäfer, P. & Scholz, J. (eds) *Bryozoan studies 2010.* Lecture Notes in Earth System Sciences **143**. Springer, Berlin.

Dick, M.H., Hirose, M., Takashima, R., Ishimura, T., Nishi, H. & Mawatari, S.F. (2008) Application of MART analysis to infer paleosea-sonality in a Pleistocene shallow marine benthic environment. Pp. 93–99 in Okada, H., Mawatari, S.F., Suzuki, N. & Gautam, P. (eds) *Origin and evolution of natural diversity*. Proceedings of International Symposium 'The Origin and Evolution of Natural Diversity, 1–5 October 2007, University of Hokkaido, Sapporo.

Dick, M.H., Lidgard, S., Gordon, D.P. & Mawatari, S.F. (2009) The origin of ascophoran bryozoans was historically contingent but likely. *Proceedings of the Royal Society, Series B* **276**, 3141–3148.

Dick, M.H. & Mawatari, S.F. (2005) Morphological and molecular concordance of *Rhynchozoon* clades (Bryozoa, Cheilostomata) from Alaska. *Invertebrate Biology* **124**, 344–354.

Dick, M.H. & Ross, J.R.P. (1985) Intertidal cheilostome bryozoans in rock-pile habitat at Narrow Strait, Kodiak, Alaska. Pp. 87–93 in Nielsen, C. & Larwood, G.P. (eds) *Bryozoa: Ordovician to Recent*. Olsen & Olsen, Fredensborg.

Dick, M.H., Sakamoto, C. & Komatsu, T. (2018) Cheilostome Bryozoa from the Upper Cretaceous Himenoura Group, Kyushu, Japan. *Paleontological Research* **22**, 239–264.

Dick, M.H., Tilbrook, K.J. & Mawatari, S.F. (2007) Diversity and taxonomy of rocky-intertidal Bryozoa on the Island of Hawaii, USA. *Journal of Natural History* **40**, 2197–2258.

Dietz, L., Dömel, J.S., Leese, F., Lehmann, T. & Melzer, R.R. (2018) Feeding ecology in sea spiders (Arthropoda: Pycnogonida): what do we know? *Frontiers in Zoology* **15**(7), 16 pp.

Dobretsov, S. & Qian, P.-Y. (2006) Facilitation and inhibition of larval attachment of the bryozoan *Bugula neritina* in association with mono-species and multi-species biofilms. *Journal of Experimental Marine Biology and Ecology* **333**, 263–274.

Dove, P.M., De Yoreo, J.J. & Weiner, S. (2003) Biomineralization. *Reviews in Mineralogy and Geochemistry* **54**, 1–381.

Driscoll, E.G., Gibson, J.W. & Mitchell, S.W. (1971) Larval selection of substrate by the Bryozoa *Discoporella* and *Cupuladria*. *Hydrobiologia* **37**, 347–359.

Dudley, J.E. (1973) Observations on the reproduction, early larval development, and colony astogeny of *Conopeum tenuissimum* (Canu). *Chesapeake Science* **14**, 270–278.

Dunn, C.W., Giribet, G., Edgecombe, G.D. & Hejnol, A. (2014) Animal phylogeny and its evolutionary implications. *Annual Review of Ecology, Evolution and Systematics* **45**, 371–395.

Durante, K.M. & Fu-Shiang, C. (1991) Epiphytism on *Agarum fimbriatum*: can herbivore preferences explain distributions of epiphytic bryozoans? *Marine Ecology Progress Series* **77**, 279–287.

Dyrynda, P.E.J. (1981) A preliminary study of patterns of polypide generation-degeneration in marine cheilostome Bryozoa. Pp. 73–81 in Larwood, G.P. & Nielsen, C. (eds) *Recent and fossil Bryozoa*. Olsen and Olsen, Fredensborg.

Dyrynda, P.E.J. (1985) Functional allelochemistry in temperate waters: chemical defences of bryozoans. Pp. 95–100 in Larwood, G.P. & Nielsen, C. (eds) *Bryozoa: Ordovician to Recent*. Olsen and Olsen, Fredensborg.

Dyrynda, P.E.J. (1986) Defensive strategies of modular organisms. *Philosophical Transactions of the Royal Society, Series B* **313**, 227–243.

Dyrynda, P.E.J., Fairall, V.R., Occipinti Ambrogi, A. & d'Hondt, J.-L. (2000) The distribution, origins and taxonomy of *Tricellaria inopinata* d'Hondt and Occhipinti Ambrogi, 1985, an invasive bryozoan new to the Atlantic. *Journal of Natural History* **34**, 1993–2006.

Dzik. J. (1975) Evolutionary relationships of the early Palaeozoic 'cyclostomatous' Bryozoa. *Palaeontology* **24**, 827–861.

Dzik. J. (1981) The origin and early phylogeny of cheilostomatous Bryozoa. *Acta Palaeontologica Polonica* **20**, 395–423.

Dzik. J. (1991) Possible solitary bryozoan ancestors from the Early Palaeozoic and the affinities of the Tentaculita. *Bulletin de la Société des Sciences Naturelle de l'Ouest de la France, Mémoire* **H.S. 1**, 121–131.

Eckman, J.E. & Okamura, B. (1998) A model of particle capture by bryozoans in turbulent flow: significance of colony form. *The American Naturalist* **152**, 861–880.

Edwards, K.F. & Schreiber, S.J. (2010) Preemption of space can lead to intransitive coexistence. *Oikos* **119**, 1201–1209.

Eggleston, D. (1972a) Factors influencing the distribution of sub-littoral ectoprocts off the south of the Isle of Man (Irish Sea). *Journal of Natural History* **6**, 247–260.

Eggleston, D. (1972b) Patterns of reproduction in the marine Ectoprocta of the Isle of Man (Irish Sea). *Journal of Natural History* **6**, 31–38.

Eldredge, N. & Gould, S.J. (1972) Punctuated equilibria: an alternative to phyletic gradualism. Pp. 82–115 in Schopf, T.J.M. (ed.) *Models in paleobiology*. Freeman, Cooper & Co., San Francisco.

El-Sorogy, A.S. (2015) Bryozoan nodules as frame-builder of bryozoan-microreef, Middle Miocene sediments, Egypt. *Journal of Earth Science* **26**, 215–258.

Elias, M.K. (1954) *Cambroporella* and *Coeloclema*, Lower Cambrian and Ordovician bryozoans. *Journal of Paleontology* **28**, 52–58.

Elias, M.K. & Condra, G.E. (1957) *Fenestella* from the Permian of West Texas. *Memoirs of the Geological Society of America* **70**, 1–158.

Engeser, T. & Taylor, P.D. (1989) Supposed Triassic bryozoans in the Klipstein Collection from the Cassian Formation of the Italian Dolomites redescribed as calcified demosponges. *Bulletin of the British Museum (Natural History), Geology Series* **45**, 39–55.

Erickson, J.M. & Bouchard, T.D. (2003) Description and interpretation of *Sanctum laurentiensis*, new ichnogenus and ichnospecies, a domichnium mined into Late Ordovician (Cincinnatian) ramose bryozoan colonies. *Journal of Paleontology* **77**, 1002–1010.

Ernst, A. (2002) Systematics and biogeography of the Permian bryozoans in Europe. Pp. 109–112 in Wyse Jackson, P.N., Buttler, C.J. & Spencer Jones, M.E. (eds) *Bryozoan studies 2001: Proceedings of the 12th International Bryozoology Association Symposium*. Balkema, Lisse.

Ernst, A. (2013) Diversity dynamics and evolutionary patterns of Devonian Bryozoa. *Palaeobiodiversity and Palaeoenvironments* **93**, 45–63.

Ernst, A. (2016a) Bryozoan fauna from the Permian (Artinskian-Kungurian) Zhongba Formation of southwestern Tibet. *Palaeontologia Electronica* **19.2.**15A, 1–59.

Ernst, A. (2016b) *Fenestrapora* (Fenestrata, Bryozoa) from the Middle Devonian of Germany. *Paläontologische Zeitschrift* **90**, 19–32.

Ernst, A. (2017) Diversity dynamics of Ordovician Bryozoa. *Lethaia* **51**, 198–206.

Ernst, A., Bogolepova, O.K., Hubmann, B., Golubkova, E. Yu., Gubanov, A.P. (2014) *Dianulites* (Trepostomata, Bryozoa) from the Early Ordovician of Severnaya Zemlya, Arctic Russia. *Geological Magazine* **151**, 328–338.

Ernst, A. & Bohaty, J. (2009) *Schischcatella* (Fenestrata, Bryozoa) from the Devonian of the Rhenish Massif, Germany. *Palaeontology* **52**, 1291–1310.

Ernst, A., Bohaty, J. & Taylor, P.D. (2014) *Botryllopora* (Cystoporata, Bryozoa) from the Middle Devonian of Canada and Germany. *Studi Trentini di Scienze Naturali* **94**, 101–109.

Ernst, A., Fernández, L.P., Fernández-Martínez, E. & Vera, C. (2012a) Description of a bryozoan fauna from mud mounds of the Lebanza Formation (Lower Devonian) in the Arauz area (Pisuerga-Carrión Province, Cantabrian Zone, NW Spain). *Geodiversitas* **34**, 693–738.

Ernst, A. & Königshof, P. (2008) The role of bryozoans in fossil reefs – an example from the Middle Devonian of the Western Sahara. *Facies* **54**, 613–620.

Ernst, A., Königshof, P. & Schäfer, P. (2009) Unusual skeletal morphology and systematic description of a new Devonian cryptostome bryozoan from the Western Sahara. *Paläontologische Zeitschrift* **83**, 449–457.

Ernst, A. & May, A. (2009) Bryozoan fauna from the Koneprusy Limestone (Pragian, Lower Devonian) of Zlaty Kun near Kineprusy (Czech Republic). *Journal of Paleontology* **83**, 767–782.

Ernst, A. & Munnecke, A. (2009) A Hirnantian (latest Ordovician) reefal bryozoan fauna from Anticosti Island, eastern Canada: taxonomy and chemostratigraphy. *Canadian Journal of Earth Sciences* **46**, 207–229.

Ernst, A., Munnecke, A. & Oswald, I. (2015) Exceptional bryozoan assemblage of a microbial-dominated reef from the early Wenlock of Gotland, Sweden. *GFF* **137**, 1–24.

Ernst, A. & Nakrem, H.-A. (2011) Late Ordovician (Sandbian) bryozoans and their depositional environment, Furuberget Formation, Mjøsa District, Oslo Region, Norway. *Bulletin of Geosciences* **87**, 21–44.

Ernst, A. & Nakrem, H.-A. (2015) Bryozoans from the lower Silurian (Wenlock) Steinsfjorden Formation of Ringerike, southern Norway. *Bulletin of Geosciences* **90**, 65–87.

Ernst, A. & Schäfer, P. (2006) Palaeozoic vs. post-Palaeozoic Stenolaemata: phylogenetic relationship or morphological convergence? *Courier Forschungsinstitut Senckenberg* **257**, 49–63.

Ernst, A., Schäfer, P. & Grant-Mackie, J.A. (2015) New Caledonian Triassic Bryozoa. *Journal of Paleontology* **89**, 730–747.

Ernst, A., Seuss, B., Taylor, P.D. & Nützel, A. (2016) Bryozoan fauna of the Boggy Formation (Deese Group, Pennsylvanian) of the Buckhorn Asphalt Quarry, Oklahoma, USA. *Palaeobiodiversity and Palaeoenvironments* **96**, 517–540.

Ernst, A., Taylor, P.D., Bohaty, J. & Wyse Jackson, P.N. (2012b) Homeomorphy in *Lunostoma*, a new Middle Devonian cryptostome bryozoan. *Paläontologische Zeitschrift* **86**, 135–145.

Ernst, A., Taylor, P.D. & Bohatý, J. (2014) A new Middle Devonian cystoporate bryozoan from Germany containing a new symbiont bioclaustration. *Acta Palaeontologica Polonica* **59**, 173–183.

Ernst, A., Taylor, P.D. & Wilson, M.A. (2007) Ordovician bryozoans from the Kanosh Formation (Whiterockian) of Utah, USA. *Journal of Paleontology* **81**, 998–1008.

Ernst, A. & Vachard, D. (2017a) The cystoporate bryozoan *Glyptopora michelinia* (Prout, 1860) in the Viséan–Serpukhovian boundary interval of the Montagne Noire (southern France). *Paläontologische Zeitschrift* **91**, 207–216.

Ernst, A. & Vachard, D. (2017b) Middle Pennsylvanian bryozoans of Cerros de Tule, Sonora, Mexico. *Neues Jahrbuch für Geologie und Paläontologie Abhandlungen* **285**, 11–38.

Ernst, A. & Voigt, E. (2002) Zooidal anatomy in Ordovician and Carboniferous trepostome bryozoans. *Paläontologische Zeitschrift* **76**, 339–345.

Ernst, A., Weidlich, O. & Schäfer, P. (2008) Stenolaemate Bryozoa from the Permian of Oman (Aseelah Unit, Batain Coast). *Journal of Paleontology* **82**, 676–716.

Erwin, D.H. (2015) Early metazoan life: divergence, environment and ecology. *Philosophical Transactions of the Royal Society, Series B* **370**, 20150036.

Evans, S. & Todd, J.A. (1997) Late Jurassic soft-bodied wood epibionts preserved by bioimmuration. *Lethaia* **30**, 185–189.

Farmer, J.D. (1977) An adaptive model for the evolution of the ectoproct life cycle. Pp. 487–517 in Woollacott, R.M. & Zimmer, R.L. (eds) *Biology of bryozoans*. Academic Press, New York.

Farmer, J.D. (1979) Morphology and function of zooecial spines in cyclostome Bryozoa. Pp. 219–246 in Larwood, G.P. & Abbott, M.B. (eds) *Advances in bryozoology*. Academic Press, London.

Farmer, J.D. & Rowell, A.J. (1973) Variation in the bryozoan *Fistulipora decora* (Moore & Dudley) from the Beil Limestone of Kansas. Pp. 377–394 in Boardman, R.S., Cheetham, A.H. & Oliver, W.A., Jr (eds) *Animal colonies*. Dowden, Hutchinson & Ross Inc., Stroudsberg.

Farmer, J.D., Valentine, J.W. & Cowen, R. (1973) Adaptive strategies leading to the ectoproct ground-plan. *Systematic Zoology* **22**, 233–239.

Farrapeira, C.M.R. (2011) Invertebrados macrobentônicos detectados na costa brasileira transportados por resíduos flutuantes sólidos abiogênicos. *Revista da Gestão Costeira Integrada* **11**, 85–96.

Federov, P.V., Koromyslova, A.V. & Martha, S.O. (2017) The oldest bryozoans of Baltoscandia from the lowermost Floian (Ordovician) of north-western Russia: two new rare, small and simple species of Revalotrypidae. *PalZ* **91**, 353–373.

Fehlauer-Ale, K.H., Mackie, J.A., Lim-Fong, G.E., Ale, E., Pie, M.R. & Waeschenbach, A. (2014) Cryptic species in the cosmopolitan *Bugula neritina* complex (Bryozoa, Cheilostomata). *Zoologica Scripta* **43**, 193–205.

Fehlauer-Ale, K.H., Winston, J.E., Tilbrook, K.J., Nascimento, K.B. & Vieira, L.M. (2015) Identifying monophyletic groups within *Bugula* sensu lato (Bryozoa, Buguloidea). *Zoologica Scripta*, **44**, 334–347.

Ferguson, N., White, C.R. & Marshall, D.J. (2013) Competition in benthic marine invertebrates: the unrecognized role of exploitative competition for oxygen. *Ecology* **94**, 126–135.

Ferretti, C., Magnino, G. & Balduzzi, A. (2007) Morphology of the larva and ancestrula of *Myriapora truncata* (Bryozoa, Cheilostomatida). *Italian Journal of Zoology* **74**, 341–350.

Figuerola, B., Angulo-Preckler, C., Núñez-Pons, L., Moles, J., Sala-Comorera, L., García-Aljaro, C., Blanch, A.R. & Avila, A. (2017a) Experimental evidence of chemical defence mechanisms in Antarctic bryozoans. *Marine Environmental Research* **129**, 68–75.

Figuerola, B. & Avila, C. (2019) The phylum Bryozoa as a promising source of anticancer drugs. *Marine Drugs* **17**(8), 477, https://doi.org/10.3390/md17080477.

Figuerola, B., Barnes, D.K.A., Brickle, P. & Brewin, P.E. (2017b) Bryozoan diversity around the Falkland and South Georgia Islands: overcoming Antarctic barriers. *Marine Environmental Research* **126**, 81–94.

Figuerola, B., Gordon, D.P. & Cristobo, J. (2018) New deep Cheilostomata (Bryozoa) species from the Southwestern Atlantic: shedding light in the dark. *Zootaxa* **4375**, 211–249.

Figuerola, B., Gordon, D.P., Polonio, V., Cristobo, J. & Avila, C. (2014) Cheilostome bryozoan diversity from the southwest Atlantic region: is

Antarctica really isolated? *Journal of Sea Research* **85**, 1–17.

Figuerola, B., Kuklinski, P., Carmona, F. & Taylor, P.D. (2017c) Evaluating potential factors influencing branch diameter and skeletal Mg-calcite using an Antarctic cyclostome bryozoan species. *Hydrobiologia* **799**, 101–110.

Figuerola, B., Kuklinski, P. & Taylor, P.D. (2015) Depth patterns in Antarctic bryozoan skeletal calcite: can they provide an analogue for future environmental changes? *Marine Ecology Progress Series* **540**, 109–120.

Fine, M. & Loya, Y. (2003) Alternate coral–bryozoan competitive superiority during coral bleaching. *Marine Biology* **142**, 989–996.

Flor, F.D. (1972) Biometrische Untersuchungen zur Autökologie ober-kretazischer Bryozoen. *Mitteilungen aus dem Geologisch-Paläontologischen Institut der Universität Hamburg* **4**, 15–128.

Flor, F.D. & Hillmer, G. (1970) Rhythmische Wachstumsvorgaenge bei *Multicrescis tuberosa* (Roemer, 1839) (Bryoz. Cycl.). *Paläontologische Zeitschrift* **44**, 171–181.

Forester, R.M., Sandberg, P.A. & Anderson, T.F. (1973) Isotopic variability of cheilostome bryozoan skeletons. Pp. 79–94 in Larwood, G.P. (ed.) *Living and fossil Bryozoa*. Academic Press, London.

Fournier, F., Montaggioni, L. & Borgomano, J. (2004) Paleoenvironments and high-frequency cyclicity from Cenozoic South-East Asian shallow-water carbonates: a case study from the Oligo-Miocene buildups of Malampaya (Offshore Palawan, Philippines). *Marine and Petroleum Geology* **21**, 1–21.

Foveau, A., Desroy, N., Dewarumez, J.M., Dauvin, J.C. & Cabioch, L. (2008) Long-term changes in the sessile epifauna of the Dover Strait pebble community. *Journal of Oceanography, Research and Data* **1**, 1–11.

Fraaije, R.H.B., Van Bakel, B.W.M., Jagt, J.W.M. & Viegas, P.A. (2018) The rise of a novel, plankton-based marine ecosystem during the Mesozoic: a bottom-up model to explain new higher-tier invertebrate morphotypes. *Boletín de la Sociedad Geológica Mexicana* **70**, 187–200.

Francis, D.R. (1997) Bryozoan statoblasts in the recent sediments of Douglas Lake, Michigan. *Journal of Paleolimnology* **17**, 255–261.

Fransen, C.H.J.M. (1986) Caribbean Bryozoa: Anasca and Ascophora Imperfecta of the inner bays of Curacao and Bonaire. *Studies on the Fauna of Curacao and other Caribbean islands* **68**, 1–119.

Franz, D.R. (1971) Population age structure, growth and longevity of the marine gastropod *Urosalpinx cinerea*. *Biological Bulletin* **140**, 63–72.

Franzén, Å. (1960) *Monobryozoon limicola* n. sp., a ctenostomatous bryozoan from the detritus layer on soft sediment. *Zoologiska Bidrag från Uppsala* **33**, 135–148.

Franzén, Å. (1998) Spermiogenesis, sperm structure and spermatozeugmata in the gymnolaematous bryozoan *Electra pilosai* [sic] (Bryozoa, Gymnolaemata). *Invertebrate Reproduction and Development* **34**, 55–63.

Frazier, J.G., Winston, J.E. & Ruckdeschel, C.A. (1992) Epizoan communities on marine turtles. III. Bryozoa. *Bulletin of Marine Science* **51**, 1–8.

Freestone, A.L., Osman, R.W. & Whitlach, R.B. (2009) Latitudinal gradients in recruitment and community dynamics in marine epifaunal communities: implications for invasion success. *Smithsonian Contributions to the Marine Sciences* **38**, 247–258.

Freitas, M.C., Cacao, M., Cancela da Fonseca, L., Caroca, C. & Galopim de Carvalho, A.M. (1994) Unusual co-occurrence of serpulids and Bryozoa in a lagoonal system (Albufeira coastal lagoon – Portugal). *Gaia* **8**, 39–46.

Fritz, M.A. (1947) Cambrian Bryozoa. *Journal of Paleontology* **21**, 434–435.

Fuchs, J., Obst, M. & Sundberg, P. (2009) The first comprehensive molecular phylogeny of Bryozoa (Ectoprocta) based on combined analysis of nuclear and mitochondrial genes. *Molecular Phylogenetics and Evolution* **52**, 225–233.

Fürsich, F.T. & Palmer, T.J. (1975) Open crustacean burrows associated with hardgrounds in the Jurassic of the Cotswolds, England.

Proceedings of the Geologists' Association **86**, 171–181.

Fürsich, F.T., Palmer, T.J. & Goodyear, K.L. (1994) Growth and disintegration of bivalve-dominated patch reefs in the Upper Jurassic of southern England. *Palaeontology* **37**, 131–171.

Galle, A. & Parsley, R.L. (2005) Epibiont relationships on hyolithids demonstrated by Ordovician trepostomes (Bryozoa) and Devonian tabulates (Anthozoa). *Bulletin of Geosciences* **80**, 125–138.

Garbelli, C., Angiolini, L. & Shen, S.Z. (2017) Biomineralization and global change: a new perspective for understanding the end-Permian extinction. *Geology* **45**, 19–22.

Gardiner, A.R. & Taylor, P.D. (1982) Computer modelling of branching growth in the bryozoan *Stomatopora*. *Neues Jahrbuch für Geologie und Paläontologie Abhandlungen* **163**, 389–416.

Gautier, T.G. (1973) Growth in bryozoans of the Order Fenestrata. Pp. 271–274 in Larwood, G.P. (ed.) *Living and fossil Bryozoa*. Academic Press, London.

Gautier, T.G., Wyse Jackson, P.N. & McKinney, F.K. (2013) *Adlatipora*, a distinctive new acanthocladiid bryozoan from the Permian of the Glass Mountains, Texas, USA, and its bearing on fenestrate astogeny and growth. *Journal of Paleontology* **87**, 444–455.

Giles, K.A. (1998) The allochthonous nature of Lower Mississippian Waulsortian Mounds in the Sacramento Mountains, New Mexico. *New Mexico Geological Society Guidebook, 49th Field Conference*, 155–160.

Gili, J.-M., Arntz, W.E., Palanques, A., Orejas, C., Clarke, A., Dayton, P.K., Isla, E., Teixidó, N., Rossi, S. & López-González, J. (2006) A unique assemblage of epibenthic sessile suspension feeders with archaic features in the high-Antarctic. *Deep-Sea Research II* **53**, 1029–1052.

Gilmour, E.H. & Snyder, E.M. (1986) *Stellahexiformis* and *Morozoviella*, two new genera of Bryozoa from the Gerster Formation, northeastern Nevada. *Contributions to Geology, University of Wyoming* **24**, 211–217.

Giribet, G., Dunn, C.W., Edgecombe, G.D., Hejnol, A., Martindale, M.Q. & Rouse, G.W. (2009) Assembling the Spiralian tree of life. Pp. 52–64 in Telford, M.J. & Littlewood, D.T.J. (eds) *Animal evolution: genomes, fossils, and trees*. Oxford University Press, Oxford.

Goldstein, M.C., Carson, H.S. & Eriksen, M. (2014) Relationship of diversity and habitat area in North Pacific plastic-associated rafting communities. *Marine Biology* **161**, 1441–1453.

Gómez, A., Hughes, R.N., Wright, P.J., Carvalho, G.R. & Lunt, D.H. (2007b) Mitochondrial DNA phylogeography and mating compatibility reveal marked genetic structuring and speciation in the NE Atlantic bryozoan *Celleporella hyalina*. *Molecular Ecology* **16**, 2173–2188.

Gómez, A., Wright, P.J., Lunt, D.H., Cancino, J.M. & Hughes, R.N. (2007a) Mating trials validate the use of DNA barcoding to reveal cryptic speciation of a marine bryozoan taxon. *Proceedings of the Royal Society, Series B* **274**, 199–207.

Goncharova, I.A. & Rostovtseva, Yu. V. (2009) Evolution of organogenic carbonate buildups in the middle through late Miocene of the Euxine–Caspian Basin (Eastern Paratethys). *Paleontological Journal* **43**, 866–876.

Gooday, A.J. & Cook, P.L. (1984) An association between komokiacean foraminifers (Protozoa) and paludicelline ctenostomes (Bryozoa) from the abyssal northeast Atlantic. *Journal of Natural History* **18**, 765–784.

Gooley, T.A., Marshall, D.J. & Monro, K. (2010) Responses to conspecific density in an arborescent bryozoan. *Marine Ecology Progress Series* **415**, 83–90.

Gordon, D.P. (1971) Colony formation in the cheilostomatous bryozoan *Fenestrulina malusii* var. *thyreophora. New Zealand Journal of Marine and Freshwater Research* **5**, 342–351.

Gordon D.P. (1972) Biological relationships of an intertidal bryozoan population. *Journal of Natural History* **6**, 503–514.

Gordon, D.P. (1973) A fine-structure study of brown bodies in the gymnolaemate *Cryptosula pallasiana* (Moll). Pp. 275–286 in Larwood, G.P. (ed.) *Living and fossil Bryozoa*. Academic Press, London.

Gordon, D.P. (1975) Ultrastructure and function of the gut in a marine bryozoan. *Cahiers de Biologie Marine* **16**, 367–382.

Gordon, D.P. (1977) The aging process in bryozoans. Pp. 335–376 in Woollacott, R.M. & Zimmer, R.L. (eds) *Biology of bryozoans*. Academic Press, New York.

Gordon, D.P. (1987) The deep-sea Bryozoa of the New Zealand region. Pp. 97–104 in Ross, J.R.P. (ed.) *Bryozoa: present and past*. Western Washington University, Bellingham.

Gordon, D.P. (1989) Intertidal bryozoans from coral reef-flat rubble in Sa'aga, Western Samoa. *New Zealand Journal of Zoology* **16**, 447–463.

Gordon, D.P. (1999) Bryozoan diversity in New Zealand and Australia. Pp. 199–204 in Ponder, W. & Lunney, D. (eds) *The other 99%: the conservation and biodiversity of invertebrates*. Royal Zoological Society of New South Wales, Mosman.

Gordon, D.P. (2000) Towards a phylogeny of cheilostomes – morphological models of frontal wall/shield evolution. Pp. 17–37 in Herrera Cubilla, A. & Jackson, J.B.C. (eds) *Proceedings of the 11th International Bryozoology Association Conference*. Smithsonian Tropical Research Institute, Balboa.

Gordon, D.P. (2012) Life on the edge: *Parachnoidea* (Ctenostomata) and *Barentsia* (Kamptozoa) on bathymodiolin mussels from an active submarine volcano in the Kermadec Volcanic Arc. Pp. 339–365 in Ernst, A., Schäfer, P. & Scholz, J. (eds) *Bryozoan studies 2010*. Lecture Notes in Earth System Sciences **143**. Springer, Berlin.

Gordon, D.P. (2014) Apprehending novel biodiversity – fifteen new genera of Zealandian Bryozoa. *Journal of the Marine Biological Association of the United Kingdom* **94**, 1597–1628.

Gordon, D.P. & Costello, M.J. (2016) Bryozoa – not a minor phylum. *New Zealand Science Review* **73**, 63–66.

Gordon, D.P. & d'Hondt, J.-L. (1991) Bryozoa: the Miocene to Recent family Petalostegidae. Systematics, affinities, biogeography. *Mémoires du Muséum National d'Histoire Naturelle, Paris* **151**, 91–103.

Gordon, D.P. & Hastings, A.B. (1979) The interzooidal communications of *Hippothoa* sensu lato (Bryozoa) and their value in classification. *Journal of Natural History* **13**, 561–579.

Gordon, D.P., Mawatari, S.F. & Kajihara, H. (2002) New taxa of Japanese and New Zealand Eurystomellidae (Phylum Bryozoa) and their phylogenetic relationships. *Zoological Journal of the Linnean Society* **136**, 199–216.

Gordon, D.P. & Parker, S.A. (1991a) An aberrant new genus and subfamily of the spiculate family Thalamoporellidae epiphytic on *Posidonia*. *Journal of Natural History* **25**, 1363–1378.

Gordon, D.P. & Parker, S.A. (1991b) The plectriform apparatus – an enigmatic structure in malacostegine Bryozoa. *Bulletin de la Société des Sciences Naturelle de l'Ouest de la France, Mémoire* **H.S. 1**, 133–145.

Gordon, D.P., Ramalho, L.V., & Taylor, P.D. (2006) An unreported invasive bryozoan that can affect livelihoods – *Membraniporopsis tubigera* in New Zealand and Brazil. *Bulletin of Marine Science* **78**, 331–342.

Gordon, D.P. & Rudman, W.B. (2006) *Integripelta acanthus* n. sp. (Bryozoa: Eurystomellidae) – a tropical prey species of *Okenia hiroi* (Nudibranchia). *Zootaxa* **1229**, 41–48.

Gordon, D.P., Stuart, I.G. & Collen, J.D. (1994) Bryozoan fauna of the Kaipuke Siltstone, northwest Nelson: a Miocene homologue of the modern Tasman Bay coralline bryozoan grounds. *New Zealand Journal of Geology and Geophysics* **37**, 239–247.

Gordon, D.P. & Taylor, P.D. (1999) Latest Paleocene to earliest Eocene bryozoans from Chatham Island, New Zealand. *Bulletin of The Natural History Museum, London, Geology Series* **55**, 1–45.

Gordon, D.P. & Taylor, P.D. (2001) New Zealand Recent Densiporidae and Lichenoporidae (Bryozoa: Cyclostomata). *Species Diversity* **6**, 243–290.

Gordon, D.P. & Taylor, P.D. (2008) Systematics of the bryozoan genus *Macropora* (Cheilostomata). *Zoological Journal of the Linnean Society* **153**, 115–146.

Gordon, D.P. & Taylor, P.D. (2010) New sea-mount- and ridge-associated cyclostome Bryozoa from New Zealand. *Zootaxa* **2533**, 43–68.

Gordon, D.P. & Taylor, P.D. (2015) Bryozoa of the Early Eocene Tumaio Limestone, Chatham Island, New Zealand. *Journal of Systematic Palaeontology* **13**, 983–1070.

Gordon, D.P., Taylor, P.D. & Bigey, F.P. (2009) Phylum Bryozoa – moss animals, sea mats, lace corals. Pp. 271–297 in Gordon, D.P. (ed.) *The New Zealand inventory of biodiversity, Volume 1: Kingdom Animalia – Radiata, Lophotrochozoa, and Deuterostomia*. Canterbury University Press, Christchurch.

Gordon, D.P. & Voigt, E. (1996) The kenozooidal origin of the ascophorine hypostegal coelom and associated frontal shield. Pp. 89–107 in Gordon, D.P., Smith, A.M. & Grant-Mackie, J.A. (eds) *Bryozoans in space and time*. NIWA, Wellington.

Gordon, D.P., Voje, K.J. & Taylor, P.D. (2017) Living and fossil Steginoporellidae (Bryozoa: Cheilostomata) from New Zealand. *Zootaxa* **4350**, 345–362.

Gorjunova, R.V. (1996) *Phylogeny of the Paleozoic bryozoans*. Nauka, Moscow, 165 pp. [in Russian].

Gorjunova, R.V. (2010) On the Paleozoic bryozoans of the genus *Ascopora* Trautschold. *Paleontological Journal* **44**, 21–35.

Gorjunova, R.V. (2011) Family Coelotubuliporidae fam. nov. and morphological parallelisms in the evolution of bryozoans. *Paleontological Journal* **45**, 510–524.

Gorodiski, A. & Balavoine, P. (1962) Bryozoaires crétacés et eocenes du Sénègal. *Bulletin du Bureau de Recherches Géologiques et Minières* **4** [for 1961], 1–15.

Gould, S.J. (2002) *The structure of evolutionary theory*. Belknap Press, Cambridge, MA.

Grabowska, M., Grezelak, K. & Kuklinski, P. (2015) Rock encrusting assemblages: structure and distribution along the Baltic Sea. *Journal of Sea Research* **103**, 24–31.

Grant, A. & Hayward, P.J. (1985) Bryozoan benthic assemblages in the English Channel. Pp. 115–124 in Nielsen, C. & Larwood, G.P. (eds) *Bryozoa: Ordovician to Recent.* Olsen & Olsen, Fredensborg.

Grave, B.J. (1930) The natural history of *Bugula flabellata* at Woods Hole, Massachusetts, including the behaviour and attachment of the larva. *Journal of Morphology and Physiology* **49**, 355–379.

Gray, C.A., McQuaid, C.D. & Davies-Coleman, M.T. (2005) A symbiotic shell-encrusting bryozoan provides subtidal whelks with chemical defence against rock lobsters. *African Journal of Marine Science* **27**, 549–556.

Greeley, R. (1969) Basally 'uncalcified' zoaria of lunulitiform Bryozoa. *Journal of Paleontology* **43**, 252–256.

Greene, S.E., Martindale, R.C., Ritterbush, K.A., Bottjer, D.J., Corsetti, F.A. & Berelson, W.M. (2012) Recognising ocean acidification in deep time: an evaluation of the evidence for acidification across the Triassic–Jurassic boundary. *Earth-Science Reviews* **113**, 72–93.

Griffiths, H.J., Barnes, D.K.A. & Linse, K. (2009) Towards a generalized biogeography of the Southern Ocean benthos. *Journal of Biogeography* **36**, 162–177.

Grischenko, A.V. & Chernyshev, A.V. (2018) Deep-water Bryozoa from the Kuril Basin, Sea of Okhotsk. *Deep-Sea Research Part II* **154**, 59–73.

Grischenko, A.V., Gordon, D.P. & Melnik, V.P. (2018) Bryozoa (Cyclostomata and Ctenostomata) from polymetallic nodules in the Russian exploration area, Clarion–Clipperton Fracture Zone, eastern Pacific Ocean – taxon novelty and implications of mining. *Zootaxa* **4484**, 1–91.

Grischenko, A.V., Gordon, D.P. & Taylor, P.D. (1998) A unique new genus of cheilostomate bryozoan with reversed-polarity zooidal budding. *Asian Marine Biology* **15**, 105–117.

Grischenko, A., Mawatari, S.F. & Taylor, P.D. (2000) Systematics and phylogeny of the cheilostome bryozoan *Doryporella. Zoologica Scripta* **29**, 247–264.

Grischenko, A.V., Taylor, P.D. & Mawatari, S.F. (2002) A new cheilostome bryozoan with gigantic zooids from the north-west Pacific. *Zoological Science* **19**, 1279–1289.

Grischenko, A.V., Taylor, P.D. & Mawatari, S.F. (2004) *Doryporella smirnovi* sp. nov. (Bryozoa: Cheilostomata) and its impact on phylogeny and classification. *Zoological Science* **21**, 327–332.

Gruhl, A., Grobe, P. & Bartolomaeus, T. (2005) Fine structure of the epistome in *Phoronis ovalis*: significance for the coelomic organization in Phoronida. *Invertebrate Biology* **124**, 332–343.

Gruhl, A. & Schwaha, T. (2016) Bryozoa (Ectoprocta). Pp. 325–340 in Schmidt-Rhaesa, A., Harzsch, S. & Purschke, G. (eds) *Structure and evolution of invertebrate nervous systems.* Oxford University Press, Oxford.

Gruhl, A., Wegener, I. & Bartolomaeus, T. (2009) Ultrastructure of the body cavities in Phylactolaemata (Bryozoa). *Journal of Morphology* **270**, 306–318.

Grünbaum, D. (1997) Hydromechanical mechanisms of colony organisation and cost of defense in an encrusting bryozoan, *Membranipora membranacea. Limnology and Oceanography* **42**, 741–752.

Grzelak, K. & Kuklinski, P. (2010) Benthic assemblages associated with rocks in a brackish environment of the southern Baltic Sea. *Journal of the Marine Biological Association of the United Kingdom* **90**, 115–124.

Guha, A.K. (2013) Tertiary Bryozoa from western Kachchh, Gujarat – a review. *Journal of the Palaeontological Society of India* **58**, 3–15.

Guha A.K. & Nathan, D.S. (1996) Bryozoan fauna of the Ariyalur Group (Late Cretaceous), Tamil Nadu and Pondicherry, India. *Palaeontologia Indica* n.s. **49**, 1–217.

Gutak, J.M., Tolokonnikova, Z.A. & Ruban, D.A. (2008) Bryozoan diversity in Southern Siberia at the Devonian–Carboniferous transition: new data confirm a resistivity to two mass extinctions. *Palaeogeography, Palaeoclimatology, Palaeoecology* **264**, 93–99.

Hageman, S.J. (1991) *Worthenopora*: an unusual cryptostome (Bryozoa) that looks like a cheilostome. *Journal of Paleontology* **65**, 648–661.

Hageman, S.J. (1995) Observed phenotypic variation in a Paleozoic bryozoan. *Paleobiology* **21**, 314–328.

Hageman, S.J. (2018) *Pywackia* is not a Cambrian bryozoan: evidence from skeletal microstructure and taphonomy. *Geological Society of America Abstracts with Programs* **50**. https://gsa.confex.com/gsa/2018AM/webprogram/Paper319226.html.

Hageman, S.J., Bayer, M.M. & Todd, C.D. (1999) Partitioning phenotypic variation: genotypic, environmental and residual components from bryozoan skeletal morphology. *Journal of Natural History* **33**, 1713–1735.

Hageman, S.J., Bone, Y., McGowran, B. & James, N.P. (1997) Bryozoan colonial growth forms as paleoenvironmental indicators: evaluation of methodology. *Palaios* **12**, 406–419.

Hageman, S.J., Bock, P.E., Bone, Y. & McGowran, B. (1998) Bryozoan growth habits: classification and analysis. *Journal of Paleontology* **72**, 418–436.

Hageman, S.J., James, N.P. & Bone, Y. (2000) Cool-water carbonate production from epizoic bryozoans on ephemeral substrates. *Palaios* **15**, 33–48.

Hageman, S.J., Lukasik, J., McGowran, B. & Bone, Y. (2003) Paleoenvironmental significance of *Celleporaria* (Bryozoa) from modern and Tertiary cool-water carbonates of southern Australia. *Palaios* **18**, 510–527.

Hageman, S.J., McKinney, F.K. & Chandler, R. (2008) *The fossils of North Carolina: Bryozoans*. The North Carolina Fossil Club, Research Triangle Park, 27 pp.

Hageman, S.J., McKinney, F.K. & Jaklin, A. (2012) Testing habitat complexity as a control over bryozoan colonial growth form and species distribution. Pp. 105–119 in Ernst, A., Schäfer, P. & Scholz, J. (eds) *Bryozoan studies 2010*. Springer, Heidelberg.

Hageman, S.J., Needham, L.L. & Todd, C.D. (2009) Threshold effects of food concentration on the skeletal morphology of the bryozoan *Electra pilosa* (Linnaeus, 1767). *Lethaia* **42**, 438–451.

Hageman, S.J. & Sawyer, J.A. (2006) Phenotypic variation in the bryozoan *Leioclema punctatum* (Hall, 1858) from Mississippian ephemeral host microcommunities. *Journal of Paleontology* **80**, 1047–1057.

Hageman, S.J. & Todd, C.D. (2014) Hierarchical (mm- to km-scale) environmental variation affecting skeletal phenotype of a marine invertebrate (*Electra pilosa*, Bryozoa): implications for fossil species concepts. *Palaeogeography, Palaeoclimatology, Palaeoecology* **396**, 213–226.

Hageman, S.J., Wyse Jackson, P.N., Abernethy, A.R. & Steinthorsdottir, M. (2011) Calendar scale, environmental variation preserved in the skeletal phenotype of a fossil bryozoan (*Rhombopora blakei* n. sp.), from the Mississippian of Ireland. *Journal of Paleontology* **85**, 853–870.

Haig, D.W., McCartain, E., Mory, A.J., Borges, G., Davydov, V.I., Dixon, M., Ernst, A., Groflin, S., Hakansson, E., Keep, M., Dos Santos, Z., Shi, G.R. & Soares, J. (2014) Postglacial Early Permian (late Sakmarian–early Artinskian) shallow-marine carbonate deposition along a 2000 km transect from Timor to West Australia. *Palaeogeography, Palaeoclimatology, Palaeoecology* **409**, 180–204.

Håkansson, E. & Thomsen, E. (1979) Distribution and types of bryozoan communities at the boundary in Denmark. Pp. 78–91 in Birkelund, T. & Bromley, R.G. (eds) *Cretaceous–Tertiary boundary events, I: the Maastrichtian and Danian of Denmark*. University of Copenhagen, Copenhagen.

Håkansson, E. & Thomsen, E. (2001) Asexual propagation in cheilostome Bryozoa. Pp. 326–347 in Jackson, J.B.C., Lidgard, S. & McKinney, F.K. (eds) *Evolutionary patterns:*

growth, form and tempo in the fossil record. University of Chicago Press, Chicago.

Håkansson, E. & Voigt, E. (1996) New free-living bryozoans from the northwest European Chalk. *Bulletin of the Geological Society of Denmark* **42**, 187–207.

Håkansson, E. & Winston, J.E. (1985) Interstitial bryozoans: unexpected life forms in a high energy environment. Pp. 125–134 in Nielsen, C. & Larwood, G.P. (eds) *Bryozoa: Ordovician to Recent*. Olsen & Olsen, Fredensborg.

Halfar, J., Godinez Orta, L., Mutti, M., Valdez Holguín, J.E. & Borges Souza, J.M. (2006) Carbonates calibrated against oceanographic parameters along a latitudinal transect in the Gulf of California, Mexico. *Sedimentology* **53**, 297–320.

Hall, S.R., Taylor, P.D., Davis, S.A. & Mann, S. (2002) Electron diffraction studies of the calcareous skeletons of bryozoans. *Journal of Inorganic Biochemistry* **88**, 410–419.

Hallam, A. (1994) *An outline of Phanerozoic biogeography*. Oxford University Press, Oxford, 246 pp.

Hampton, G.L. (1979) Stratigraphy and archaeocyathans of Lower Cambrian strata of Old Douglas Mountain, Stevens County, Washington. *Geology Studies, Brigham Young University* **26**, 27–39.

Hao, J., Li, C., Sun, X. & Yang, Q. (2005) Phylogeny and divergence time estimation of cheilostome bryozoans based on mitochondrial 16S rRNA sequences. *Chinese Science Bulletin* **50**, 1205–1211.

Hara, U. (2001) Bryozoans from the Eocene of Seymour Island, Antarctic Peninsula. *Palaeontologia Polonica* **60**, 33–156.

Hara, U. (2007) Biogeographical relationship of the South America–Antarctic Cenozoic bryozoan biota: the example of austral genus *Aspidostoma. US Geological Survey and The National Academies; USGS OF-2007-1047*, Extended Abstract **214**, 6 pp.

Hara, U. (2015) Bryozoan internal moulds from the La Meseta Formation (Eocene) of Seymour Island, Antarctic Peninsula. *Polish Polar Research* **36**, 25–49.

Hara, U. & Jasionowski, M. (2012) The Early Sarmatian bryozoan *Celleporina medoborensis* sp. nov. from the Medobory reefs of western Ukraine (Central Paratethys). *Geological Quarterly* **56**, 895–906.

Hardie, L.A. (1996) Secular variation in seawater chemistry: an explanation for the coupled secular variation in the mineralogies of marine limestones and potash evaporites over the past 600 my. *Geology* **24**, 279–283.

Harmelin, J.-G. (1973) Morphological variations and ecology of the Recent cyclostome bryozoan 'Idmonea' atlantica from the Mediterranean. Pp. 95–106 in Larwood, G.P. (ed.) *Living and fossil Bryozoa*. Academic Press, London.

Harmelin, J.G. (1974) A propos d'une forme stomatoporienne typique, *Stomatopora gingrina* Jullien, 1882 (Bryozoaires Cyclostomes), et de son gonozoïde. *Journal of Natural History* **8**, 1–9.

Harmelin, J.-G. (1976) Le sous-ordre des Tubuliporina (Bryozoaires Cyclostomes) en Méditerranée. *Memoires de l'Institut Oceanographique, Monaco* **10**, 1–326.

Harmelin, J.-G. (1990) Deep-water crisiids (Bryozoa: Cyclostomata) from the northeast Atlantic Ocean. *Journal of Natural History* **24**, 1597–1616.

Harmelin, J.-G. (1997) Diversity of bryozoans in a Mediterranean sublittoral cave with bathyal-like conditions: role of dispersal processes and local factors. *Marine Ecology Progress Series* **153**, 139–152.

Harmelin, J.-G. (2000) Ecology of cave and cavity-dwelling bryozoans. Pp. 38–53 in Herrera Cubilla, A. & Jackson, J.B.C. (eds) *Proceedings of the 11th International Bryozoology Association Conference*. Smithsonian Tropical Research Institute, Balboa.

Harmelin, J.-G., Bitar, G. & Zibrowius, H. (2016) High xenodiversity versus low native diversity in the south-eastern Mediterranean: bryozoans from the coastal zone of Lebanon. *Mediterranean Marine Science* **17**, 417–439.

Harmelin, J.-G., Boury-Esnault, N. & Vacelet, J. (1994) A bryozoan-sponge symbiosis: the association between *Smittina cervicornis* and *Halisarca* cf. *dujardini* in the Mediterranean. Pp. 69–74 in Hayward, P.J., Ryland, J.S. & Taylor, P.D. (eds) *Biology and palaeobiology of bryozoans*. Olsen & Olsen, Fredensborg.

Harmelin, J.-G., Vieira, L.M., Ostrovsky, A.N., Cáceres-Chamizo, J.P. & Sanner J. (2012) *Scorpiodinipora costulata* (Canu & Bassler, 1929) (Bryozoa, Cheilostomata), a taxonomic and biogeographic dilemma: complex of cryptic species or human-mediated cosmopolitan colonizer? *Zoosystema* **34**, 123–138.

Harmer, S.F. (1893) On the occurrence of embryonic fission in cyclostomatous Polyzoa. *Quarterly Journal of Microscopical Science* **34**, 199–241.

Harmer, S.F. (1900) A revision of the genus *Steganoporella*. *Quarterly Journal of Microscopical Science* **43**, 225–297.

Harper, D.A.T. (2006) The Ordovician biodiversification: setting an agenda for marine life. *Palaeogeography, Palaeoclimatology, Palaeoecology* **232**, 148–166.

Hartikainen, H., Fontes, I. & Okamura, B. (2013) Parasitism and phenotypic change in colonial hosts. *Parasitology* **140**, 1403–1412.

Hartikainen, H., Humphries, S. & Okamura, B. (2014) Form and metabolic scaling in colonial animals. *The Journal of Experimental Biology* **217**, 779–786.

Harvell, C.D. (1984) Why nudibranchs are partial predators: intracolonial variation in bryozoan palatability. *Ecology* **65**, 716–724.

Harvell, C.D. (1986) The ecology and evolution of inducible defences in a marine bryozoan: cues, costs, and consequences. *American Naturalist* **128**, 810–823.

Harvell, C.D., Caswell, H. & Simpson, P. (1990) Density effects in a colonial monoculture: experimental studies with a marine bryozoan (*Membranipora membranacea* L.). *Oecologia* **82**, 227–237.

Harvell, C.D. & Padilla, D.K. (1990) Inducible morphology, heterochrony, and size hierarchies in a colonial inveretebrate monoculture. *Proceedings of the National Academy of Sciences, USA* **87**, 508–512.

Harzhauser, M., Kroh, A., Mandic, O., Piller, W.E., Göhlich, U., Reuter, M. & Berning, B. (2007) Biogeographic responses to geodynamics: a key study all around the Oligo-Miocene Tethyan Seaway. *Zoologischer Anzeiger* **246**, 241–256.

Hauck, J., Gerdes, D., Hillenbrand, C.-G., Hoppema, M., Kuhn, G., Nehrke, G., Völker, C. & Wolf-Gladrow, D.A. (2011) Distribution and mineralogy of carbonate sediments on Antarctic shelves. *Journal of Marine Systems* **90**, 77–87.

Hautmann, M., Benton, M.J. & Tomasovych, A. (2008) Catastrophic ocean acidification at the Triassic–Jurassic boundary. *Neues Jahrbuch für Geologie und Paläontologie Abhandlungen* **249**, 119–127.

Haygood, M.G. & Davidson, S.K. (1997) Small-subunit rRNA genes and in situ hybridization with oligonucleotides specific for the bacterial symbionts in the larvae of the bryozoan *Bugula neritina* and proposal of 'Candidatus Endobugula sertula'. *Applied and Environmental Microbiology* **63**, 4612–4616.

Hayward, P.J. (1981) The Cheilostomata (Bryozoa) of the deep sea. *Galathea Report* **15**, 21–68.

Hayward, P.J. (1983) Biogeography of *Adeonella* (Bryozoa, Cheilostomata): a preliminary account. *Bulletin of Marine Science* **33**, 582–596.

Hayward, P.J. (1985) Ctenostome bryozoans. *Synopses of the British Fauna* n.s. **33**, 1–169.

Hayward, P.J. (1995) *Antarctic cheilostomatous Bryozoa*. Oxford University Press, Oxford, 355 pp.

Hayward, P.J. & Ryland, J.S. (1985) Cyclostome bryozoans. *Synopses of the British Fauna* n.s. **34**, 1–147.

Hayward, P.J. & Taylor, P.D. (1984) Fossil and Recent Cheilostomata (Bryozoa) from the Ross Sea, Antarctica. *Journal of Natural History* **18**, 71–94.

Hayward, P.J. & Winston, J.E. (2011) Bryozoa collected by the United States Antarctic

Research Program: new taxa and new records. *Journal of Natural History* **45**, 2259–2338.

Hausdorf, B., Helmkampf, M., Meyer, A., Witek, A., Herlyn, H., Bruchhaus, I., Hankeln, T., Struck, T.H. & Lieb, B. (2007) Spiralian phylogenomics supports the resurrection of Bryozoa comprising Ectoprocta and Entoprocta. *Molecular Biology and Evolution* **24**, 2723–2729.

Heindl, H., Wiese, J., Thiel, V. & Imhoff, J.F. (2010) Phylogenetic diversity and antimicrobial activities of bryozoan-associated bacteria isolated from Mediterranean and Baltic Sea habitats. *Systematic and Applied Microbiology* **33**, 94–104.

Helm, C. (1999) Astogenese von *Aulopora* cf. *enodis* Klaamann 1966 (Visby-Mergel, Silur von Gotland). *Paläontologische Zeitschrift* **73**, 241–246.

Helm, C. & Schülke, I. (2006) Patch reef development in the *florigemma*-Bank Member (Oxfordian) from the Deister Mts (NW Germany): a type example for Late Jurassic coral thrombolite thickets. *Facies* **52**, 441–467.

Hengherr, S. & Schill, R.O. (2011) Dormant stages in freshwater bryozoans – an adaptation to transcend environmental constraints. *Journal of Insect Physiology* **57**, 595–601.

Henrich, R., Freiwald, A., Betzler, C., Bader, B., Schäfer, P., Samtleben, C., Brachert, T.C., Wehrmann, A., Zankl, H. & Kühlmann, D.H.H. (1995) Controls on modern carbonate sedimentation on warm-temperate to Arctic coasts, shelves and seamounts in the northern hemisphere: implications for fossil counterparts. *Facies* **32**, 71–108.

Herrera, A. & Jackson, J.B.C. (1996) Life history variation among 'dominant' encrusting cheilostomate Bryozoa. Pp. 117–123 in Gordon, D.P., Smith, A.M. & Grant-Mackie, J.A. (eds) *Bryozoans in space and time*. NIWA, Wellington.

Hickey, D.R. (1987) Skeletal structure, development and elemental composition of the Ordovician trepostome bryozoan *Peronopora*. *Palaeontology* **30**, 691–716.

Hillebrand, H. (2004) On the generality of the latitudinal diversity gradient. *American Naturalist* **163**, 192–211.

Hillmer, G. (1991) A 300-million-year gap in the bryozoan fossil record: modern features in the Ordovician. *Naturwissenschaften* **78**, 123–125.

Hillmer, G., Gautier, T.G. & McKinney, F.K. (1975) Budding by intrazooecial fission in the stenolaemate bryozoans *Stenoporella*, *Reptomulticava* and *Canalipora*. *Mitteilungen aus dem Geologisch-Paläontologisches Institut der Universität Hamburg* **44**, 123–132.

Hinds, R.W. (1973) Intrazooecial structures in some tubuliporinid cyclostome Bryozoa. Pp. 299–306 in Larwood, G.P. (ed.) *Living and fossil Bryozoa*. Academic Press, London.

Hinds, R.W. (1975) Growth mode and homeomorphism in cyclostome Bryozoa. *Journal of Paleontology* **49**, 875–910.

Hirose, M. (2011) Orientation and righting behaviour of the sand-dwelling bryozoan *Conescharellina catella*. *Invertebrate Biology* **130**, 282–290.

Hirose, M., Dick, M.H. & Mawatari, S.F. (2008) Molecular phylogenetic analysis of phylactolaemate bryozoans based on mitochondrial gene sequences. *Virginia Museum of Natural History, Special Publication* **15**, 65–74.

Hirose, M., Dick, M.H. & Mawatari, S.F. (2011) Are plumatellid statoblasts in freshwater bryozoans phylogenetically informative? *Zoological Science* **28**, 318–326.

Hladllová, S., Zágorsek, K. & Ziegler, V. (2004) A Middle Miocene gastropod shell with epifauna from the locality Buituri (Transylvanian Basin, Romania). *Fragmenta Palaeontologia Hungarica* **22**, 99–105.

Hoare, R.D. & Steller, D.L. (1967) A Devonian brachiopod with epifauna. *Ohio Journal of Science* **67**, 291–297.

Hoffmeister, J.E., Stockmann, K.W. & Multer, H.G. (1967) Miami Limestone of Florida and its Recent Bahamian counterpart. *Geological Society of America Bulletin* **78**, 175–190.

Hollingworth, N. (1991) A well preserved, Upper Permian, Zechstein Cycle 1 reef fauna from

North-East England. *Zentralblatt für Geologie und Paläontologie, Teil I, Geologie* **H.4**, 839–852.

d'Hondt, J.-L. (1976) Bryozoaires cténostomes bathyaux et abyssaux de l'Atlantique Nord. *Documents des laboratoires de géologie de la Faculté des sciences de Lyon* **H.S. 3**(2), 311–333.

d'Hondt, J.-L. (1983) Tabular keys for identification of the Recent ctenostomatous Bryozoa. *Mémoires de l'Institut océanographique* **14**, 1–134.

d'Hondt, J.-L. (2016) Biosystématique actualisée des Bryozoaires Eurystomes. *Bulletin de la Société zoologique de France* **141**, 15–23.

d'Hondt, J.-L. & Horowitz, A.S. (2007) Matériaux pour une revision systématique des Bryozoaires Cténostomes fossiles. *Bulletin de la Société Linnéenne de Bordeaux* **35**, 265–280.

Hong, J., Oh, J.-R., Lee, J.-H., Choh, S.-J. & Lee, D.-L. (2014) The earliest evolutionary link of metazoan bioconstruction: Laminar stromatoporoid–bryozoan reefs from the Middle Ordovician of Korea. *Palaeogeography, Palaeoclimatology, Palaeoecology* **492**, 126–133.

Horowitz, A.S. & Pachut, J.F. (2000) The fossil record of bryozoan species diversity. Pp. 245–248 in Herrera Cubilla, A. & Jackson, J.B.C. (eds) *Proceedings of the 11th International Bryozoology Association Conference.* Smithsonian Tropical Research Institute, Balboa.

Hu, Z.-X. (1984) Triassic Bryozoa from Xizang (Tibet) with reference to their biogeographical provincialism in the world. *Acta Palaeontologica Sinica* **23**, 568–577.

Hughes, D.J. (1989) Variation in reproductive strategy among clones of the bryozoan *Celleporella hyalina* (L.). *Ecological Monographs* **59**, 387–403.

Hughes, D.J. & Hughes, R.N. (1986) Life history variation in *Celleporella hyalina* (Bryozoa). *Proceedings of the Royal Society, Series B* **228**, 127–132.

Hughes, R.N. (2005) Lessons in modularity: the evolutionary ecology of colonial invertebrates. *Scientia Marina* **69** (Supplement 1), 169–179.

Hughes, R.N., D'Amato, M.E., Bishop, J.D.D., Carvalho, G.R., Craig, S.F., Hansson, L.J., Harley, M.A. & Pemberton, A.J. (2005) Paradoxical polyembryony? Embryonic cloning in an ancient order of marine bryozoans. *Biology Letters* **1**, 178–180.

Hughes, R.N., Manriquez, P.H. & Bishop, J.D.D. (2002) Female investment is retarded pending reception of allosperm in a hermaphroditic colonial invertebrate. *Proceedings of the National Academy of Sciences, USA* **99**, 14884–14886.

Hughes, R.N., Manriquez, P.H., Bishop, J.D.D. & Burrows, M.T. (2003) Stress promotes maleness in hermaphroditic modular animals. *Proceedings of the National Academy of Sciences, USA* **100**, 10326–10330.

Hughes, R.N., Manriquez, P.H., Morley, S., Craig, S.F. & Bishop, J.D.D. (2004) Kin or self-recognition? Colonial fusibility of the bryozoan *Celleporella hyalina. Evolution and Development* **6**, 431–437.

Huntley, J. & De Baets, K. (2015) Trace fossil evidence of trematode–bivalve parasite–host interactions in deep time. *Advances in Parasitology* **90**, 201–231.

Hurd, C.L., Durante, K.M., Chia, F.S. & Harrison, P.J. (1994) Effect of bryozoan colonisation on inorganic nitrogen acquisition by the kelps *Agarum fimbriatum* and *Macrocystis integrifolia. Marine Biology* **12**, 167–173.

Hylleberg, J. (1970) On the ecology of the sipunculan *Phascolion strombi* (Montagu). Pp. 241–250 in Rice, M.E. & Todorovic, M. (eds) *Proceedings of the International Conference on the Biology of the Sipuncula and Echiura.* Institute for Biological Research, Belgrade, and Smithsonian Institution, Washington DC.

Hyman, L.H. (1959) *The invertebrates: smaller coelomate groups.* McGraw-Hill, New York.

Igawa, M., Hata, H. & Kato, M. (2017) Reciprocal symbiont sharing in the lodging mutualism between walking corals and sipunculans. *PLoS ONE* **12**, e0169825.

Ishii, T. & Saito, Y. (1995) Colony specificity in the marine bryozoan *Dakaria subovoidea*. *Zoological Science* **12**, 435–441.

Ito, M., Onishi, T. & Dick, M.H. (2015) *Cribrilina mutabilis* n. sp., an eelgrass-associated bryozoan (Gymnolaemata: Cheilostomata) with large variation in zooid morphology related to life history. *Zoological Science* **32**, 485–497.

Ivany, L.C., Portell, R.W. & Jones, D.S. (1990) Animal–plant relationships and paleobiogeography of an Eocene seagrass community from Florida. *Palaios* **5**, 244–258.

Jablonski, D., Lidgard, S. & Taylor, P.D. (1997) Comparative ecology of bryozoan radiations: origin of novelties in cyclostomes and cheilostomes. *Palaios* **12**, 505–523.

Jackson, J.B.C. (1979a) Morphological strategies of sessile animals. Pp. 499–555 in Larwood, G. & Rosen, B.R. (eds) *Biology and systematics of colonial organisms*. Academic Press, London.

Jackson, J.B.C. (1979b) Overgrowth competition between encrusting cheilostome ectoprocts in a Jamaican cryptic reef environment. *Journal of Animal Ecology* **48**, 805–823.

Jackson, J.B.C. & Cheetham, A.H. (1990) Evolutionary significance of morphospecies: a test with cheilostome Bryozoa. *Science* **248**, 579–583.

Jackson, J.B.C. & Cheetham, A.H. (1994) Phylogeny reconstruction and the tempo of speciation in cheilostome Bryozoa. *Paleobiology* **20**, 407–423.

Jackson, J.B.C. & Coates, A.G. (1986) Life cycles and evolution of clonal (modular) animals. *Philosophical Transactions of the Royal Society, Series B* **313**, 7–22.

Jackson, J.B.C. & Herrera Cubilla, A. 2000. Adaptation and constraint as determinants of zooid and ovicell size among encrusting ascophoran cheilostome Bryozoa from opposite sides of the Isthmus of Panama. Pp. 249–258 in Herrera Cubilla, A. & Jackson, J.B.C. (eds) *Proceedings of the 11th International Bryozoology Association Conference*. Smithsonian Tropical Research Institute, Balboa.

Jackson, J.B.C. & Wertheimer, S.P. (1985) Patterns of reproduction in five common species of Jamaican reef-associated bryozoans. Pp. 161–168 in Nielsen, C. & Larwood, G.P. (eds) *Bryozoa: Ordovician to Recent*. Olsen & Olsen, Fredensborg.

Jackson, J.B.C. & Winston, J.E. (1981) Modular growth and longevity in bryozoans. Pp. 121–126 in Larwood, G.P. & Nielsen, C. (eds) *Recent and fossil Bryozoa*. Olsen & Olsen, Fredensborg.

Jackson, J.B.C., Winston, J.E. & Coates, A.G. (1985) Niche breadth, geographic range, and extinction of Caribbean reef-associated cheilostome Bryozoa and Scleractinia. *Proceedings of the 5th International Coral Reef Congress* **4**, 151–158.

Jacob, D.E., Ruthensteiner, B., Trimby, P., Henry, H., Martha, S.O., Leitner, J., Otter, L.M. & Scholz, J. (2019) Architecture of *Anoteropora latirostris* (Bryozoa, Cheilostomata) and implications for their biomineralization. *Scientific Reports* **9**, 11439.

Jaeckle, W.B. (1994) Rates of energy consumption and acquisition by lecithotrophic larvae of *Bugula neritina* (Bryozoa, Cheilostomata). *Marine Biology* **119**, 517–523.

James, D.W., Foster, M.S. & O'Sullivan, J. (2006) Bryoliths (Bryozoa) in the Gulf of California. *Pacific Science* **60**, 117–124.

James, N.P. (1997) The cool-water carbonate depositional realm. *SEPM Special Publication* **56**, 1–20.

James, N.P., Feary, D.A., Betzler, C., Bone, Y., Holbourn, A.E., Li, Q., Machiyama, H., Toni Simo, J.A. & Surlyk, F. (2004) Origin of late Pleistocene bryozoan reef mounds; Great Australian Bight. *Journal of Sedimentary Research* **74**, 20–48.

James, N.P., Feary, D.A., Surlyk, F., Simo, J.A., Betzler, C., Holbourn, A.E., Li, Q., Matsuda, H., Machiyama, H., Brooks, G.R., Andres, M.S., Hine, A.C. & Malone, M.J. (2000) Quaternary bryozoan reef-mounds in cool-water, upper-slope environments, Great Australian Bight. *Geology* **28**, 647–650.

James, N.P., Jones, B., Nelson, C.S., Campbell, H.J. & Titjen, J. (2011) Cenozoic temperate and sub-tropical carbonate sedimentation on an oceanic volcano – Chatham Islands, New Zealand. *Sedimentology* **58**, 1007–1029.

Jaramillo-Vogel, D., Bover-Arnal, T. & Strasser, A., Frijia, G. (2016) Bryozoan beds in northern Italy as a shallow-water expression of environmental changes during the Oligocene isotope event 1. *Sedimentary Geology* **331**, 148–161.

Jaramillo-Vogel, D., Strasser, A., Frijia, G. & Spezzaferri, S. (2013) Neritic isotope and sedimentary records of the Eocene–Oligocene greenhouse–icehouse transition: the Calcare di Nago Formation (northern Italy) in a global context. *Palaeogeography, Palaeoclimatology, Palaeoecology* **369**, 361–376.

Jasionowski, M. (2006) Facies and geochemistry of Lower Sarmatian reefs along the northern margins of the Paratethys in Roztocze (Poland) and Medobory (Ukraine) regions: paleoenvironmental implications. *Przeglad Geologiczny* **54**, 445–454. [In Polish, with English abstract.]

Jebram, D. (1969) Bryozoen als Holzschädlinge im Brackwasser. *Kieler Meeresforschungen* **25**, 224–231.

Jebram, D. (1973) The importance of different growth directions in the Phylactolaemata and Gymnolaemata for reconstructing the phylogeny of the Bryozoa. Pp. 565–576 in Larwood, G.P. (ed.) *Living and fossil Bryozoa*. Academic Press, London.

Jebram, D. (1975) Dauerknospen ('Hibernacula') bei den Bryozoa Ctenostomata in mesohalinen und vollmarinen Gewässern. *Marine Biology* **31**, 129–137.

Jebram, D. (1992) The polyphyletic origin of the 'Cheilostomata' (Bryozoa). *Zeitschrift für zoologische Systematik und Evolutionsforschung* **30**, 46–52.

Jenkins, H.L., Bishop, J.D.D. & Hughes, R.N. (2015) Prudent female allocation by modular hermaphrodites: female investment is promoted by the opportunity to outcross in cyclostome bryozoans. *Biological Journal of the Linnean Society* **116**, 593–602.

Jenkins, H.L. & Taylor, P.D. (2014) New or poorly known skeletal features in the early astogeny of some cyclostome bryozoans. *Studi Trentini di Scienze Naturali* **94**, 125–130.

Jenkins, H.L. & Taylor, P.D. (2017) Ancestrular morphology in cyclostome bryozoans and the quest for phylogenetically informative skeletal characters. *Journal of Natural History* **51**, 2849–2861.

Jenkins, H.L., Waeschenbach, A., Okamura, B., Hughes, R.N. & Bishop, J.D.D. (2017) Phylogenetically widespread polyembryony in cyclostome bryozoans and the protracted asynchronous release of clonal brood-mates. *PLoS ONE*, **12**(1), e0170010.

Jenner, R.A. & Littlewood, D.T.J. (2008) Problematica old and new. *Philosophical Transactions of the Royal Society, Series B* **363**, 1503–1512.

Jiménez-Sánchez, A., Anstey, R.L. & Azanza, B. (2010) Description and phylogenetic analysis of *Iberostomata fombuenensis* new genus and species (Bryozoa, Ptilodictyina). *Journal of Paleontology* **84**, 695–708.

Jiménez-Sánchez, A., Taylor, P.D. & Gómez, J.B. (2013) Palaeogeographical patterns in Late Ordovician bryozoan morphology as proxies for temperature. *Bulletin of Geosciences* **88**, 417–426.

Jiménez-Sánchez, A., Vennin, E. & Villas, E. (2015) Trepostomate bryozoans from the upper Katian (Upper Ordovician) of Morocco: gigantism in high latitude Gondwana platforms. *Journal of Paleontology* **89**, 195–221.

Jiménez-Sánchez, A. & Villas, E. (2009) The bryozoan dispersion into the Mediterranean margin of Gondwana during the pre-glacial Late Ordovician. *Palaeogeography, Palaeoclimatology, Palaeoecology* **294**, 220–231.

Johnson, C. & Wendt, D. (2007) Availability of dissolved organic matter offsets metabolic costs of a protracted larval period for *Bugula neritina* (Bryozoa). *Marine Biology* **151**, 301–311.

Johnson, C.H. (2010) Effects of selfing on offspring survival and reproduction in a colonial simultaneous hermaphrodite (*Bugula stolonifera*, Bryozoa). *Biological Bulletin* **219**, 27–37.

Johnson, M.E. (1992) Studies in ancient rocky shores: brief history and annotated bibliography. *Journal of Coastal Research* **8**, 797–812.

Kaandorp, J.A. (1999) Morphological analysis of growth forms of branching marine sessile organisms along environmental gradients. *Marine Biology* **134**, 295–306.

Kahle, J., Liebezeit, G. & Gerdes, G. (2003) Growth aspects of *Flustra foliacea* (Bryozoa, Cheilostomata) in laboratory culture. *Hydrobiologia* **503**, 237–244.

Kaneko, N. (1993) Bryozoan fossils from the Morozaki Group, Chiba Peninsula, Central Japan. *The Tokai Fossil Society* **4**, 91–94.

Karagodina, N.P., Vishnyakov, A.E., Kotenko, O.N., Maltseva, A.L. & Ostrovsky, A.N. (2018) Ultrastructural evidence for nutritional relationships between a marine colonial invertebrate (Bryozoa) and its bacterial symbionts. *Symbiosis* **75**, 155–164.

Karklins, O.L. (1983) Introduction to the Suborder Ptilodictyina. Pp. 453–488 in Boardman, R.S., Cheetham, A.H., Blake, D.B., Utgaard, J., Karklins, O.L., Cook, P.L., Sandberg, P.A., Lutaud, G. & Wood, T.S. (eds) *Treatise on invertebrate paleontology, part G, Bryozoa, revised*. Geological Society of America and University of Kansas, Boulder and Lawrence.

Karlson, R.H. & Buss, L.W. (1984) Competition, disturbance and local diversity patterns of substratum-bound clonal organisms: a simulation. *Ecological Modelling* **23**, 243–255.

Kaselowsky, J., Scholz, J., Mawatari, S.F., Probert, P.K., Gerdes, G., Kadagies, N. & Hillmer, G. (2005) Bryozoans and microbial communities of cool-temperate to subtropical latitudes – paleoecological implication, I: growth morphologies of shallow-water bryozoans settling on bivalve shells (Japan and New Zealand). *Facies* **50**, 349–361.

Kátcha, P. & Saric, R. (2009) Host preferences in Late Ordovician (Sandbian) epibenthic bryozoans: example from the Zahořany Formation of Prague Basin. *Bulletin of Geosciences* **84**, 169–178.

Kaufmann, K.W. (1971) The form and functions of the avicularia of *Bugula* (Phylum Ectoprocta). *Postilla* **151**, 1–26.

Keij, A.J. (1972) *Sylonika* and *Kylonisa*, two new Palaeogene bryozoan genera (Cheilostomata, Skyloniidae). *Scripta Geologica* **11**, 1–15.

Kelly, S.M. & Horowitz, A.S. (1987) Growth forms and palaeoecology of Mississippian bryozoans: critical applications of Stach's 1936 model, eastern United States. Pp. 137–144 in Ross, J.R.P. (ed.) *Bryozoa: present and past*. Western Washington University, Bellingham.

Kelmo, F., Attrill, M.J., Gomes, R.C.T. & Jones, M.B. (2004) El Niño induced local extinction of coral reef bryozoan species from Northern Bahia, Brazil. *Biological Conservation* **118**, 609–617.

Keough, M.J. (1984) Kin-recognition and the spatial distribution of larvae of the bryozoan *Bugula neritina* (L.). *Evolution* **38**, 142–147.

Keough, M.J. (1986) The distribution of a bryozoan on seagrass blades: settlement, growth, and mortality. *Ecology* **67**, 846–857.

Keough, M.J. (1989) Variation on growth rate and reproduction of the bryozoan *Bugula neritina*. *Biological Bulletin* **177**, 277–286.

Kershaw, S. (1980) Cavities and cryptic faunas beneath non-reef stromatoporoids. *Lethaia* **13**, 327–338.

Kesling, R.V., Hoare, R.D. & Sparks, D.K. (1980) Epizoans of the Middle Devonian brachiopod *Paraspirifer bownockeri*: their relationships to one another and to their host. *Journal of Paleontology* **54**, 1141–1154.

Key, M.M., Jr (1987) Partitioning of morphologic variation across stability gradients in Upper Ordovician trepostomes. Pp. 144–152 in Ross, J.R.P. (ed.) *Bryozoa: present and past*. Western Washington University, Bellingham.

Key, M.M., Jr (1990) Intracolony variation in skeletal growth rates in Paleozoic ramose trepostome bryozoans. *Paleobiology* **16**, 483–491.

Key, M.M., Jr & Barnes, D.K.A. (1999) Bryozoan colonization of the marine isopod *Glyptonotus antarcticus* at Signy Island, Antarctica. *Polar Biology* **21**, 48–55.

Key, M.M., Jr, Hollenbeck, P.M., O'Dea, A. & Patterson, W.P. (2013) Stable isotope profiling in modern marine bryozoan colonies across the Isthmus of Panama. *Bulletin of Marine Science* **89**, 837–856.

Key, M.M., Jr, Hyžný, M., Khosravi, E., Hudáčkova, N., Robin, N. & Ataabadi, M.M. (2017) Bryozoan epibiosis on fossil crabs: a rare occurrence from the Miocene of Iran. *Palaios* **32**, 491–505.

Key, M.M., Jr, Jeffries, W.B. & Voris, H.K. (1995) Epizoic bryozoans, sea snakes and other nektonic substrates. *Bulletin of Marine Science* **56**, 462–474.

Key, M.M., Jr, Jeffries, W.B., Voris, H.K. & Yang, C.M. (1996) Epizoic bryozoans, horseshoe crabs, and other mobile benthic substrates. *Bulletin of Marine Science* **58**, 368–384.

Key, M.M., Jr, Jeffries, W.B., Voris, H.K. & Yang, C.M. (2000) Bryozoan fouling pattern on the horseshoe crab *Tachypleus gigas* (Müller) from Singapore. Pp. 265–271 in Herrera Cubilla, A. & Jackson, J.B.C. (eds) *Proceedings of the 11th International Bryozoology Association Conference*. Smithsonian Tropical Research Institute, Balboa.

Key, M.M., Jr, Knauff, J.B. & Barnes, D.K.A. (2012) Epizoic bryozoans on predatory pycnogonids from the South Orkney Islands, Antarctica: 'If You Can't Beat Them, Join Them'. Pp. 137–153 in Ernst, A., Schäfer, P. & Scholz, J. (eds) *Bryozoan studies 2010*. Springer, Heidelberg.

Key, M.M., Jr, Rossi, R.K., Smith, A.M., Hageman, S.J. & Patterson, W.P. (2018) Stable isotope profiles of skeletal carbonate validate annually-produced growth checks in the bryozoan *Melicerita chathamensis* from Snares Platform, New Zealand. *Bulletin of Marine Science* **94**, 1447–1464.

Key, M.M., Jr, Schuhmacher, G.A., Babcock, L.E., Frey, R.C., Heimbrock, W.P., Felton, S.H.,

Cooper, D.L., Gibson, W.B., Scheid, D.G. & Schuhmacher, S.A. (2010) Paleoecology of commensal epizoans fouling *Flexicalymene* (Trilobita) from the Upper Ordovician, Cincinnati arch region, USA. *Journal of Paleontology* **84**, 1121–1134.

Key, M.M., Jr, Thrane, L. & Collins, J.A. (2001) Space-filling problems in ramose trepostome bryozoans as exemplified in a giant colony from the Permian of Greenland. *Lethaia* **34**, 125–135.

Key, M.M., Jr, Thrane, L. & Collins, J.A. (2002) Functional morphology of maculae in a giant ramose bryozoan from the Permian of Greenland. Pp. 163–170 in Wyse Jackson, P.N., Buttler, C.J. & Spencer Jones, M.E. (eds) *Bryozoan studies 2001: Proceedings of the 12th International Bryozoology Association Symposium*. Balkema, Lisse.

Key, M.M., Jr, Winston, J.E., Volpe, J.W., Jeffries, W.B. & Voris, H.K. (1999) Bryozoan fouling of the blue crab *Callinectes sapidus* at Beaufort, North Carolina. *Bulletin of Marine Science* **64**, 513–533.

Key, M.M., Wyse Jackson, P.N. & Felton, S.H. (2016) Intracolony variation in colony morphology in reassembled fossil ramose stenolaemate bryozoans from the Upper Ordovician (Katian) of the Cincinnati Arch region, USA. *Journal of Paleontology* **90**, 400–412.

Key, M.M., Jr, Wyse Jackson, P.N., Håkansson, E., Patterson, W.P. & Moore, M.D. (2005) Gigantism in Permian trepostomes from Greenland: testing the algal symbiosis hypothesis using δ13C and δ18O values. Pp. 141–151 in Moyano G., H.I., Cancino, J.M. & Wyse Jackson, P.N. (eds) *Bryozoan studies 2004*. Balkema, Leiden.

Key, M.M., Jr, Wyse Jackson, P.N., Miller, K.E. & Patterson, W.P. (2008) A stable isotope test for the origin of fossil brown bodies in trepostome bryozoans from the Ordovician of Estonia. *Virginia Museum of Natural History, Special Publication* **15**, 75–84.

Key, M.M., Jr, Wyse Jackson, P.N. & Vitiella, L.J. (2011) Stream channel network analysis applied to colony-wide feeding structures in a Permian bryozoan from Greenland. *Paleobiology* **37**, 287–302.

Key, M.M., Jr, Zagorsek, K. & Patterson, W.P. (2013) Paleoenvironmental reconstruction of the Early to Middle Miocene Central Paratethys using stable isotopes from bryozoan skeletons. *International Journal of Earth Sciences* **102**, 305–318.

Keynes, R.D. (2003) From bryozoans to tsunami: Charles Darwin's findings on the *Beagle*. *Proceedings of the American Philosophical Society* **147**, 103–127.

Kidwell, S.M. & Gyllenhaal, E.D. (1998) Symbiosis, competition, and physical disturbance in the growth histories of Pliocene cheilostome bryoliths. *Lethaia* **31**, 221–239.

Kiessling, T., Gutow, L. & Thiel, M. (2015) Marine litter as habitat and dispersal vector. Pp. 141–181 in Bergmann, M., Gutow, L., & Klages, M. (eds) *Marine anthropogenic litter*. Springer, Cham.

Kiessling, W. & Simpson, C. (2011) On the potential for ocean acidification to be a general cause of ancient reef crises. *Global Change Biology* **17**, 56–67.

Kipp, R., Bailey, S.A., MacIsaac, H.J. & Ricciardi, A. (2010) Transoceanic ships as vectors for nonindigenous freshwater bryozoans. *Diversity and Distributions* **16**, 77–83.

Kirkpatrick, R. & Metzelaar, J. (1922) On an instance of commensalism between a hermit crab and a polyzoon. *Journal of Zoology* **92**, 983–990.

Kittelmann, S. & Harder, T. (2005) Species- and site-specific bacterial communities associated with four encrusting bryozoans from the North Sea, Germany. *Journal of Experimental Marine Biology and Ecology* **327**, 201–209.

Klicpera, A., Michel, J. & Westphal, H. (2015) Facies patterns of a tropical heterozoan carbonate platform undereutrophic conditions: the Banc d'Arguin, Mauritania. *Facies* **421**, 420–444.

Klicpera, A., Taylor, P.D. & Westphal, H. (2013) Bryoliths constructed by bryozoans in symbiotic associations with hermit crabs in a tropical heterozoan carbonate system, Golfe d'Arguin, Mauritania. *Marine Biodiversity* **43**, 429–444.

Klompmaker, A.A., Portella, R.W., Lad, S.E. & Kowelawski, M. (2015) The fossil record of drilling predation on barnacles. *Palaeogeography, Palaeoclimatology, Palaeoecology* **426**, 95–111.

Knight, S., Gordon, D.P. & Lavery, S.D. (2011) A multi-locus analysis of phylogenetic relationships within cheilostome bryozoans supports multiple origins of ascophoran frontal shields. *Molecular Phylogenetics and Evolution* **61**, 351–362.

Knoerich, A.C. & Mutti, M. (2003) Controls of facies and sediment composition on the diagenetic pathway of shallow-water heterozoan carbonates: the Oligocene of the Maltese Islands. *International Journal of Earth Sciences* **92**, 494–510.

Knoll, A.H. (2003) Biomineralization and evolutionary history. *Reviews in Mineralogy and Geochemistry* **54**, 329–356.

Knoll, A.H., Bambach, R.K., Payne, J.L., Pruss, S. & Fischer, W.W. (2007) Paleophysiology and end-Permian mass extinction. *Earth and Planetary Science Letters* **256**, 295–313.

Knowles, T. (2008) The cheilostome bryozoan *Floridina* from Plio-Pleistocene deposits of the coastal plain of North America. *Virginia Museum of Natural History Special Publication* **15**, 85–92.

Knowles, T., Leng, M.J., Williams, M., Taylor, P.D., Sloane, H.J. & Okamura, B. (2010) Interpreting seawater temperature range using oxygen isotopes and zooid size variation in *Pentapora foliacea* (Bryozoa). *Marine Biology* **157**, 1171–1180.

Knowles, T., Taylor, P.D., Williams, M., Haywood, A.M. & Okamura, B. (2009) Pliocene seasonality across the North Atlantic inferred from cheilostome bryozoans. *Palaeogeography, Palaeoclimatology, Palaeoecology* **277**, 226–235.

Kobluk, D.R., Cuffey, R.J., Fonda, S.S. & Lysenko, M.A. (1988) Cryptic Bryozoa, leeward fringing reef of Bonaire, Netherlands Antilles, and their paleoecological application. *Journal of Paleontology* **62**, 427–439.

Koçak, F., Balduzzi, A. & Benli, H.A. (2002) Epiphytic bryozoan community of *Posidonia oceanica* (L.) Delile meadow in the northern Cyprus (Eastern Mediterranean). *Indian Journal of Marine Sciences* **31**, 235–238.

Kocsis, A.T., Reddin, C.J. & Kiessling, W. (2018) The stability of coastal benthic biogeography over the last 10 million years. *Global Ecology and Biogeography* **27**, 1106–1120.

Kohring, R. & Pint, A. (2005) Fossile Süßwasserbryozoen – vorkommen, überlief-erung, fundbedingungen. *Denisia* **16**, 95–102.

Koletic, N., Novosel, M., Rajevic, N. & Franjevic, D. (2014) Bryozoans are returning home: recolonization of freshwater ecosystems inferred from phylogenetic relationships. *Ecology and Evolution* **5**, 255–264.

Kollár, P., Rajchard, J., Balounová, Z. & Pazourek, J. (2014) Marine natural products: bryostatins in preclinical and clinical studies. *Pharmaceutical Biology* **52**, 237–242.

Koromyslova, A.V. (2011) Bryozoans of the Latorp and Volkhov Horizons (Lower-Middle Ordovician) of the Leningrad Region. *Paleontological Journal* **45**, 887–980.

Koromyslova, A.V. (2014a) The earliest calcified opercula of bryozoans of the order Cheilostomata. *Paleontological Journal* **48**, 587–593.

Koromyslova, A.V. (2014b) Morphological features and systematic position of the bryozo-ans *Onychocella rowei* and *O. mimosa* (Cheilostomata) from a Campanian erratic block (Belarus). *Paleontological Journal* **48**, 275–286.

Koromyslova, A.V., Baraboshkin, E. Yu. & Martha, S.O. (2018) Late Campanian to late Maastrichtian bryozoans encrusting on belem-nite rostra from the Aktolagay Plateau in western Kazakhstan. *Geobios* **51**, 307–333.

Koromyslova, A.V., Martha, S.O. & Pakhnevich, A.V. (2018) The internal morphology of *Acoscinopleura* Voigt, 1956 (Cheilostomata, Bryozoa) from the Campanian–Maastrichtian of Central and Eastern Europe. *PalZ* **92**, 241–266.

Koromyslova, A.V. & Pakhnevich, A.V. (2016) Species of *Pachydermopora* Gordon, 2012 and *Beisselina* Canu, 1913 (Bryozoa: Cheilostomida) from a Campanian erratic block (Belarus) and their micro-CT investigation. *Paleontological Journal* **50**, 41–53.

Kouchi, N., Nakaoka, M. & Mukai, H. (2006) Effects of temoral dynamics and vertical structure of the seagrass *Zostera caulescens* on distribution and recruitment of the epifaunal encrusting Bryozoa *Microporella trigonellata*. *Marine Ecology* **27**, 145–153.

Kowalewski, M., Dulai, A. & Fürsich, F.T. (1998) A fossil record full of holes: the Phanerozoic history of drilling predation. *Geology* **26**, 1091–1094.

Krause, F.F., Scotese, C.R., Nieto, C., Sayegh, S.G., Hopkins, J.C. & Meyer, R.O. (2004) Paleozoic stromatactis and zebra carbonate mud-mounds: global abundance and paleo-geographic distribution. *Geology* **32**, 181–184.

Kröger, B., Descrochers, A. & Ernst, A. (2017) The reengineering of reef habitats during the Great Ordovician Biodiversification Event. *Palaios* **32**, 584–599.

Krzeminska, M., Kuklinski, P., Najorka, J. & Iglikowska, A. (2016) Skeletal mineralogy patterns of Antarctic Bryozoa. *The Journal of Geology* **124**, 411–422.

Kuklinski, P. (2009) Ecology of stone-encrusting organisms in the Greenland Sea – a review. *Polar Research* **28**, 222–237.

Kuklinski, P. & Bader, B. (2007) Diversity, structure and interactions of encrusting litho-phyllic macrofaunal assemblages from Belgica Bank, East Greenland. *Polar Biology* **30**, 709–717.

Kuklinski, P. & Barnes, D.K.A. (2005) Microhabitat diversity of Svalbard Bryozoa. *Journal of Natural History* **39**, 539–554.

Kuklinski, P. & Barnes, D.K.A. (2010) First bipolar benthic brooder. *Marine Ecology Progress Series* **410**, 15–20.

Kuklinski, P., Barnes, D.K.A. & Taylor, P.D. (2006) Latitudinal patterns of diversity and abundance in North Atlantic intertidal boulder-fields. *Marine Biology* **149**, 1577–1583.

Kukliński, P., Barnes, D.K.A. & Wlodarska-Kowalczuk, M. (2008) Gastropod shells, hermit crabs and Arctic bryozoan richness. *Virginia Museum of Natural History, Special Publication* **15**, 93–100.

Kuklinski, P., Gulliksen, B., Lønne, O.J. & Weslawski, J.M. (2006) Substratum as a structuring influence on assemblages of Arctic bryozoans. *Polar Biology* **29**, 652–661.

Kukliński, P. & Porter, J.S. (2004) *Alcyonidium disciforme*: an exceptional Arctic bryozoan. *Journal of the Marine Biological Association of the United Kingdom* **84**, 267–275.

Kuklinski, P., Sokolowski, A., Ziolkowska, M., Balazy, P., Novosel, M. & Barnes, D.K.A. (2012) Growth rate of selected sheet-encrusting bryozoan colonies along a latitudinal transect: preliminary results. Pp. 155–167 in Ernst, A., Schäfer, P. & Scholz, J. (eds) *Bryozoan studies 2010*. Springer, Heidelberg.

Kuklinski, P. & Taylor, P.D. (2006) Unique life history strategy in a successful Arctic bryozoan, *Harmeria scutulata*. *Journal of the Marine Biological Association of the United Kingdom* **86**, 1305–1314.

Kuklinski, P. & Taylor, P.D. (2008) Are bryozoans adapted for living in the Arctic? *Virginia Museum of Natural History, Special Publication* **15**, 101–100.

Kuklinski, P. & Taylor, P.D. (2009) Mineralogy of Arctic bryozoan skeletons in a global context. *Facies* **55**, 489–500.

Kuklinski, P., Taylor, P.D., Denisenko, N.V. & Berning, B. (2013) Atlantic origin of the Arctic biota? Evidence from phylogenetic and biogeographical analysis of the cheilostome bryozoan genus *Pseudoflustra*. *PLoS ONE* **8**, e59152.

Lagaaij, R. (1963a) *Cupuladria canariensis* (Busk) – portrait of a bryozoan. *Palaeontology* **6**, 172–217.

Lagaaij, R. (1963b) New additions to the bryozoan fauna of the Gulf of Mexico. *Institute of Marine Science, Texas, Publication* **9**, 181–236.

Lagaaij, R. (1968) Fossil Bryozoa reveal long-distance sand transport along the Dutch coast.
Proceedings of the Koninklijke Nederlandse Akademie van Wetenschappen, Series B **71**, 31–50.

Lagaaij, R. & Gautier, Y.V. (1965) Bryozoan assemblages from marine sediments of the Rhone delta, France. *Micropaleontology* **11**, 39–58.

Landing, E., Antcliffe, J.B., Brasier, M.D. & English, A.B. (2015) Distinguishing Earth's oldest known bryozoan (*Pywackia*, late Cambrian) from pennatulacean octocorals (Mesozoic–Recent). *Journal of Paleontology* **89**, 292–317.

Landing, E., English, A. & Keppie, J.D. (2010) Cambrian origin of all skeletalized metazoan phyla – discovery of Earth's oldest bryozoans (Upper Cambrian, southern Mexico). *Geology* **38**, 547–550.

Lang, W.D. (1919) The Pelmatoporinae, an essay on the evolution of a group of Cretaceous Polyzoa. *Philosophical Transactions of the Royal Society, Series B* **209**, 191–228.

Larsen, N. & Håkansson, E. (2000) Microfacies mosaics across latest Maastrichtian bryozoan mounds in Denmark. Pp. 272–281 in Herrera Cubilla, A. & Jackson, J.B.C. (eds) *Proceedings of the 11th International Bryozoology Association Conference*. Smithsonian Tropical Research Institute, Balboa.

Larsen, P.S. & Riisgård, H.U. (2002) On ciliary sieving and pumping in bryozoans. *Journal of Sea Research* **48**, 181–195.

Larwood, G.P. & Taylor, P.D. (1979) Early structural and ecological diversification in the Bryozoa. Pp. 209–234 in House, M.R. (ed.) *The origin of major invertebrate groups*. Academic Press, London.

Larwood, G.P. & Taylor, P.D. (1981) Mesozoic bryozoan evolution: response to increasing predation pressure? Pp. 312–313 in Larwood, G.P. & Nielsen, C. (eds) *Recent and fossil Bryozoa*. Olsen & Olsen, Fredensborg.

Larwood, G.P., Voigt, E. & Scholz, J. (2008) Palaeoecological, morphological and taxonomic of the pelmatoporinid genus *Ubaghsia* Jullien from the Maastricht Chalk. *Virginia Museum of Natural History, Special Publication* **15**, 111–121.

Laubier, L. (1959) Contribution a la faunistique de coraligene. III. Deux spionidiens inquilins sur des bryozoaires chilostomides. *Vie et Milieu* **10**, 347–349.

Laumer, C.E., Fernández, Rosa, Lemer, S., Combosch, D., Kocot, K.M., Riesgo, A., Andrade, S.C.S., Sterrer, W., Sørensen, M.V. & Giribet, G. (2019) Revisiting metazoan phylogeny with genomic sampling of all phyla. *Proceedings of the Royal Society, Series B* **286**, 20190831.

Lee, D.E., Scholz, J. & Gordon, D.P. (1997) Paleoecology of a Late Eocene mobile rock-ground from North Otago, New Zealand. *Palaios* **12**, 568–581.

Lees, A. (1964) The structure and origin of the Waulsortian (Lower Carboniferous) 'reefs' of west-central Eire. *Philosophical Transations of the Royal Society, Series B* **247**, 483–531.

Lamarti-Sefian, N., Pouyet, S., El Hajjaji, K., André, J.P. & Ben Moussa, A. (1997) Les bryozoarites du Miocène supérieur du bassin de Charf el Akab (Maroc nord occidental): sédimentologie et paléoenvironnements. *Géologie Méditerranéenne* **24**, 161–173.

Lescinsky, H.L. (1997) Epibiont communities: recruitment and competition on North American Carboniferous brachiopods. *Journal of Paleontology* **71**, 34–53.

Lescinsky, H.L., Ledesma-Vázquez, J. & Johnson, M.E. (1991) Dynamics of Late Cretaceous Rocky Shores (Rosario Formation) from Baja California, Mexico. *Palaios* **6**, 126–141.

Lev, S.M., Key, M.M., Jr & Lighthart, A. (1993) A paleobiologic test for diastems using the internal stratigraphy of trepostome bryozoans. *Journal of the Pennsylvania Academy of Science* **67**, 32–37.

Levinsen, G.M.R. (1912) Studies of the Cyclostomata Operculata. *Danske kgl Vidensk Selsk Skr* **7**, 1–52.

Liddell, W.D. & Brett, C.E. (1982) Skeletal overgrowths among epizoans from the Silurian (Wenlockian) Waldron Shale. *Paleobiology* **8**, 67–78.

Lidgard, S. (1981) Water flow, feeding, and colony form in an encrusting cheilostome. Pp. 135–142 in Larwood, G.P. & Nielsen, C. (eds) *Recent and fossil Bryozoa*. Olsen and Olsen, Fredensborg.

Lidgard, S. (1985) Budding processes and geometry in encrusting cheilostome bryozoans. Pp. 175–182 in Nielsen, C. & Larwood, G.P. (eds) *Bryozoa: Ordovician to Recent*. Olsen and Olsen, Fredensborg.

Lidgard, S. (1986) Ontogeny in animal colonies: a persistent trend in the bryozoan fossil record. *Science* **232**, 230–232.

Lidgard, S. (1996) Zooid skeletal morphogenesis of some Australian and New Zealand *Adeonellopsis* (Cheilostomatida). Pp. 167–177 in Gordon, D.P., Smith, A.M. & Grant-Mackie, J.A. (eds) *Bryozoans in space and time*. NIWA, Wellington.

Lidgard, S. (2008) Predation on marine bryozoan colonies: taxa, traits and trophic groups. *Marine Ecology Progress Series* **359**, 117–131.

Lidgard, S., Carter, M.C., Dick, M.H., Gordon, D.P. & Ostrovsky, A.N. (2012) Division of labor and recurrent evolution of polymorphisms in a group of colonial animals. *Evolutionary Ecology* **26**, 233–257.

Lidgard, S. & Jackson, J.B.C. (1989) Growth in encrusting cheilostome bryozoans: I. Evolutionary trends. *Paleobiology* **15**, 255–282.

Lidgard, S., McKinney, F.K. & Taylor, P.D. (1993) Competition, clade replacement, and a history of cyclostome and cheilostome bryozoan diversity. *Paleobiology* **19**, 352–371.

Lindquist, N. & Hay, M.E. (1996) Palatability and chemical defense of marine invertebrate larvae. *Ecological Monographs* **66**, 431–450.

Lindskog, A., Eriksson, M.E., Bergström, S.M., Terfelt, F. & Marone, F. (2017) Palaeozoic 'conodont pearls' and other phosphatic microspherules. *Lethaia* **50**, 26–40.

Linneman, J., Paulus, D., Lim-Fong, G. & Lopanik, N.B. (2014) Latitudinal variation of a defensive symbiosis in the *Bugula neritina* (Bryozoa) sibling species complex. *PLoS ONE* **9**(10), e108783.

Liow, L.H., Di Martino, E., Krzeminska, M., Ramsfjell, M., Rust, S., Taylor, P.D. & Voje, K.L. (2017) Relative size predicts competitive outcome through 2 million years. *Ecology Letters*, https://doi.org/10.1111/ele.12795.

Liow, L.H., Di Martino, E., Voje, K.L., Rust, S. & Taylor, P.D. (2016) Interspecific interactions through 2 million years: are competitive outcomes predictable? *Proceedings of the Royal Society, Series B* **283**, 20160981.

Liow, L.H., Reitan, T., Voje, K.L., Taylor, P.D. & Di Martino, E. (2019) Size, weapons, and armor as predictors of competitive outcomes in fossil and contemporary marine communities. *Ecological Monographs* **89**(2), e01354.

Liow, L.H. & Taylor, P.D. (2019) Cope's Rule in a modular organism: directional evolution without an overarching macroevolutionary trend. *Evolution* **73**, 1863–1872.

Litherland, M. (1975) Organic remains and traces from the Dalradian of Benderloch, Argyll. *Scottish Journal of Geology* **11**, 47–50.

Liu, X.-X. (1992) On the genus *Membranipora* (Anasca: Cheilostomata: Bryozoa) from southern Chinese Seas. *Raffles Bulletin of Zoology* **40**, 103–144.

Liuzzi, M.G. & López Gappa, J. 2008. The distribution of colonies of the bryozoan *Antarctothoa bougainvillei* on the red alga *Hymenena laciniata*. *Hydrobiologia* **605**, 65–73.

Loeb, M.J. & Walker, G. (1977) Origin, composition, and function of secretions from pyriform organs and internal sacs of four settling cheiloctenostome bryozoan larvae. *Marine Biology* **42**, 37–46.

Lombardi, C., Cocito, S., Gambi, M.C., Cisterna, B., Flach, F., Taylor, P.D., Keltie, K., Freer, A. & Cusack, M. (2011) Effects of ocean acidification on growth, organic tissue and protein profile of the Mediterranean bryozoan *Myriapora truncata*. *Aquatic Biology* **13**, 251–262.

Lombardi, C., Cocito, S., Hiscock, K., Occhipinti-Ambrogi, A., Setti, M. & Taylor, P.D. (2008) Influence of seawater temperature on growth bands, mineralogy and carbonate production in a bioconstructional bryozoan. *Facies* **54**, 333–342.

Lombardi, C., Taylor, P.D. & Cocito, S. (2010) Systematics of the Miocene–Recent bryozoan genus *Pentapora* (Cheilostomata). *Zoological Journal of the Linnean Society* **160**, 17–39.

Lombardi, C., Taylor, P.D. & Cocito, S. (2014) Bryozoan constructions in a changing Mediterranean Sea. Pp. 373–384 in Goffredo, S. & Zubinsky, Z. (eds) *The Mediterranean Sea: its history and present challenges*. Springer, Dordrecht.

Lombardi, C., Taylor, P.D., Cocito, S., Bertolini, C. & Calosi, P. (2017) Low pH conditions impair module capacity to regenerate in a calcified colonial invertebrate, the bryozoan *Cryptosula pallasiana*. *Marine Environmental Research* **125**, 110–117.

Long, E.R. & Rucker, J.B. (1970) Offshore marine cheilostome Bryozoa from Fort Lauderdale, Florida. *Marine Biology* **6**, 18–25.

Lopanik, N., Lindquist, N. & Targett, N. (2004) Potent cytotoxins produced by a microbial symbiont protect host larvae from predation. *Oecologia* **139**, 131–139.

López Gappa, J.J. (1989) Overgrowth competition in an assemblage of encrusting bryozoans settled on artificial substrata. *Marine Ecology Progress Series* **51**, 121–130.

López Gappa, J. (2000) Species richness of marine Bryozoa in the continental shelf and slope off Argentina (south-west Atlantic). *Diversity and Distributions* **6**, 15–27.

López Gappa, J., Carranza, A., Gianuca, N.M. & Scarabino, F. (2010) *Membraniporopsis tubigera*, an invasive bryozoan in sandy beaches of southern Brazil and Uruguay. *Biological Invasions* **12**, 977–982.

López Gappa, J.J. & Landoni, N.A. (2009) Space utilisation patterns of bryozoans on the Patagonian scallop *Psychrochlamys patagonica*. *Scientia Marina* **73**, 161–171.

López Gappa, J.J. & Liuzzi, M.G. (2008) A new Antarctic *Osthimosia* (Bryozoa, Cheilostomata, Celleporidae) with dimorphic zooids. *Polar Biology* **32**, 47–51.

Lörz, A-N., Myers, A. & Gordon, D.P. (2014) An inquiline deep-water bryozoan/amphipod association from New Zealand, including the description of a new genus and species of Chevaliidae. *European Journal of Taxonomy* **72**, 1–17.

Lowenstam, H.A. (1981) Minerals formed by organisms. *Science* **211**, 1126–1131.

Loxton, J., Kuklinski, P., Barnes, D.K.A., Najorka, J., Spencer Jones, M. & Porter, J.S. (2014a) Variability of Mg-calcite in Anatrctic bryozoan skeletons across spatial scales. *Marine Ecology Progress Series* **507**, 169–180.

Loxton, J., Kuklinski, P., Mair, J.M., Spencer Jones, M. & Porter, J.S. (2012) Patterns of magnesium-calcite distribution in the skeleton of some polar bryozoan species. Pp. 169–185 in Ernst, A., Schäfer, P. & Scholz, J. (eds) *Bryozoan studies 2010*. Lecture Notes in Earth System Sciences **143**. Springer, Berlin.

Loxton, J., Kuklinski, P., Najorka, J., Spencer Jones, M. & Porter, J.S. (2014b) Variability in the skeletal mineralogy of temperate bryozoans: the relative influence of environmental and biological factors. *Marine Ecology Progress Series* **510**, 45–57.

Loxton, J., Spencer Jones, M., Najorka, J., Smith, A.M. & Porter, J.S. (2018) Skeletal carbonate mineralogy of Scottish bryozoans. *PLoS ONE* **13**, e0197533.

Luci, L. & Cichowolski, M. (2014) Encrustation in nautilids: a case study in the Cretaceous species *Cymatoceras perstriatum*, Neuquén Basin, Argentina. *Palaios* **29**, 101–120.

Luci, L., Cichowolski, M. & Aguirre-Urreta, M.B. (2016) Sclerobionts, shell morphology and biostratinomy on ammonites: two early Cretaceous cases from the Neuquén Basin, Argentina. *Palaios* **31**, 41–54.

Lutaud, G. (1985) Preliminary experiments on interzooidal metabolic transfer in anascan bryozoans. Pp. 183–191 in Nielsen, C. & Larwood, G.P. (eds) *Bryozoa: Ordovician to Recent*. Olsen & Olsen, Fredensborg.

Ma, J., Buttler, C.J. & Taylor, P.D. (2014) Cladistic analysis of the 'trepostome' Suborder Esthonioporina and the systematics of Palaeozoic bryozoans. *Studi Trentini di Scienze Naturali* **94**, 153–161.

Ma, J.-Y., Taylor, P.D. & Feng-Sheng, X. (2014) New observations on the skeletons of the earliest bryozoans from the Fenhsiang Formation (Tremadocian, Lower Ordovician), Yichang, China. *Palaeoworld* **23**, 25–30.

Ma, J., Taylor, P.D., Fengsheng, X. & Renbin, Z. (2015) The oldest known bryozoan: *Prophyllodictya* (Cryptostomata) from the Lower Tremadocian (Lower Ordovician) of Liujiachang, south-western Hubei, central China. *Palaeontology* **58**, 925–934.

Machiyama, H., Yamada, T., Kaneko, N., Iryu, Y., Odawara, K., Asami, R., Matsuda, H., Mawatari, S.F., Bone, Y. & James, N.P. (2003) Carbon and oxygen isotopes of cool-water bryozoans from the Great Australian Bight, and their paleoenvironmental significance. *Proceedings of the Ocean Drilling Programme, Scientific Results* **182**, 1–29.

Mackie, G.O. (1986) From aggregates to integrates: physiological aspects of modularity in colonial animals. *Philosophical Transactions of the Royal Society, Series B* **313**, 175–196.

MacLeod, N., Rawson, P.F., Forey, P.L., Banner, F.T., Boudagher-Fadel, M.K., Bown, P.R., Burnett, J.A., Chambers, P., Culver, S., Evans, S.E., Jeffrey, C., Kaminski, M.A., Lord, A.R., Milner, A.C., Milner, A.R., Morris, N., Owen, E., Rosen, B.R., Smith, A.B., Taylor, P.D., Urquhart, E. & Young, J.R. (1997) The Cretaceous–Tertiary biotic transition. *Journal of the Geological Society* **154**, 265–292.

Madsen, L. (1987) Growth and polypide morphology in some ramose trepostome bryozoans from the Permo-Carboniferous of the Arctic. Pp. 169–176 in Ross, J.R.P. (ed.) *Bryozoa: present and past*. Western Washington University, Bellingham.

Madsen, L. & Håkansson, E. (1989) Upper Palaeozoic bryozoans from the Wandel Sea Basin, north Greenland. *Rapport Grønlands Geologiske Undersøgelse* **144**, 43–52.

Madurell, T., Zabala, M., Dominguez-Carrió, C. & Gili, J.M. (2013) Bryozoan faunal composition and community structure from the continental shelf off Cap de Creus (Northwestern Mediterranean). *Journal of Sea Research* **83**, 123–136.

Mahler, D.L., Weber, M.G., Wagner, C.E. & Ingram, T. (2017) Pattern and process in the comparative study of convergent evolution. *The American Naturalist* **190**(S1), S13–S28.

Mallela, J. (2013) Calcification by reef-building sclerobionts. *PLoS ONE* **8**, e60010.

Manríquez, P.H. & Cancino, J.M. (1996) Bryozoan-macroalgal interactions: do epibionts benefit? *Marine Ecology Progress Series* **138**, 189–197.

Manríquez, P.H., Hughes, R.N. & Bishop, J.D.D. (2001) Age-dependent loss of fertility in water-borne sperm of the bryozoan *Celleporella hyalina*. *Marine Ecology Progress Series* **224**, 87–92.

Marcus, E. (1941) Sôbre o desenvolvimento do bryozoario *Synnotum aegyptiacum*. *Arquivos de Cirurgia Clinica e Experimental* **5**, 227–234.

Markham, J.B. & Ryland, J.S. (1987) Function of the gizzard in Bryozoa. *Journal of Experimental Marine Biology and Ecology* **107**, 21–37.

Mariani, S. (2003) Recruitment in invertebrates with short-lived larvae: the case of the bryozoan *Disporella hispida* (Fleming). *Helgolander Marine Research* **57**, 47–53.

Martha, S.O. & Taylor, P.D. (2017) The oldest erect cheilostome bryozoan: *Jablonskipora* gen. nov. from the Upper Albian of south-west England. *Papers in Palaeontology* **4**, 55–66.

Martha, S.O., Taylor, P.D. & Rader, W.L. (2019) Early Cretaceous gymnolaemate bryozoans from the early to middle Albian of the Glen Rose and Walnut formations of Texas, USA. *Journal of Paleontology* **93**, 260–277.

Martin, R.E. (1996) Secular increase in nutrient levels through the Phanerozoic: implications for productivity, biomass, and diversity of the marine biosphere. *Palaios* **11**, 209–219.

Mawatari, S. (1951) The natural history of a common fouling bryozoan, *Bugula neritina* (Linnaeus). *Miscellaneous Reports of the Research Institute for Natural Resources, Tokyo* **19–21**, 47–54.

Mawatari, S. (1952) On *Watersipora cucullata* (Busk). 2. Anatomical study. *Miscellaneous Reports of the Research Institute for Natural Resources, Tokyo* **28**, 17–27.

Mawatari, S. (1968) A pedunculate ctenostome from Antarctic region. *Proceedings of the Japanese Society of Systematic Zoology* **4**, 42–45.

Mawatari, S.F. (1986) A new genus and species of celleporid bryozoan with ancestrular triad from Hokkaido, Japan. *Journal of Natural History* **20**, 193–202.

Maynard Smith, J. & Brown, R.L.W. (1986) Competition and body size. *Theoretical Population Biology* **30**, 166–179.

McClelland, H.L.O., Taylor, P.D., O'Dea, A. & Okamura, B. (2014) Revising and refining the bryozoan zs-MART seasonality proxy. *Palaeogeography, Palaeoclimatology, Palaeoecology* **410**, 412–420.

McCoy, V.E. & Anstey, R.L. (2010) Biogeographic associations of Silurian bryozoan genera in North America, Baltica and Siberia. *Palaeogeography, Palaeoclimatology, Palaeoecology* **297**, 420–427.

McCuller, M.I. & Carlton, J.T. (2018) Transoceanic rafting of Bryozoa (Cyclostomata, Cheilostomata, and Ctenostomata) across the North Pacific Ocean on Japanese tsunami marine debris. *Aquatic Invasions* **13**, 137–162.

McGhee, G.R., Jr & McKinney, F.K. (2000) A theoretical morphologic analysis of convergently evolved erect helical colony form in the Bryozoa. *Paleobiology* **26**, 556–577.

McKinney, F.K. (1977a) Autozooecial budding patterns in dendroid Paleozoic bryozoans. *Journal of Paleontology* **51**, 303–329.

McKinney, F.K. (1977b) Functional interpretation of lyre-shaped Bryozoa. *Paleobiology* **3**, 90–97.

McKinney, F.K. (1981a) Intercolony fusion suggests polyembryony in Paleozoic fenestrate bryozoans. *Paleobiology* **7**, 247–251.

McKinney, F.K. (1981b) Planar branch systems in colonial suspension feeders. *Paleobiology* **7**, 344–354.

McKinney, F.K. (1983) Asexual colony multiplication by fragmentation: an important mode of genet longevity in the Carboniferous bryozoan *Archimedes*. *Paleobiology* **9**, 35–43.

McKinney, F.K. (1986a) Evolution of erect marine bryozoan faunas: repeated success of unilaminate species. *The American Naturalist* **128**, 795–809.

McKinney, F.K. (1986b) Historical record of erect bryozoan growth forms. *Proceedings of the Royal Society, Series B* **228**, 133–149.

McKinney, F.K. (1987) Paleobiological interpretation of some skeletal characters of Lower Devonian fenestrate Bryozoa, Prague Basin, Czechoslovakia. Pp. 161–168 in Ross, J.R.P. (ed.) *Bryozoa: present and past*. Western Washington University, Bellingham.

McKinney, F.K. (1988) Elevation of lophophores by exposed introverts in Bryozoa: a gymnolaemate character recorded in some stenolaemate species. *Bulletin of Marine Science* **43**, 317–322.

McKinney, F.K. (1990) Feeding and associated colonial morphology in marine bryozoans. *Reviews in Aquatic Sciences* **2**, 255–280.

McKinney, F.K. (1991) Colonial feeding currents of *Exidmonea atlantica* (Cyclostomata). *Bulletin de la Société des Sciences Naturelle de l'Ouest de la France, Mémoire* **H.S. 1**, 263–270.

McKinney, F.K. (1992) Competitive interactions between related clades: evolutionary implications of overgrowth interactions between encrusting cyclostome and cheilostome bryozoans. *Marine Biology* **114**, 645–652.

McKinney, F.K. (1993) A faster-paced world? Contrasts in biovolume and process rates in cyclostome (Class Stenolaemata) and cheilostome (Class Gymnolaemata) bryozoans. *Paleobiology* **19**, 335–351.

McKinney, F.K. (1994) The bryozoan genera *Lyropora* and *Lyroporidra* (Order Fenestrida, Family Polyporidae) in Upper Mississippian (Chesterian) Rocks of eastern North America. *American Museum Novitates* **3111**, 1–31.

McKinney F.K. (1995a) One hundred million years of competitive interactions between bryozoan clades: asymmetrical but not escalating. *Biological Journal of the Linnean Society* **56**, 465–481.

McKinney F.K. (1995b) Taphonomic effects and preserved overgrowth relationships among encrusting marine organisms. *Palaois* **10**, 279–282.

McKinney, F.K. (1996) Encrusting organisms on co-occurring disarticulated valves of two marine bivalves: comparison of living assemblages and skeletal residues. *Paleobiology* **22**, 543–567.

McKinney, F.K. (1998) Avicularia-like structures in a Paleozoic fenestrate bryozoan. *Journal of Paleontology* **72**, 819–826.

McKinney, F.K. (2000) Phylloporinids and the phylogeny of the Fenestrida. Pp. 54–65 in Herrera Cubilla, A. & Jackson, J.B.C. (eds) *Proceedings of the 11th International Bryozoology Association Conference*. Smithsonian Tropical Research Institute, Balboa.

McKinney, F.K. (2009) Bryozoan-hydroid symbiosis and a new ichnogenus, *Caupokeras*. *Ichnos* **16**, 193–201.

McKinney, F.K., Broadhead, T.W. & Gibson, M.A. (1990) Coral-bryozoan mutualism: structural innovation and greater resource exploitation. *Science* **248**, 466–468.

McKinney, F.K. & Burdick, D.W. (2001) A rare, larval-founded colony of the bryozoan *Archimedes* from the Carboniferous of Alabama. *Palaeontology* **44**, 855–859.

McKinney, F.K. & Hageman, S.J. (2006) Paleozoic to modern marine ecological shift displayed in the northern Adriatic Sea. *Geology* **34**, 881–884.

McKinney, F.K., Hageman, S.J. & Jaklin, A. (2007) Crossing the ecological divide: Paleozoic to modern marine ecosystem in the Adriatic Sea. *The Sedimentary Record* **5**(2), 4–8.

McKinney, F.K. & Jackson, J.B.C. (1989) *Bryozoan evolution*. Unwin Hyman, Boston.

McKinney, F.K. & Jaklin, A. (1993) Living populations of free-lying bryozoans: implications for post-Paleozoic decline of the growth habit. *Lethaia* **26**, 171–179.

McKinney, F.K. & Jaklin, A. (2000) Spatial niche partitioning in the *Cellaria* meadow epibiont association, northern Adriatic Sea. *Cahiers de Biologie Marine* **41**, 1–17.

McKinney, F.K., Lidgard, S., Sepkoski, J.J., Jr & Taylor, P.D. (1998) Decoupled temporal patterns of evolution and ecology in two post-Paleozoic clades. *Science* **281**, 807–809.

McKinney, F.K., Lidgard, S. & Taylor, P.D. (2001) Macroevolutionary trends: perception depends on the measure used. Pp. 348–385 in Jackson, J.B.C., Lidgard, S. & McKinney, F.K. (eds) *Evolutionary patterns: growth, form and tempo in the fossil record in honor of Alan Cheetham*. University of Chicago Press, Chicago.

McKinney, F.K., Listokin, M.R.A. & Phifer, C.D. (1986) Flow and polypide distribution in the cheilostome bryozoan *Bugula* and their inference in *Archimedes*. *Lethaia* **19**, 81–93.

McKinney, F.K. & McGhee, G.R., Jr (2003) Evolution of erect helical colony form in the Bryozoa: phylogenetic, functional, and ecological factors. *Biological Journal of the Linnean Society* **80**, 235–260.

McKinney, F.K., McKinney, M.J. & Listokin, M.R.A. (1987) Erect bryozoans are more than baffling: enhanced sedimentation rate by a living unilaminate branched bryozoan and possible implications for fenestrate bryozoan mudmounds. *Palaios* **2**, 41–47.

McKinney, F.K. & Raup, D.M. (1982) A turn in the right direction: simulation of erect spiral growth in the bryozoans *Archimedes* and *Bugula*. *Paleobiology* **8**, 101–112.

McKinney, F.K. & Taylor, P.D. (1997) Life histories of some Mesozoic encrusting cyclostome bryozoans. *Palaeontology* **40**, 515–556.

McKinney, F.K. & Taylor, P.D. (2001) Bryozoan generic extinctions and originations during the last one hundred million years. *Palaeontologia Electronica* **4**(1), Article 3, 26 pp. http://palaeo-electronica.org/2001_1/bryozoan/issue1_01.htm/.

McKinney, F.K. & Taylor, P.D. (2003) Palaeoecology of free-lying domal bryozoan colonies from the Upper Eocene of southeastern USA. *Acta Palaeontologica Polonica* **48**, 447–462.

McKinney, F.K. & Taylor, P.D. (2006) Encrusting community on an Upper Cretaceous erect bryozoan assemblage, eastern North America. *Courier Forschungsinstitut Senckenberg* **257**, 93–102.

McKinney, F.K., Taylor, P.D. & Lidgard, S. (2003) Predation on bryozoans and its reflection in the fossil record. Pp. 239–261 in Kelley, P.H., Kowalewski, M. & Hansen, T.A. (eds) *Predator–prey interactions in the fossil record*. Kluwer Academic/Plenum Publishers, New York.

McKinney, F.K., Taylor, P.D. & Zullo, V.A. (1993) Lyre-shaped hornerid bryozoan colonies: homeomorphy in colony form between Paleozoic Fenestrata and Cenozoic Cyclostomata. *Journal of Paleontology* **67**, 343–354.

McKinney, M.J. (1997) Fecal pellet disposal in marine bryozoans. *Invertebrate Biology* **116**, 151–160.

McKinney, M.J. & Dewel, R.A. (2002) The ctenostome collar – an enigmatic structure. Pp. 191–197 in Wyse Jackson, P.N., Buttler, C.J. & Spencer-Jones, M.E. (eds) *Bryozoan studies 2001: Proceedings of the 12th International Bryozoology Association Symposium*. Balkema, Lisse.

McNamara, K.J. (1978) Symbiosis between gastropods and bryozoans in the Late Ordovician of Cumbria, England. *Lethaia* **11**, 25–40.

McShea, D.W. & Venit, E.P. (2002) Testing for bias in the evolution of coloniality: a demonstration in cyclostome bryozoans. *Paleobiology* **28**, 308–327.

Medd, A.W. (1964) On the musculature of some Cretaceous membranimorph Polyzoa. *Annals and Magazine of Natural History, Series 13*, **7**, 185–187.

Menon, N.R. (1972) Heat tolerance, growth and regeneration in three North Sea bryozoans exposed to different constant temperatures. *Marine Biology* **15**, 1–11.

Menon, N.R. (1973) A note on the occurrence of *Pelagonema obtusicauda*, a free-living nematode inside bryozoans. *Helgolander Meeresunters* **25**, 170–172.

Menon, N.R. & Nair, N.B. (1974) On the nature of tolerance to salinity in two euryhaline intertidal bryozoans *Victorella pavida* Kent and *Electra crustulenta* Pallas. *Proceedings of the*

Indian National Science Academy, Series B **8**, 414–424.

Mercado, J.M., Carmona, R., & Niell, F.X. (1998) Bryozoans increase available CO_2 for photosynthesis in *Gelidium sesquipedale* (Rhodophyceae). *Journal of Phycology* **34**, 925–927.

Mergl, M. (2004) The earliest brachiopod–bryozoan dominated community in the Ordovician of peri-Gondwana and its ancestors: a case study from the Klabava Formation (Arenigian) of the Barrandian, Bohemia. *Journal of the Czech Geological Society* **49**, 127–136.

Mesentseva, O.P. (2008) Trepostomids (Bryozoa) from the Devonian of Salair, Kuznetsky Basin, Gorny and Rudny Altai, Russia. *Bulletin of Geosciences* **83**, 449–460.

Metcalfe, K., Gordon, D.P. & Hayward, E. (2007) An amphibious bryozoan from living mangrove leaves – *Amphibiobeania* new genus (Beaniidae). *Zoological Science* **24**, 563–570.

Miles, J.S., Harvell, C.D., Griggs, C.M. & Eisner, S. (1995) Resource translocation in a marine bryozoan: quantification and visualization of ^{14}C and ^{35}S. *Marine Biology* **122**, 439–445.

Min, B.S., Seo, J.E., Grischenko, A.V. & Gordon, D.P. (2017) Intertidal Bryozoa from Korea – new additions to the fauna and a new genus of Bitectiporidae (Cheilostomata) from Baengnyeong Island, Yellow Sea. *Zootaxa* **4226**, 451–470.

Moissette, P. (1996) The cheilostome bryozoan *Batopora rosula* (Reuss, 1848): a paleobathymetric indicator in the Mediterranean Neogene. Pp. 193–198 in Gordon, D.P., Smith, A.M. & Grant-Mackie, J.A. (eds) *Bryozoans in space and time*. NIWA, Wellington.

Moissette, P. (2000a) Changes in bryozoan assemblages and bathymetric variations. Examples from the Messinian of northwest Algeria. *Palaeogeography, Palaeoclimatology, Palaeoecology* **155**, 305–326.

Moissette, P. (2000b) The use of Neogene bryozoans for a better understanding of the ecology of some Recent species. Pp. 291–297 in Herrera Cubilla, A. & Jackson, J.B.C. (eds) *Proceedings of the 11th International Bryozoology Association Conference*. Smithsonian Tropical Research Institute, Balboa.

Moissette, P. (2012) Seagrass-associated bryozoan communities from the late Pliocene of the Island of Rhodes (Greece). Pp. 187–201 in Ernst, A., Schäfer, P. & Scholz, J. (eds) *Bryozoan studies 2010*. Springer, Heidelberg.

Moissette, P., Dulai, A., Escarguel, G., Kazmer, M., Mueller, P. & Saint Martin, J.-P. (2007a) Mosaic of environments recorded by bryozoan faunas from the Middle Miocene of Hungary. *Palaeogeography, Palaeoclimatology, Palaeoecology* **252**, 530–556.

Moissette, P., Dulai, A. & Muller, P. (2006) Bryozoan faunas in the Middle Miocene of Hungary: biodiversity and biogeography. *Palaeogeography, Palaeoclimatology, Palaeoecology* **233**, 300–314.

Moissette, P., Koskeridou, E., Cornée, J.-J., Guillocheau, F. & Lécuyer, C. (2007b) Spectacular preservation of seagrasses and seagrass-associated communities from the Pliocene of Rhodes, Greece. *Palaios* **22**, 200–211.

Moissette, P. & Pouyet, S. (1991) Bryozoan masses in the Miocene-Pliocene and Holocene of France, North Africa and the Mediterranean. *Bulletin de la Societe des Sciences Naturelles de l'Ouest de la France Mémoire* **H.S. 1**, 271–279.

Moissette, P., Saint Martin, J-P., André, J-P. & Pestrea, S. (2002) L'association microbialite-bryozoaires dans le Messinien de Sicile et de Sardaigne. *Geodiversitas* **24**, 611–623.

Moissette, P. & Spjeldnaes, N. (1995) Plio-Pleistocene deep-water bryozoans from Rhodes, Greece. *Palaeontology* **38**, 771–799.

Morgado, E.H. & Tanaka, M.O. (2001) The macrofauna associated with the bryozoan *Schizoporella errata* (Waters) in southeastern Brazil. *Scientia Marina* **65**, 173–181.

Morozova, I.P. (1966) A new suborder of Late Paleozoic bryozoans of the order Cryptostomata. *Paleontologicheskii Zhurnal* **1966**(2), 33–41.

Morris, P.A. (1976) Middle Pliocene temperature implications based on the Bryozoa *Hippothoa* (Cheilostomata-Ascophora). *Journal of Paleontology* **50**, 1143–1149.

Morris, P.A., von Bitter, P.H., Schenk, P.E. & Wentworth, S.J. (2002) Interactions of bryozoans and microbes in a chemosynthetic hydrothermal vent system: Big Cove Formation (Lower Codroy Group, Lower Carboniferous, Middle Viseann/Arundian), Port au Port Peninsula, western Newfoundland, Canada. Pp. 181–186 in Wyse Jackson, P.N., Buttler, C.J. & Spencer Jones, M.E. (eds) *Bryozoan studies 2001*. Balkema, Lisse.

Morris, P.J., Linsley, R.M. & Cottrell, J.F. (1991) A Middle Devonian symbiotic relationship involving a gastropod, a trepostomatous bryozoan, and an inferred secondary occupant. *Lethaia* **24**, 55–67.

Morrison, S.J. & Anstey, R.L. (1979) Ultrastructure and composition of brown bodies in some Ordovician trepostome bryozoans. *Journal of Paleontology* **53**, 943–949.

Moyano G., H.I. (1999) Magellan Bryozoa: a review of the diversity and of the Subantarctic and Antarctic zoogeographical links. *Scientia Marina* **63**, 219–226.

Moyano G., H.I. (2005a) Bryozoa de la expedición Chilena Cimar 5 Islas Oceánicas I: el género *Jellyella* Taylor & Monks 1997 (Bryozoa, Cheilostomatida) en la Isla de Pascua. *Ciencia y Tecnología del Mar* **28**, 97–90.

Moyano G., H.I. (2005b) Scotia Arc bryozoans from the LAMPOS expedition: a narrow bridge between two different faunas. *Scientia Marina* **69** (Supplement 2), 103–112.

Moyano G., H.I. (2006) Holocne bryozoan links between Australia, New Zealand, southern South America, and Antarctica – a preliminary evaluation. Pp. 207–219 in Gordon, D.P., Smith, A.M. & Grant-Mackie, J.A. (eds) *Bryozoans in space and time*. NIWA, Wellington.

Moysiuyk, J., Smith, M.R. & Caron, J.-B. (2017) Hyoliths are Palaeozoic lophophorates. *Nature* **541**, 394–397.

Mukai, H. (1999) Comparative morphological studies on the statoblasts of lower phylactolaemate bryozoans, with discussion on the systematics of the Phylactolaemata. *Science Reports of the Faculty of Education, Gunma University*, **46**, 51–91.

Müller, P., Hahn, G. & Bohaty, J. (2013) Agelacrinitid Edrioasteroidea (Echinodermata) from the Middle Devonian of the Eifel (Rhenish Massif, Germany). *Paläontologische Zeitschrift* **87**, 455–472.

Mutti, M. & Hallock, P. (2003) Carbonate systems along nutrient and temperature gradients: some sedimentological and geochemical constraints. *International Journal of Earth Science* **92**, 465–475.

Nakrem, H.A. (1994) Environmental distribution of bryozoans in the Permian of Spitsbergen. Pp. 133–137 in Hayward, P.J., Ryland, J.S. & Taylor, P.D. (eds) *Biology and palaeobiology of bryozoans*. Olsen & Olsen, Fredensborg.

Nakrem, H.A. & Spjeldnaes, N. (1995) *Ramipora hochstetteri* Toula, 1875 (Bryozoa, Cystoporata), from the Permian of Svalbard. *Journal of Paleontology* **69**, 831–838.

Nebelsick, J.H., Schmid, B. & Stachowitsch, M. (1997) The encrustation of fossil and recent sea-urchin tests: ecological and taphonomic significance. *Lethaia* **30**, 271–284.

Nelson, C.S., Hyden, F.M., Keane, S.L., Leask, W.L. & Gordon, D.P. (1988) Application of bryozoan zoarial growth-form studies in facies analysis of non-tropical carbonate deposits in New Zealand. *Sedimentary Geology* **60**, 301–322.

Nelson, C.S., Keane, S.L. & Head, P.S. (1988) Non-tropical carbonate deposits on the modern New Zealand shelf. *Sedimentary Geology* **60**, 71–94.

Nesnidal, M.P., Helmkampf, M., Meyer, A., Witek, A., Bruchhaus, I., Ebersberger, I., Hankeln, T., Lieb, B., Struck, T.H. & Hausdorf, B. (2013) New phylogenomic data support the monophyly of Lophophorata and an Ectoproct-Phoronid clade and indicate that Polyzoa and Kryptrochozoa are caused by systematic bias. *BMC Evolutionary Biology* **13**, 253.

Nielsen, C. (1970) On metamorphosis and ancestrula formation in cyclostomatous bryozoans. *Ophelia* **7**, 217–256.

Nielsen, C. (1981) On morphology and reproduction of '*Hippodiplosia*' *insculpta* and *Fenestrulina malusii* (Bryozoa, Cheilostomata). *Ophelia* **20**, 91–125.

Nielsen, C. (1985) Ovicell formation in *Tegella* and four cellularioids (Bryozoa, Cheilostomata). Pp. 213–220 in Nielsen, C. & Larwood, G.P. (eds) *Bryozoa: Ordovician to Recent*. Olsen & Olsen, Fredensborg.

Nielsen, C. (1987) Structure and function of metazoan ciliary bands and their phylogenetic significance. *Acta Zoologica* **68**, 205–262.

Nielsen, C. (2013) The triradiate sucking pharynx in animal phylogeny. *Invertebrate Biology* **132**, 1–13.

Nielsen, C. & Pedersen, K.J. (1979) Cystid structure and protrusion of the polypide in *Crisia* (Bryozoa, Cyclostomata). *Acta Zoologica Stockholm* **60**(2), 65–88.

Nielsen, C. & Riisgård, H.U. (1998) Tentacle structure and filter-feeding in *Crisia eburnea* and other cyclostomatous bryozoans, with a review of upstream-collecting mechanisms. *Marine Ecology Progress Series* **168**, 163–186.

Nielsen, C. & Worsaae, K. (2010) Structure and occurrence of cyphonautes larvae (Bryozoa, Ectoprocta). *Journal of Morphology* **271**, 1094–1109.

Nielsen, K.S.S. & Nielsen, J.K. (2001) Bioerosion in Pliocene to Late Holocene tests of benthic and planktonic foraminiferans, with a revision of the ichnogenera *Oichnus* and *Tremichnus*. *Ichnos* **8**, 99–116.

Nielsen, S.N. (2009) Pliocene balanuliths from northern Chile: the first report of fossil balanuliths. *Palaios* **24**, 334–335.

Nikulina, E.A., Hanel, R. & Schäfer, P. (2007) Cryptic speciation and paraphyly in the cosmopolitan bryozoan *Electra pilosa* – impact of the Tethys closing on species evolution. *Molecular Phylogenetics and Evolution* **45**, 765–776.

Nikulina, E.A. & Schäfer, P. (2006) Bryozoans of the Baltic Sea (Bryozoen in der Ostsee). *Meyniana* **58**, 75–95.

Novak, V., Santodomingo, N., Rösler, A., Di Martino, E., Braga, J.C., Taylor, P.D., Johnson, K.G. & Renema, W. (2013) Environmental reconstruction of a late Burdigalian (Miocene) patch reef in deltaic deposits (East Kalimantan, Indonesia). *Palaeogeography, Palaeoclimatology, Palaeoecology* **374**, 110–122.

Novosel, M. (2005) Bryozoans of the Adriatic Sea. *Denisia* **16**, 231–246.

Numakunai, T. (1960) An observation on the budding of the stolon of a bryozoan, *Bugula neritina* Linné. *Bulletin of the Marine Biological Station of Asamushi* **10**, 99–101.

Nye, O.B., Jr & Lemone, D.V. (1978) Multilaminar growth in *Reptomulticava texana*, a new species of cyclostome Bryozoa. *Journal of Paleontology* **52**, 830–845.

O'Dea, A. (2003) Seasonality and zooid size variation in Panamanian encrusting bryozoans. *Journal of the Marine Biological Association of the United Kingdom* **83**, 1107–1108.

O'Dea, A. (2005) Zooid size parallels contemporaneous oxygen isotopes in a large colony of *Pentapora foliacea* (Bryozoa). *Marine Biology* **146**, 1075–1081.

O'Dea, A. (2006) Asexual propagation in the marine bryozoan *Cupuladria exfragminis*. *Journal of Experimental Marine Biology and Ecology* **335**, 312–322.

O'Dea, A. (2009) Relation of form to life habit in free-living cupuladriid bryozoans. *Aquatic Biology* **7**, 1–18.

O'Dea, A., Håkansson, E., Taylor, P.D. & Okamura, B. (2011) Environmental change prior to the K-T boundary inferred from temporal variation in the morphology of cheilostome bryozoans. *Palaeogeography, Palaeoclimatology, Palaeoecology* **308**, 502–512.

O'Dea, A., Herrera Cubilla, A., Fortunata, H. & Jackson, J.B.C. (2004) Life history variation in cupuladriid bryozoans from either side of the Isthmus of Panama. *Marine Ecology Progress Series* **280**, 145–161.

O'Dea, A. & Jackson, J.B.C. (2002) Bryozoan growth mirrors contrasting seasonal regimes across the Isthmus of Panama. *Palaeogeography, Palaeoclimatology, Palaeoecology* **185**, 77–94.

O'Dea, A. & Jackson, J.B.C. (2009) Environmental change drove macroevolution in cupuladriid bryozoans. *Proceedings of the Royal Society, Series B* **279**, 3629–3634.

O'Dea, A., Jackson, J.B.C., Taylor, P.D. & Rodriguez, F. (2008) Modes of reproduction in Recent and fossil cupuladriid bryozoans. *Palaeontology* **51**, 847–864.

O'Dea, A. & Okamura, B. (1999) Infuence of seasonal variation in temperature, salinity and food availability on module size and colony growth of the estuarine bryozoan *Conopeum seurati*. *Marine Biology* **135**, 581–588.

O'Dea, A. & Okamura, B. (2000a) Cheilostome bryozoans as indicators of seasonality in the Neogene epicontinental seas of western Europe. Pp. 316–320 in Herrera Cubilla, A. & Jackson, J.B.C. (eds) *Proceedings of the 11th International Bryozoology Association Conference*. Smithsonian Tropical Research Institute, Balboa.

O'Dea, A. & Okamura, B. (2000b) Intracolony variation in zooid size in cheilostome bryozoans as a new technique for investigating palaeoseasonality. *Palaeogeography, Palaeoclimatology, Palaeoecology* **162**, 319–332.

Oakley, K.P. (1934) Phosphatic calculi in Silurian Polyzoa. *Proceedings of the Royal Society, Series B* **116**, 296–314.

Ochi Agostini, V., Ritter, M. do N., Macedo, A.J., Muxagata, E. & Erthal, F. (2017) What determines sclerobiont colonization on marine mollusk shells? *PLoS ONE* **12**(9), e0184745.

Occhipinti Ambrogi, A. (1985) The zonation of bryozoans along salinity gradients in the Venice Lagoon (northern Adriatic). Pp. 221–231 in Nielsen, C. & Larwood, G.P. (eds) *Bryozoa: Ordovician to Recent*. Olsen & Olsen, Fredensborg.

Okamura, B. & Bishop, J.D.D. (1988) Zooid size in cheilostome bryozoans as an indicator of relative palaeotemperature. *Palaeogeography, Palaeoclimatology, Palaeoecology* **66**, 145–152.

Okamura, B., Harmelin, J.-G. & Jackson, J.B.C. (2001) Refuges revisited: enemies versus flow and feeding as determinant of sessile animal distribution and form. Pp. 61–93 in Jackson, J.B.C., Lidgard, S. & McKinney, F.K. (eds) *Evolutionary patterns: growth, form and tempo in the fossil record in honor of Alan Cheetham*. University of Chicago Press, Chicago.

Okamura, B., Hartikainen, H. & Trew, J. (2019) Waterbird-mediated dispersal and freshwater biodiversity: general insights from bryozoans. *Frontiers in Ecology and Evolution*, https://doi.org/10.3389/fevo.2019.00029.

Okamura, B., O'Dea, A. & Knowles, T. (2011) Bryozoan growth and environmental reconstruction by zooid size variation. *Marine Ecology Progress Series* **430**, 133–146.

Okamura, B., O'Dea, A., Taylor, P.D. & Taylor, A. (2013) Evidence of El Niño/La Niña–Southern Oscillation variability in the Neogene–Pleistocene of Panama revealed by a new bryozoan assemblage-based proxy. *Bulletin of Marine Science* **89**, 857–876.

Okamura, B. & Partridge, J.C. (1999) Suspension feeding adaptations to extreme flow environments in a marine bryozoan. *Biological Bulletin* **196**, 205–215.

Okuyama, M., Wada, H. & Ishii, T. (2006) Phylogenetic relationships of freshwater bryozoans (Ectoprocta, Phylactolaemata) inferred from mitochondrial ribosomal DNA sequences. *Zoologica Scripta* **35**, 243–249.

Olempska, E. (2012) Exceptional soft-tissue preservation in boring ctenostome bryozoans and associated 'fungal' borings from the Early Devonian of Podolia, Ukraine. *Acta Palaeontologica Polonica* **57**, 925–940.

Olempska E. (2015) Decay and mineralization of soft tissue in Early Devonian boring ctenostome bryozoans. *Lethaia* **49**, 421–432.

Olempska, E. & Rakowicz, L. (2014) Affinities of Palaeozoic encrusting ascodictyid 'pseudobryozoans'. *Journal of Systematic Palaeontology* **12**, 983–999.

Oren, U., Benayahu, Y., Lubinevsky, H. & Loya, Y. (2001) Colony integration during regeneration in the stony coral *Favia favus. Ecology* **82**, 802–813.

Orpin, A.R. (1991) The chimney revolution. *University of Otago Marine Science Bulletin* **3**, 1–2.

Orr, R.J.S., Waeschenbach, A., Enevoldsen, E.L.G., Boeve, J.P., Haugen, M.N., Voje, K.L., Porter, J.S., Zágoršek, K., Smith, A.M., Gordon, D.P. & Liow, L.H. (2018) Bryozoan genera *Fenestrulina* and *Microporella* no longer confamilial; multi-gene phylogeny supports separation. *Zoological Journal of the Linnean Society* **186**, 190–199.

Osborne, S. (1984) Bryozoan interactions: observations on stolonal outgrowths. *Australian Journal of Marine and Freshwater Research* **35**, 453–462.

Oschmann, W. (1990) Dropstones – rocky miniislands in high-latitude pelagic soft substrate environments. *Senckenbergiana Maritima* **21**, 55–75.

Osman, R.W. & Haugsness, J.A. (1981) Mutualism among sessile invertebrates: a mediator of competition and predation. *Science* **211**, 846–848.

Ostrovsky, A.N. (1998) Comparative studies of ovicell anatomy and reproductive patterns in *Cribrilina annulata* and *Celleporella hyalina* (Bryozoa: Cheilostomatida). *Acta Zoologica* **79**, 287–318.

Ostrovsky, A.N. (2013) *Evolution of sexual reproduction in marine invertebrates: example of gymnolaemate bryozoans.* Springer, Dordrecht, 356 pp.

Ostrovsky, A.N., Grischenko, A.V., Taylor, P.D., Bock, P. & Mawatari, S.F. (2006) Comparative anatomical study of internal brooding in three anascan bryozoans (Cheilostomata) and its taxonomic and evolutionary implications. *Journal of Morphology* **267**, 739–749.

Ostrovsky, A.N., Nielsen, C., Vavra, N. & Yagunova, E.B. (2009) Diversity of the brooding structures in calloporid bryozoans (Gymnolaemata: Cheilostomata): comparative anatomy and evolutionary trends. *Zoomorphology* **128**, 13–35.

Ostrovsky, A.N., O'Dea, A. & Rodríguez, F. (2009) Comparative anatomy of internal incubational sacs in cupuladriid bryozoans and the evolution of brooding in free-living cheilostomes. *Journal of Morphology* **270**, 1413–1430.

Ostrovsky, A.N. & Taylor, P.D. (2004) Systematics of Upper Cretaceous calloporid bryozoans with primitive spinose ovicells. *Palaeontology* **47**, 775–793.

Ostrovsky, A.N. & Taylor, P.D. (2005a) Brood chambers constructed from spines in fossil and Recent cheilostome bryozoans. *Zoological Journal of the Linnean Society* **144**, 317–361.

Ostrovsky, A.N. & Taylor, P.D. (2005b) Ovicell development in the early calloporid *Wilbertopora* Cheetham, 1954 (Bryozoa: Cheilostomata) from the mid-Cretaceous of the USA. Pp. 223–230 in Moyano G., H.I., Cancino, J.M. & Wyse Jackson, P.N. (eds) *Bryozoan studies 2004.* Balkema, Leiden.

Oswald, R.C., Telford, N., Seed, R. & HappeyWood, C.M. (1984) The effect of encrusting bryozoans on the photosynthetic activity of *Fucus serratus* L. *Estuarine and Coastal Shelf Science* **19**, 697–702.

Pachut, J.F. & Anstey, R.L. (2007) Inferring evolutionary order and durations using both stratigraphy and cladistics in a fossil lineage (Bryozoa: *Peronopora*). *Palaios* **22**, 476–488.

Pachut, J.F. & Anstey, R.L. (2012) Rates of anagenetic evolution and selection intensity in Middle and Upper Ordovician species of the bryozoan genus *Peronopora. Paleobiology* **38**, 403–423.

Pachut, J.F. & Fisherkeller, P. (2010) Inferring larval type in fossil bryozoans. *Lethaia* **43**, 396–410.

Pachut, J.F. & Fisherkeller, P. (2011) Patterns of early colony development (astogeny) in four genera of trepostome bryozoans from the Upper Ordovician. *Journal of Paleontology* **85**, 744–756.

Padilla, D.K., Harvell, C.D., Marks, J. & Helmuth, B. (1996) Inducible aggression and intraspecific competition for space in a marine bryozoan, *Membranipora membranacea. Limnology and Oceanography* **41**, 505–512.

Palau, M., Cornet, C., Riera, T. & Zabala, M. (1991) Planktonic gradients along a Mediterranean sea cave. *Oecologia aquatica* **10**, 299–316.

Palinska, K.A., Scholz, J., Sterflinger, K., Gerdes, G. & Bone, Y. (1999) Microbial mats associated with bryozoans (Coorong Lagoon, South Australia). *Facies* **41**, 1–14.

Palmer, T.J. & Fürsich, F.T. (1974) The ecology of a Middle Jurassic hardground and crevice fauna. *Palaeontology* **17**, 507–524.

Palmer, T.J. & Fürsich, F.T. (1981) Ecology of sponge reefs from the Upper Bathonian of Normandy. *Palaeontology* **24**, 1–23.

Palmer, T.J. & Hancock, C.D. (1973) Symbiotic relationships between ectoprocts and gastropods and between ectoprocts and hermit crabs in the French Jurassic. *Palaeontology* **16**, 563–566.

Palmer, T.J. & Palmer, C.D. (1977) Faunal distribution and colonization strategy in a Middle Ordovician hardground community. *Lethaia* **10**, 179–199.

Palmer, T.J., Taylor, P.D. & Todd, J.A. (1993) Epibiont shadowing: a hitherto unrecognized way of preserving soft-bodied fossils. *Terra Nova* **5**, 568–572.

Palmer, T.J. & Wilson, M.A. (1988) Parasitism of Ordovician bryozoans and the origin of pseudo-borings. *Palaeontology* **31**, 939–949.

Palmer, T.J. & Wilson, M.A. (1990) Growth of ferruginous oncoliths in the Bajocian (Middle Jurassic) of Europe. *Terra Nova* **2**, 142–147.

Palumbi, S.R. & Jackson, J.B.C. (1983) Ageing in modular organisms: ecology of zooid senescence in *Steginoporella* sp. (Bryozoa: Cheilostomata). *Biological Bulletin* **164**, 267–278.

Parrass, A. & Casadio, S. (2006) The oyster *Crassostrea?* *hatcheri* (Ortmann, 1897), a physical ecosystem engineer from the Upper Oligocene–Lower Miocene of Patagonia, southern Argentina. *Palaios* **21**, 168–186.

Patzkowsky, M.E. (1987) Inferred water flow patterns in the fossil *Fistulipora* M'Coy (Cystoporata, Bryozoa). Pp. 213–219 in Ross, J.R.P. (ed.) *Bryozoa: present and past*. Western Washington University, Bellingham.

Patzkowsky, M.E. (1995) A hierarchical branching model of evolutionary radiations. *Palaeobiology* **21**, 440–460.

Pätzold, J., Ristedt, H. & Wefer, G. (1987) Rate of growth and longevity of a large colony of *Pentapora foliacea* (Bryozoa) recorded in their oxygen isotope profiles. *Marine Biology* **96**, 535–538.

Peck, L.S., Hayward, P.J. & Spencer Jones, M.E. (1995) A pelagic bryozoan from Antarctica. *Marine Biology* **123**, 757–762.

Pemberton, A.J., Hansson, L.J., Craig, S.F., Hughes, R.N. & Bishop, J.D.D. (2007) Microscale genetic differentiation in a sessile invertebrate with cloned larvae: investigating the role of polyembryony. *Marine Biology* **153**, 71–82.

Pemberton, A.J., Hansson, L.J. & Bishop, J.D.D. (2011) Does sperm supply limit the number of broods produced by a polyembryonous bryozoan? *Marine Ecology Progress Series* **430**, 113–119.

Perez, F.M. & Banta, W.C. (1996) How does *Cellaria* get out of its box? A new cheilostome hydrostatic mechanism (Bryozoa: Cheilostomata). *Invertebrate Biology* **115**, 162–169.

Pérez, L.M., Griffin, M., Pastorino, G., López-Gappa, J. & Manceñido, M.O. (2015) Redescription and palaeoecological significance of the bryozoan *Hippoporidra patagonica* (Pallaroni, 1920) in the San Julián Formation (late Oligocene) of Santa Cruz province, Argentina. *Alcheringa* **39**, 1–7.

Pérez, L.M., López-Gappa, J. & Griffin, M. (2015) Los géneros *Aspidostoma* y *Melychocella* (Bryozoa: Cheilostomata: Aspidostomatidae) en el Mioceno de la Patagonia Argentina. *Ameghiniana* **53**(4), 36R–37R.

Pillans, B. (2017) Quaternary stratigraphy of Whanganui Basin – a globally significant archive. Pp. 141–170 in Shulmeister, J. (ed.) *Landscape and Quaternary environmental change in New Zealand*. Atlantis Press.

Pistevos, J.C.A., Calosi, P., Widdicombe, S. & Bishop, J.D.D. (2011) Will variation among genetic individuals influence species responses to global climate change? *Oikos* **120**, 675–689.

Pitt, L.J. & Taylor, P.D. (1990) Cretaceous Bryozoa from the Faringdon Sponge Gravel (Aptian) of Oxfordshire. *Bulletin of the British Museum (Natural History), Geology Series* **46**, 61–152.

Plaziat, J.-C. (1970) Huitres de mangrove et peuplements littoraux de l'Eocene Inférieur des Corbières. *Geobios* **3**, 7–27.

Pohle, A., Klug, C., Toom, U. & Kröger, B. (2019) Conch structures, soft-tissue imprints and taphonomy of the Middle Ordovician cephalopod *Tragoceras falcatum* from Estonia. *Fossil Imprint* **75**, 70–78.

Pohowsky, R.A. (1973) A Jurassic cheilostome from England. Pp. 447–461 in Larwood, G.P. (ed.) *Living and fossil Bryozoa*. Academic Press, London.

Pohowsky, R.A. (1974) Notes on the study and nomenclature of boring Bryozoa. *Journal of Paleontology* **48**, 557–564.

Pohowsky, R.A. (1978) The boring ctenostomate Bryozoa: taxonomy and paleobiology based on cavities in calcareous substrata. *Bulletin of American Paleontology* **73**, 1–192.

Poluzzi, A. (1980) I briozoi membraniporiformi del delta settertrionale del Po. *Atti della Società italiana di scienze naturali, e del Museo civico di storia naturale, Milano* **121**, 101–120.

Poluzzi, A. & Sabelli, B. (1985) Polymorphic zooids in deltaic species populations of *Conopeum seurati* (Canu, 1928) (Bryozoa, Cheilostomata). *Marine Ecology* **6**, 265–284.

Poluzzi, A. & Sartori, N. (1975) Report on the carbonate mineralogy of Bryozoa. *Documents des Laboratoires de Géologie de la Faculté des Sciences de Lyon, Hors Série* **3**(1), 193–210.

Porter, J.S. & Hayward, P.J. (2004) Species of *Alcyonidium* (Bryozoa: Ctenostomata) from Antarctica and Magellan Strait, defined by morphological, reproductive and molecular characters. *Journal of the Marine Biological Association of the United Kingdom* **84**, 253–265.

Porter, S.M. (2010) Calcite and aragonite seas and the de novo acquisition of carbonate skeletons. *Geobiology* **8**, 256–277.

Pouyet, S. (1978) Révision de quatre espèces actuelles de *Celleporaria* (Bryozoa, Cheilostomata) déscrites par Lamarck en 1816. *Geobios* **11**, 611–621.

Pouyet, S. & David, L. (1979) Revision systematique du genre *Steginoporella* Smitt, 1873 (Bryozoa, Cheilostomata). *Geobios* **12**, 763–817.

Powers, C.M. & Bottjer, D.J. (2007) Bryozoan paleoecology indicates mid-Phanerozoic extinctions were the product of long-term environmental stress. *Geology* **35**, 995–998.

Powers, C.M. & Bottjer, D.J. (2009) Behavior of lophophorates during the end-Permian mass extinction and recovery. *Journal of Asian Earth Sciences* **36**, 413–419.

Powers, C.M. & Pachut, J.F. (2008) Diversity and distribution of Triassic bryozoans in the aftermath of the end-Permian mass extinction. *Journal of Paleontology* **82**, 362–371.

Pratt, M.C. (2005) Consequences of coloniality: influence of colony form and size on feeding success in the bryozoan *Membranipora membranacea*. *Marine Ecology Progress Series* **303**, 153–165.

Probert, P.K. & Batham, E.J. (1979) Epibenthic macrofauna off southeastern New Zealand and mid-shelf bryozoan dominance. *New Zealand Journal of Marine and Freshwater Research* **13**, 379–392.

Pröts, P., Wanninger, A. & Schwaha, T. (2019) Life in a tube: morphology of the ctenostome bryozoan *Hypophorella expansa*. *Zoological Letters* **5**, 28.

Puce, S., Bavestrello, G., Di Camillo, C.G. & Boero, F. (2007) Symbiotic relationships between hydroids and bryozoans. *Symbiosis* **44**, 137–143.

Pugaczewska, H. (1965) Les organismes sédentaires sur les rostres des bélemnites du Crétacé supérieur. *Acta Palaeontologica Polonica* **10**, 73–95.

Pushkin, V.I. & Popov, L.E. (2005) Two enigmatic bryozoans from the Middle Ordovician of the East Baltic. *Palaeontology* **48**, 1065–1074.

Quine, M. & Bosence, D. (1991) Stratal geometries, facies and sea-floor erosion in Upper Cretaceous Chalk, Normandy, France. *Sedimentology* **38**, 1113–1152.

Ramalho, L.V., Taylor, P.D., Morales, F.C., Moura, R., Amado-Filho, G.M. & Bastos, A.C. (2018) Bryozoan framework composition in the oddly shaped reefs from Abrolhos Bank, Brazil, southwestern Atlantic: taxonomy and ecology. *Zootaxa* **4483**, 155–186.

Rao, C.P. (1993) Carbonate minerals, oxygen and carbon isotopes in modern temperate Bryozoa, eastern Tasmania, Australia. *Sedimentary Geology* **88**, 123–135.

Rao, K.S. & Ganapati, P.N. (1980) Epizoic fauna of *Thalamoporella gothica* var. *indica* and *Pherusella tubulosa* (Bryozoa). *Bulletin of Marine Science* **30**, 34–44.

Raup, D.M. & Sepkoski, J.J., Jr (1982) Mass extinctions in the marine fossil record. *Science* **215**, 1501–1503.

Rech, S., Thiel, M., Borrell Pichs, Y.J. & García-Vazquez, E. (2018) Travelling light: fouling biota on macroplastics arriving on beaches of remote Rapa Nui (Easter Island) in the South Pacific Subtropical Gyre. *Marine Pollution Bulletin* **137**, 119–128.

Reed, C.G. (1988) The reproductive biology of the gymnolaemate bryozoan *Bowerbankia gracilis* (Ctenostomata: Vesiculariidae). *Ophelia* **29**, 1–23.

Reed, C.G. (1991) Bryozoa. Pp. 85–245 in Giese, A.C., Pearse, J.S. & Pearse, V.B. (eds) *Reproduction of marine invertebrates*. Boxwood Press, Pacific Grove.

Reguant, S. (1993) Semi-multilamellar growth in *Reptomulticava alhamensis*, a new cyclostome (Bryozoa) species from the Tortonian of Alhama de Granada (S. Spain). *Revista Española de Paleontología* **8**, 147–152.

Reid, C.M. (2003) Permian Bryozoa of Tasmania and New South Wales: systematics and their use in Tasmanian biostratigraphy. *Memoirs of the Association of Australasian Palaeontologists* **28**, 1–133.

Reid, C.M. (2010) Environmental controls on the distribution of Late Paleozoic bryozoan colony morphotypes: an example from the Permian of Tasmania, Australia. *Palaios* **25**, 692–702.

Reid, C.M. (2012) Large sediment encrusting trepostome bryozoans from the Permian of Tasmania, Australia. Pp. 237–249 in Ernst, A., Schäfer, P. & Scholz, J. (eds) *Bryozoan studies 2010*. Springer, Heidelberg.

Reid, C.M. (2014) Growth and calcification rates in polar bryozoans from the Permian of Tasmania, Australia. *Studi Trentini di Scienze Naturali* **94**, 189–197.

Reid, C.M. & James, N.P. (2010) Permian higher latitude bryozoan biogeography. *Palaeogeography, Palaeoclimatology, Palaeoecology* **298**, 31–41.

Reid, C.M., James, N.P., Beauchamp, B. & Kyser, T.K. (2007) Faunal turnover and changing oceanography: Late Palaeozoic warm-to-cool water carbonates, Sverdrup Basin, Canadian Arctic Archipelago. *Palaeogeography, Palaeoclimatology, Palaeoecology* **249**, 128–159.

Reijmer, J.J.G., Bauch, T. & Schäfer, P. (2012) Carbonate facies patterns in surface sediments of upwelling and non-upwelling shelf environments (Panama, East Pacific). *Sedimentology* **59**, 32–56.

Renema, W., Bellwood, D.R., Braga, J.C., Bromfield, K., Hall, R., Johnson, K.G., Lunt, P., Meyer, C.P., McMonagle, L.B., Morley, R.J., O'Dea, A., Todd, J.A., Wesselingh, F.P., Wilson, M.E.J. & Pandolfi, J.M. (2008) Hopping hotspots: global shifts in marine biodiversity. *Science* **321**, 654–657.

Reolid, M. (2007) Taphonomy of the Oxfordian-lowermost Kimmeridgian siliceous sponges of the Prebetic Zone (Southern Iberia). *Journal of Taphonomy* **5**, 71–90.

Reverter-Gil, O. & Fernández-Pulpeiro, E. (2005) A new genus of cyclostome bryozoan from the European Atlantic coast. *Journal of Natural History* **39**, 2379–2387.

Reymond, C.E., Zihrul, K.-S., Halfar, J., Riegl, B., Humphreys, A. & Westphal, H. (2016) Heterozoan carbonates from the equatorial rocky reefs of the Galápagos Archipelago. *Sedimentology* **63**, 940–958.

Riding, R. (1991) Classification of microbial carbonates. Pp. 21–51 in Riding, R. (ed.) *Calcareous algae and stromatolites.* Springer-Verlag, Berlin.

Riegel, W. (2008) The Late Palaeozoic phytoplankton blackout – artefact or evidence of global change? *Review of Palaeobotany and Palynology* **148**, 73–90.

Rigby, J.K. (1957) Relationships between *Acanthocladia guadalupensis* and *Solenopora texana* and the bryozoan-algal consortium hypothesis. *Journal of Paleontology* **31**, 603–606.

Riisgård, H.U. & Manríquez, P.H. (1997) Filter-feeding in fifteen marine ectoprocts (Bryozoa): particle capture and water pumping. *Marine Ecology Progress Series* **154**, 223–239.

Riisgård, H.U., Okamura, B. & Funch, P. (2009) Particle capture in ciliary filter-feeding gymnolaemate and phylactolaemate bryozoans – a comparative study. *Acta Zoologica* **91**, 416–425.

Rinkevich, B. & Weissman, I.L. (1987) Chimeras in colonial invertebrates: a synergistic symbiosis or somatic- and germ-cell parasitism? *Symbiosis* **4**, 117–134.

Ristedt, H. (1991) Ancestrula and early astogeny of some anascan Bryozoa: their taxonomic importance and possible phylogenetic implications. *Bulletin de la Societe des Sciences Naturelles de l'Ouest de la France, Mémoire* **H.S. 1**, 371–382.

Ristedt, H. (1996) Initial frontal budding in some nodular cheilostomate Bryozoa. Pp. 237–242 in Gordon, D.P., Smith, A.M. & Grant-Mackie, J.A. (eds) *Bryozoans in space and time.* NIWA, Wellington.

Ristedt, H. & Schuhmacher, H. (1985) The bryozoan *Rhynchozoon larreyi* (Audouin, 1826) a successful competitor in coral reef communities of the Red Sea. *P.S.Z.N.I, Marine Ecology* **6**, 167–179.

Robertson, A. (1903) Embryology and embryonic fission in the genus *Crisia*. *University of California Publications in Zoology* **1**, 115–156.

Rodgers, P.J. & Woollacott, R.M. (2006) Systematics, variation, and developmental instability: analysis of spine patterns in ancestrulae of a common bryozoan. *Journal of Natural History* **40**, 1351–1368.

Rodland, D.L., Kowalewski, M., Carroll, M. & Simões, M.G. (2006) The temporal resolution of epibiont assemblages: are they ecological snapshots or overexposures? *The Journal of Geology* **114**, 313–324.

Rodland, D.L., Simões, M.G., Krause, R.A., Jr & Kowalewski, M. (2014) Stowing away on ships that pass in the night: sclerobiont assemblages on individually dated bivalve and brachiopod shells from a subtropical shelf. *Palaios* **29**, 170–183.

Rogick, M.D. & Croasdale, H. (1949) Studies on marine Bryozoa, III: Woods Hole region Bryozoa associated with algae. *Biological Bulletin* **96**, 32–69.

Rosen, B.R. (1986) Modular growth and form of corals: a matter of metamers? *Philosophical Transactions of the Royal Society, Series B* **313**, 115–142.

Rosenberg. G.D. & Hughes, W.W. (1991) A metabolic model for the determination of shell composition in the bivalve mollusc, *Mytilus edulis*. *Lethaia* **24**, 83–96.

Ross, J.P. (1964) Champlainian cryptostome Bryozoa from New York State. *Journal of Paleontology* **38**, 1–32.

Ross, J.P. (1967) Champlainian Ectoprocta (Bryozoa), New York State. *Journal of Paleontology* **41**, 632–648.

Ross, J.R.P. (1978) Biogeography of Permian ectoproct Bryozoa. *Palaeontology* **21**, 341–356.

Ross, J.R.P. (1981) Biogeography of Carboniferous ectoproct Bryozoa. *Palaeontology* **24**, 313–341.

Rosso, A. (2008) *Leptichnus tortus* sp. nov., a new etching trace and remarks on other bryozoan-produced fossil traces. *Studi Trentini di Scienze Naturali, Acta Geologica* **83**, 75–85.

Rosso, A. (2009) The first catenicellid (Bryozoa, Ascophora) from Mediterranean shallow waters: a hidden resident or a new immigrant? *Journal of Natural History* **43**, 2209–2226.

Rosso, A. & Di Geronimo, I. (1998) Deep-sea Pleistocene Bryozoa of southern Italy. *Geobios* **30**, 303–317.

Rosso, A. & Di Martino, E. (2016) Bryozoan diversity in the Mediterranean Sea: an update. *Mediterranean Marine Science* **17**, 567–607.

Rosso, A., Di Martino, E., Sanfilippo, R. & Di Martino, V. (2012) Bryozoan communities and thanatocoenoses from submarine caves in the Plemmirio Marine Protected Area (SE Sicily). Pp. 251–269 in Ernst, A., Schäfer, P. & Scholz, J. (eds) *Bryozoan studies 2010*. Springer, Heidelberg.

Rosso, A. & Sanfilippo, R. (2005) Bryozoans and serpuloideans in skeletobiont communities from the Pleistocene of Sicily: spatial utilisation and competitive interactions. *Annali dell'Università degli Studi di Ferrara, Museologia Scientifica e Naturalistica, Volume Speciale* **2005**, 109–124.

Rowden, A.A., Warwick, R.M. & Gordon, D.P. (2004) Bryozoan biodiversity in the New Zealand region and implications for marine conservation. *Biodiversity and Conservation* **13**, 2695–2721.

Rubin, J.A. (1982) The degree of intransitivity and its measurement in an assemblage of encrusting cheilostome Bryozoa. *Journal of Experimental Marine Biology and Ecology* **60**, 119–128.

Rucker, J.B. (1967) Palaeocological analysis of cheilostome Bryozoa from Venezuela–British Guiana shelf sediments. *Bulletin of Marine Science* **17**, 787–839.

Rucker, J.B. & Carver, R.E. (1969) A survey of the carbonate mineralogy of the cheilostome Bryozoa. *Journal of Paleontology* **43**, 791–799.

Runnegar, B. (1984) Crystallography of the foliated calcite shell layers of bivalve molluscs. *Alcheringa* **8**, 273–290.

Rust, S. & Gordon, D.P. (2011) Plio-Pleistocene bryozoan faunas of the Wanganui Basin, New Zealand: stratigraphic distribution and diversity. *New Zealand Journal of Geology and Geophysics* **54**, 151–165.

Ryland, J.S. (1970) *Bryozoans*. Hutchinson, London.

Ryland, J.S. (1979) Structural and physiological aspects of coloniality in Bryozoa. Pp. 211–242 in Larwood, G. & Rosen, B.R. (eds) *Biology and systematics of colonial organisms*. Academic Press, London.

Ryland, J.S. (1981) Colonies, growth and reproduction. Pp. 221–226 in Larwood, G.P. & Nielsen, C. (eds) *Recent and fossil Bryozoa*. Olsen & Olsen, Fredensborg.

Ryland, J.S. (2001) Convergent colonial organization and reproductive function in two bryozoan species epizoic on gastropod shells. *Journal of Natural History* **35**, 1085–1101.

Ryland, J.S., Holt, R., Loxton, J., Spencer Jones, M.E. & Porter, J.S. (2014) First occurrence of the non-native bryozoan *Schizoporella japonica* Ortmann (1890) in Western Europe. *Zootaxa* **3780**, 481–502.

Ryland, J.S. & Warner, G.F. (1986) Growth and form in modular animals: ideas on the size and arrangement of zooids. *Philosophical Transactions of the Royal Society, Series B* **313**, 53–76.

Saier, B. & Chapman, A.S. (2004) Crusts of the alien bryozoan *Membranipora membranacea* can negatively impact spore output from native kelps (*Laminaria longicruris*). *Botanica Marina* **47**, 265–271.

Sakagami, S. (1985) Paleogeographic distribution of Permian and Triassic Ectoprocta (Bryozoa). Pp. 171–183 in Nakazawa, K. & Dickins, J.M. (eds) *The Tethys: her paleogeography and paleobiogeography from Paleozoic to Mesozoic*. Tokai University Press, Tokyo.

Sánchez-Beristain, F. & Reitner, J. (2017) *Reptonoditrypa cautica*, briozoo incrustante de los olistolitos Cipit de la Formación San Cassiano (Triásico, Ladiniano/Carniano; NE de Italia) y sus implicaciones paleoecológicas. *Boletín de la Sociedad Geológica Mexicana* **69**, 409–420.

Sandberg, P.A. (1977) Ultrastructure, mineralogy, and development of bryozoan skeletons. Pp. 143–181 in Woollacott, R.M. & Zimmer, R.L. (eds) *Biology of bryozoans*. Academic Press, New York.

Sandberg, P.A. (1983) An oscillating trend in Phanerozoic non-skeletal carbonate mineralogy. *Nature* **305**, 19–22.

Sanders, D. & Baron-Szabo, R. (2008) Palaeoecology of solitary corals in soft-substrate habitats: the example of *Cunnolites* (upper Santonian, Eastern Alps). *Lethaia* **41**, 1–14.

Sanders, H.C., Geary, D.H. & Byers, C.W. (2002) Paleoecology and sedimentology of the *Prasopora* zonule in the Dunleith Formation (Ordovician), Upper Mississipi Valley. *Geoscience Wisconsin* **17**, 11–20.

Sandford, F. & Brown, C. (1997) Shell substrates of the Florida hermit-crab sponge *Spongosorites suberitoides*, from the Gulf of Mexico. *Bulletin of Marine Science* **61**, 225–223.

Sanfilippo, R., Rosso, A., Basso, D., Violanti, D., Di Geronimo, I., Benzoni, F. & Robba, E. (2011) Cobbles colonization pattern from a tsunami-affected coastal area (SW Thailand, Andaman Sea). *Facies* **57**, 1–13.

Santagata, S. & Banta, W.C. (1996) Origin of brooding and ovicells in cheilostome bryozoans: interpretive morphology of *Scrupocellaria ferox*. *Invertebrate Biology* **115**, 170–180.

Saunders, M.I. & Metaxas, A. (2009) Effects of temperature, size, and food on the growth of *Membranipora membranacea* in laboratory and field studies. *Marine Biology* **156**, 2267–2276.

Savoie, L., Miron, G. & Biron, M. (2007) Fouling community of the snow crab *Chionoecetes opilio* in Sydney Bight, Canada: preliminary observations in relation to sampling period and depth/geographical location. *Cahiers Biologie Marine* **48**, 347–359.

Schack, C.R., Gordon, D.P. & Ryan, K.G. (2018a) Classification of cheilostome polymorphs. Pp. 85–134 in Wyse Jackson, P.N. & Spencer Jones, M. (eds) *Annals of bryozoology 6*. International Bryozoology Association, Dublin.

Schack, C.R., Gordon, D.P. & Ryan, K.G. (2018b) Modularity is the mother of invention: a review of polymorphism in bryozoans. *Biological Reviews* **94**, 773–809.

Schäfer, P. (1985) Significance of soft part morphology in the classification of Recent tubuliporoid cyclostomes. Pp. 273–284 in Nielsen, C. & Larwood, G.P. (eds) *Bryozoa: Ordovician to Recent*. Olsen & Olsen, Fredensborg.

Schäfer, P. & Bader, B. (2008) Geochemical composition and variability in the skeleton of the bryozoan *Cellaria sinuosa* (Hassall): biological versus environmental control. *Virginia Museum of Natural History Special Publication* **15**, 269–279.

Schäfer, P., Bader, B. & Blaschek, H. (2006) Morphology and function of the flexible nodes in the cheilostome bryozoan *Cellaria sinuosa* (Hassall). *Courier Forschungsinstitut Senckenberg* **257**, 119–131.

Schäfer, P., Cuffey, R.J. & Young, A.R. (2003) New trepostome Bryozoa from the Early Triassic (Smithian/Spathian) of Nevada. *Paläontologische Zeitschrift* **77**, 323–340.

Schäfer, P. & Fois-Erickson, E. (1986) Triassic Bryozoa and the evolutionary crisis of Paleozoic Stenolaemata. *Lecture Notes in Earth Sciences* **8**, 251–255.

Schäfer, P. & Grant-Mackie, J. (1998) Revised systematics and palaeobiogeography of some Late Triassic colonial invertebrates from the Pacific Region. *Alcheringa* **22**, 87–122.

Schäfer, P., Herrera Cubilla, A. & Bader, B. (2012) Distribution and zoogeography of cheilostomate Bryozoa along the Pacific Coast of Panama: comparison between the Gulf of Panama and Gulf of Chiriquí. Pp. 303–319 in Ernst, A., Schäfer, P. & Scholz, J. (eds) *Bryozoan studies 2010*. Lecture Notes in Earth System Sciences **143**. Springer, Berlin.

Schäfer, P., Senowbari-Daryan, B. & Hamedani, A. (2003) Stenolaemate bryozoans from the Upper Triassic (Norian–Rhaetian) Nayband Formation, central Iran. *Facies* **49**, 135–150.

Schaumberg, G. (1979) Neue Nachweise von Bryozoen und Brachiopoden Nahrung des permischen Holocephalen, *Janassa bituminosa* (Schlotheim), *Philippia* **4**, 3–11.

Schlager, W. (2003) Benthic carbonate factories of the Phanerozoic. *International Journal of Earth Science* **92**, 445–464.

Schlögl, J., Michalík, J., Zágorsek, K. & Atrops, F. (2008) Early Tithonian serpulid-dominated cavity-dwelling fauna, and the recruitment pattern of the serpulid larvae. *Journal of Paleontology* **82**, 351–361.

Schmidt, R. (2007) Australian Cenozoic Bryozoa, 2: Free-living Cheilostomata of the Eocene St. Vincent Basin, SA, including *Bonellina* gen. nov. *Alcheringa* **31**, 67–84.

Schmidt, R. & Bone, Y. (2004) Australian Cainozoic Bryozoa, 1: *Nudicella* gen. nov. (Onychocellidae, Cheilostomata): taxonomy, palaeoenvironments and biogeography. *Alcheringa* **28**, 185–203.

Schneider, S., Jäger, M., Kroh, A., Mitterer, A., Niebuhr, B., Vodrázka, R., Wilmsen, M., Wood, C.J. & Zágorsek, K. (2013) Silicified sea life – Macrofauna and palaeoecology of the Neuburg Kieselerde Member (Cenomanian to Lower Turonian Wellheim Formation, Bavaria, southern Germany). *Acta Geologica Polonica* **63**, 555–610.

Scholz, J. & Hillmer, G. (1995) Reef-bryozoans and bryozoan micro-reefs: control factor evidence from the Philippines and other regions. *Facies* **32**, 109–143.

Scholz, J. & Krumbein, W.E. (2006) Microbial mats and biofilms associated with bryozoans. Pp. 283–298 in Gordon, D.P., Smith, A.M. & Grant-Mackie, J.A. (eds) *Bryozoans in space and time*. NIWA, Wellington.

Scholz, J. & Levit, G.S. (2003) Bryozoan morphoprocesses. Pp. 181–195 in Krumbein, W.E., Paterson, D.M. & Zavarzin, G.A. (eds) *Fossil and Recent biofilms*. Springer, Dordrecht.

Schopf, T.J.M. (1969) Paleoecology of ectoprocts (bryozoans). *Journal of Paleontology* **43**, 234–244.

Schopf, T.J.M. (1970) Taxonomic diversity gradients of ectoprocts and bivalves and their geologic implications. *Bulletin of the Geological Society of America* **81**, 3765–3768.

Schopf, T.J.M. (1979) The role of biogeographic provinces in regulating marine faunal diversity through geologic time. Pp. 449–457 in Gray, J. & Boucot, A.J. (eds) *Historical biogeography, plate tectonics, and the changing environment.* Oregon State University Press, Corvallis.

Schopf, T.J.M., Collier, K.O. & Bach, B.O. (1980) Relation of the morphology of stick-like bryozoans to bottom currents and suspended matter and depth at Friday Harbor, Washington. *Paleobiology* **6**, 466–476.

Schopf, T.J.M. & Dutton, A.R. (1976) Parallel clines in morphologic and genetic differentiation in a coastal zone marine invertebrate: the bryozoan *Schizoporella errata. Paleobiology* **2**, 255–264.

Schwaha, T., Bernhard, J.M., Edgcomb, V.P. & Todaro, M.A. (2019) *Aethozooides uraniae*, a new deep-sea genus and species of solitary bryozoan from the Mediterranean Sea, with a revision of the Aethozoidae. *Marine Biodiversity* **49**, 1843–1856.

Schwaha, T., Handschuh, S., Ostrovsky, A.N. & Wanninger, A. (2018) Morphology of the bryozoan *Cinctipora elegans* (Cyclostomata, Cinctiporidae) with first data on its sexual reproduction and the cyclostome neuro-muscular system. *BMC Evolutionary Biology* **18**(92), 1–28.

Schwaha, T., Wood, T.S. & Wanninger, A. (2011) Myoanatomy and serotonergic nervous system of the ctenostome *Hislopia malayensis*: evolutionary trends in bodyplan patterning of Ectoprocta. *Frontiers in Zoology* **8**(11), 1–16.

Schwaninger, H.R. (1999) Population structure of the widely dispersing marine bryozoan *Membranipora membranacea* (Cheilostomata): implications for population history, biogeography, and taxonomy. *Marine Biology* **135**, 411–423.

Schwaninger, H.R. (2008) Global mitochondrial DNA phylogeography and biogeographic history of the antitropically and longitudinally disjunct marine bryozoan *Membranipora membranacea* L. (Cheilostomata): another cryptic marine sibling species complex? *Molecular Phylogenetics and Evolution* **49**, 893–908.

Scoffin, T.P. (1971) The conditions of growth of the Wenlock reefs of Shropshire (England). *Sedimentology* **17**, 173–219.

Seed, R., Elliott, M.N., Boaden, P.J.S. & O'Connor, R.J. (1981) The composition and seasonal changes amongst the epifauna associated with *Fucus serratus* L. in Strangford Lough,

Northern Ireland. *Cahiers de Biologie Marine* **22**, 243–266.

Sendino, C., Suárez Andrés, J.L. & Wilson, M.A. (2019) A rugose coral–bryozoan association from the Lower Devonian of NW Spain. *Palaeogeography, Palaeoclimatology, Palaeoecology* **530**, 271–280.

Sepkoski, J.J., Jr (1984) A kinetic model of Phanerozoic taxonomic diversity. III. Post-Paleozoic families and mass extinctions. *Paleobiology* **10**, 246–267.

Sepkoski, J.J., Jr, McKinney, F.K., & Lidgard, S. (2000) Competitive displacement among post-Paleozoic cyclostome and cheilostome bryozoans. *Paleobiology* **26**, 718.

Serra-Kiel, J. & Reguant, S. (1991) Biofaciès de plate-forme aphotique silico-clastique avec bryozoaires et spongiares (Éocène Moyen, sector oriental du Bassin Sud-Pyrénéen). *Geobios* **24**, 33–40.

Servais, T., Harper, D.A.T., Munnecke, A., Owen, A.W. & Sheehan, P.M. (2009) Understanding the Great Ordovician Biodiversification Event (GOBE): influences of paleogeography, paleoclimate, or paleoecology? *GSA Today* **19**(4/5), 4–10.

Shanks, A.L. (2009) Pelagic larval duration and dispersal distance revisited. *Biological Bulletin* **216**, 373–385.

Shapiro, D.F. (1996) Size-dependent neural integration between genetically different colonies of a marine bryozoan. *The Journal of Experimental Biology* **199**, 1229–1239.

Sharples, A.G.W.D., Huuse, M., Hollis, C., Totterdell, J.M. & Taylor, P.D. (2014) Giant middle Eocene bryozoan reef mounds in the Great Australian Bight. *Geology* **42**, 683–686.

Sheehan, P.M. (2001) The Late Ordovician Mass Extinction. *Annual Review of Earth and Planetary Sciences* **29**, 331–364.

Shunatova, N.N. & Ostrovsky, A.N. (2001) Individual autozooidal behaviour and feeding in marine bryozoans. *Sarsia* **86**, 113–142.

Shunatova, N.N. & Ostrovsky, A.N. (2002) Group autozooidal behaviour and chimneys in marine bryozoans. *Marine Biology* **140**, 503–518.

Silén, L. (1972) Fertilization in the Bryozoa. *Ophelia* **10**, 27–34.

Silén, L. (1977) Polymorphism. Pp. 183–231 Woollacott, R.M. & Zimmer, R.L. (eds) *Biology of bryozoans*. Academic Press, New York.

Silén, L. (1981) Colony structure in *Flustra foliacea* (Linnaeus) (Bryozoa, Cheilostomata). *Acta Zoologica* **62**, 219–232.

Silén, L. (1982) Multizooidal budding in *Parasmittina trispinosa* (Johnston) (Bryozoa, Cheilostomata). *Acta Zoologica* **63**, 25–32.

Silén, L. (1987) Colony growth pattern in *Electra pilosa* (Linnaeus) and comparable encrusting cheilostome bryozoans. *Acta Zoologica* **68**, 17–34.

Silén, L. & Harmelin, J.-G. (1974) Observations on living Diastoporidae (Bryozoa, Cyclostomata), with special regard to polymorphism. *Acta Zoologica* **55**, 81–96.

Silén, L. & Harmelin, J.-G. (1976) *Haplopoma sciaphilum* sp.n., a cave-living bryozoan from the Skagerrak and the Mediterranean. *Zoologica Scripta* **5**, 61–66.

Simpson, C. (2011) The evolutionary history of division of labour. *Proceedings of the Royal Society, Series B* **279**, 116–121.

Simpson, C., Jackson, J.B.C. & Herrera Cubilla, A. (2017) Evolutionary determinants of morphological polymorphism in colonial animals. *The American Naturalist* **190**, 17–28.

Skovsted, C.B., Holmer, L.E., Larsson, C.M., Högström, A.E.S., Brock, G.A., Topper, T.P., Balthasar, U., Petterson Stolk, S. & Paterson J.R. (2009) The scleritome of *Paterimitra*: an Early Cambrian stem group brachiopod from South Australia. *Proceedings of the Royal Society, Series B* **276**, 1651–1656.

Smith, A.M. (2014) Growth and calcification of marine bryozoans in a changing ocean. *Biological Bulletin* **226**, 203–210.

Smith, A.M. & Clark, D.E. (2010) Skeletal carbonate mineralogy of bryozoans from Chile: an independent check of phylogenetic patterns. *Palaios* **25**, 229–233.

Smith, A.M. & Girvan, E. (2010) Understanding a bimineralic bryozoan: skeletal structure and carbonate mineralogy of *Odontionella cyclops* (Foveolariidae: Cheilostomata: Bryozoa) in New Zealand. *Palaeogeography, Palaeoclimatology, Palaeoecology* **289**, 113–122.

Smith, A.M. & Lawton, E.I. (2010) Growing up in the temperate zone: age, growth, calcification and carbonate mineralogy of *Melicerita chathamensis* (Bryozoa) in southern New Zealand. *Palaeogeography, Palaeoclimatology, Palaeoecology* **298**, 271–277.

Smith, A.M. & Key, M.M., Jr (2004) A detailed stable isotope record in a single bryozoan skeleton: controls, variation, and a record of climate change. *Quaternary Research* **61**, 123–133.

Smith, A.M., Key, M.M. & Gordon, D.P. (2006) Skeletal mineralogy of bryozoans: taxonomic and temporal patterns. *Earth-Science Reviews* **78**, 287–306.

Smith, A.M. & Nelson, C.S. (1994) Selectivity in sea-floor processes: taphonomy of bryozoans. Pp. 177–180 in Hayward, P.J., Ryland, J.S. & Taylor, P.D. (eds) *Biology and palaeobiology of bryozoans*. Olsen & Olsen, Fredensborg.

Smith, A.M. & Nelson, C.S. (2003) Effects of early sea-floor processes on the taphonomy of temperate shelf skeletal carbonate deposits. *Earth-Science Reviews* **63**, 1–31.

Smith, A.M. & Nelson, C.S. (2006) Differential abrasion of bryozoan skeletons: taphonomic implications for paleoenvironmental interpretation. Pp. 305–313 in Gordon, D.P., Smith, A.M. & Grant-Mackie, J.A. (eds) *Bryozoans in space and time*. NIWA, Wellington.

Smith, A.M., Nelson, C.S., Key, M.M., Jr & Patterson, W.P. (2004) Stable isotope values in modern bryozoan carbonate from New Zealand and implications for paleoenvironmental interpretation. *New Zealand Journal of Geology and Geophysics* **47**, 809–821.

Smith, A.M., Nelson, C. & Spencer, H. (1998) Skeletal carbonate mineralogy of New Zealand bryozoans. *Marine Geology* **151**, 27–46.

Smith, A.M., Taylor, P.D. & Spencer, H.G. (2008) Resolution of taxonomic issues in the Horneridae (Bryozoa: Cyclostomata). Pp. 359–411

in Wyse Jackson, P.N. & Spencer Jones, M.E. (eds) *Annals of bryozoology 2*. International Bryozoloogy Association, Dublin.

Smrecak, T.A. & Brett, C.E. (2014) Establishing patterns in sclerobiont distribution in a Late Ordovician (Cincinnatian) depth gradient: toward a sclerobiofacies model. *Palaios* **29**, 74–85.

Smyth, M.J. (1988) *Penetrantia clionoides*, sp. nov. (Bryozoa), a boring bryozoan in gastropod shells from Guam. *Biological Bulletin* **174**, 276–286.

Sogot, C.E., Harper, E.M. & Taylor, P.D. (2013) Biogeographical and ecological patterns in bryozoans across the Cretaceous–Paleogene boundary: implications for the phytoplankton collapse hypothesis. *Geology* **41**, 631–634.

Sogot, C.E., Harper, E.M. & Taylor, P.D. (2014) The Lilliput effect in colonial organisms: cheilostome bryozoans at the Cretaceous–Paleogene mass extinction. *PLOS One* **9**(2): e87048.

Sokolover, N., Ostrovsky, A.N. & Ilan, M. (2018) *Schizoporella errata* (Bryozoa, Cheilostomata) in the Mediterranean Sea: abundance, growth rate, and reproductive strategy. *Marine Biology Research* **14**, 868–882.

Song, H., Wignall, P.B., Tong, J.N. & Yin, H.F. (2012) Two pulses of extinction during the Permian–Triassic crisis. *Nature Geoscience* **6**, 52–56.

Sørensen, A.M., Håkansson, E. & Stemmerik, L. (2007) Faunal migration into the Late Permian Zechstein Basin – evidence from bryozoan palaeobiogeography. *Palaeogeography, Palaeoclimatology, Palaeoecology* **251**, 198–209.

Soták, J. (2010) Paleoenvironmental changes across the Eocene–Oligocene boundary: insights from the Central-Carpathian Paleogene Basin. *Geologica Carpathica* **61**, 393–418.

Soule, D.F. (1973) Morphogenesis of giant avicularia and ovicells in some Pacific Smittinidae. Pp. 485–495 in Larwood, G.P. (ed.) *Living and fossil Bryozoa*. Academic Press, London.

Soule, D.F. & Soule, J.D. (1969) Systematics and biogeography of burrowing bryozoans. *American Zoologist* **9**, 791–802.

Soule, D.F., Soule, J.D. & Chaney, H.W. (1992) The genus *Thalamoporella* worldwide (Bryozoa, Anasca), morphology, evolution and speciation. *Irene McCulloch Foundation Monograph Series* **1**, 1–93.

Sparks, D.K., Hoare, R.D. & Kesling, R.V. (1980) Epizoans on the brachiopod *Paraspirifer bownockeri* (Stewart) from the Middle Devonian of Ohio. *University of Michigan, Papers on Paleontology* **23**, 1–105.

Spjeldnaes, N. (2000) Cryptic bryozoans from West Africa. Pp. 385–391 in Herrera Cubilla, A. & Jackson, J.B.C. (eds) *Proceedings of the 11th International Bryozoology Association Conference.* Smithsonian Tropical Research Institute, Balboa.

Spjeldnaes, N. (2006) Bryozoan colonies as indicators of bottom conditions in the Lower Ordovician. Pp. 315–319 in Gordon, D.P., Smith, A.M. & Grant-Mackie, J.A. (eds) *Bryozoans in space and time.* NIWA, Wellington.

Sprinkle, J. & Rodgers, J.C. (2010) Competition between a Pennsylvanian (Late Carboniferous) edrioasteroid and a bryozoan for living space on a brachiopod. *Journal of Paleontology* **84**, 356–359.

Stach, L.W. (1936) Correlation of zoarial form with habitat. *Journal of Geology* **44**, 60–65.

Stach, L.W. (1938) Observations on *Carbasea indivisa* Busk (Bryozoa). *Proceedings of the Zoological Society of London* **108**, 389–399.

Stanley, S.M. (2006) Influence of seawater chemistry on biomineralization throughout Phanerozoic time: paleontological and experimental evidence. *Palaeogeography, Palaeoclimatology, Palaeoecology* **232**, 214–236.

Starcher, R.W. & McGhee, G.R., Jr (2000) Fenestrate theoretical morphology: geometric constraints on lophophore shape and arrangement in extinct Bryozoa. *Paleobiology* **26**, 116–136.

Stebbing, A.R.D. (1971) Growth of *Flustra foliacea* (Bryozoa). *Marine Biology* **9**, 267–273.

Stebbing, A.R.D. (1973) Observations on colony overgrowth and spatial competition. Pp. 173–183 in Larwood, G.P. (ed.) *Living and fossil Bryozoa.* Academic Press, London.

Steger, K.K. & Smith, A.M. (2005) Carbonate mineralogy of free-living bryozoans (Bryozoa: Otionellidae), Otago Shelf, southern New Zealand. *Palaeogeography, Palaeoclimatology, Palaeoecology* **218**, 195–203.

Steinthorsdottir, M. & Håkansson, E. (2017) Endo- and epilithic faunal succession in a Pliocene–Pleistocene cave on Rhodes, Greece: record of a transgression. *Palaeontology* **60**, 663–681.

Steinthorsdottir, M., Lidgard, S. & Håkansson, E. (2006) Fossils, sediments, tectonics. *Facies* **52**, 361–380.

Stephens, G.C. & Schinske, R.A. (1961) Uptake of amino acides by marine invertebrates. *Limnology and Oceanography* **6**, 175–181.

Stepień, A., Kukliński, P., Włodarska-Kowalczuk, M., Krzemińska, M. & Gudmundsson, G. (2017) Bryozoan zooid size variation across a bathymetric gradient: a case study from the Icelandic shelf and continental slope. *Marine Biology* **164**, https://doi.org/10.1007/s00227-017-3231-9.

Stilwell, J.D. & Håkansson, E. (2012) Survival, but …! New tales of 'Dead Clade Walking' from Austral and Boreal post-K-T assemblages. Pp. 795–810 in Talent, J.A. (ed.) *Earth and life: global biodiversity, extinction intervals and biogeographic perturbations through time.* Springer, Dordrecht.

Ström, R. (1977) Brooding patterns of bryozoans. Pp. 23–89 in Woollacott, R.M. & Zimmer, R.L. (eds) *Biology of bryozoans.* Academic Press, New York.

Suárez Andrés, J.L. (2014) Bioclaustration in Devonian fenestrate bryozoans. The ichnogenus *Caupokeras* McKinney, 2009. *Spanish Journal of Palaeontology* **29**, 5–14.

Suárez Andrés, J.L. & McKinney, F.K. (2010) Revision of the Devonian fenestrate bryozoan genera *Cyclopelta* Bornemann, 1884 and *Pseudoisotrypa* Prantl, 1932, with description of a rare fenestrate growth habit. *Revista Española de Paleontología* **25**, 123–138.

Suárez Andrés, J.L. & Wyse Jackson, P.N. (2015a) *Ernstipora mackinneyi*, a new unique fenestrate bryozoan genus and species with an encrusting growth habit from the Emsian (Devonian) of NW Spain. *Neues Jahrbuch für Geologie und Paläontologie, Abhandlungen* **271**, 229–242.

Suárez Andrés, J.L. & Wyse Jackson, P.N. (2015b) Feeding currents: a limiting factor for disparity of Palaeozoic fenestrate bryozoans. *Palaeogeography, Palaeoclimatology, Palaeoecology* **433**, 219–232.

Suárez Andrés, J.L. & Wyse Jackson, P.N. (2018) First report of a Palaeozoic fenestrate bryozoan with an articulated growth habit. *Journal of Iberian Geology* **44**, 273–283.

Sun, M., Wu, Z., Shen, X., Ren, J., Liu, X., Liu, H. & Liu, B. (2009) The complete mitochondrial genome of *Watersipora subtorquata* (Bryozoa, Gymnolaemata, Ctenostomata) with phylogenetic consideration of Bryozoa. *Gene* **439**, 17–24.

Surlyk, F. (1997) A cool-water carbonate ramp with bryozoan mounds: Late Cretaceous–Danian of the Danish Basin. *SEPM Special Publication* **56**, 293–307.

Surlyk, F., Damholt, T. & Bjerager, M. (2006) Stevns Klint, Denmark: uppermost Maastrichtian chalk, Cretaceous–Tertiary boundary, and lower Danian bryozoan mound complex. *Bulletin of the Geological Society of Denmark* **54**, 1–48.

Swezey, D.S., Bean, J.R., Ninokawa, A.T., Hill, T.M., Gaylord, B. & Sanford, E. (2017) Interactive effects of temperature, food and skeletal mineralogy mediate biological responses to ocean acidification in a widely distributed bryozoan. *Proceedings of the Royal Society, Series B* **284**: 20162349.

Tamberg, Y. & Shunatova, N. (2016) Feeding behavior in freshwater bryozoans: function, form, and flow. *Invertebrate Biology* **135**, 138–149.

Tapanila, L. (2005) Palaeoecology and diversity of endosymbionts in Palaeozoic marine invertebrates: trace fossil evidence. *Lethaia* **38**, 89–99.

Tavener-Smith, R. (1965) A new fenestrate bryozoan from the Lower Carboniferous of County Fermanagh. *Palaeontology* **8**, 478–491.

Tavener-Smith, R. (1969) Skeletal structure and growth in the Fenestellidae (Bryozoa). *Palaeontology* **12**, 281–309.

Tavener-Smith, R. (1973a) Fenestrate Bryozoa from the Visean of County Fermanagh, Ireland. *Bulletin of the British Museum (Natural History), Geology Series* **23**, 391–493.

Tavener-Smith, R. (1973b) Some aspects of skeletal organization in Bryozoa. Pp. 349–359 in Larwood, G.P. (ed.) *Living and fossil Bryozoa*. Academic Press, London.

Tavener-Smith, R. (1975) The phylogenetic affinities of fenestelloid bryozoans. *Palaeontology* **18**, 1–17.

Tavener-Smith, R. & Williams, A. (1970) Structure of the compensation sac in two ascophoran bryozoans. *Proceedings of the Royal Society, Series B* **175**, 235–254.

Tavener-Smith, R. & Williams, A. (1972) The secretion and structure of the skeleton of living and fossil Bryozoa. *Philosophical Transactions of the Royal Society, Series B* **264**, 97–159.

Taylor, P.D. (1976) Multilamellar growth in two Jurassic cyclostomatous Bryozoa. *Palaeontology* **19**, 293–306.

Taylor, P.D. (1978) The spiral bryozoan *Terebellaria* from the Jurassic of southern England and Normandy. *Palaeontology* **21**, 357–391.

Taylor, P.D. (1979a) The inference of extrazooidal feeding currents in fossil bryozoan colonies. *Lethaia* **12**, 47–56.

Taylor, P.D. (1979b) Palaeoecology of the encrusting epifauna of some British Jurassic bivalves. *Palaeogeography, Palaeoclimatology, Palaeoecology* **28**, 241–262.

Taylor, P.D. (1981) Functional morphology and evolutionary significance of differing modes of tentacle eversion in marine bryozoans. Pp. 235–247 in Larwood, G.P. & Nielsen, C. (eds) *Recent and fossil Bryozoa*. Olsen and Olsen, Fredensborg.

Taylor, P.D. (1982) Probable predatory borings in late Cretaceous bryozoans. *Lethaia* **15**, 67–74.

Taylor, P.D. (1984a) Adaptations for spatial competition and utilization in Silurian encrusting bryozoans. *Special Papers in Palaeontology* **32**, 197–210.

Taylor, P.D. (1984b) *Marcusodictyon* Bassler from the Lower Ordovician of Estonia: not the earliest bryozoan but a phosphatic problematicum. *Alcheringa* **8**, 177–186.

Taylor, P.D. (1985) Polymorphism in melicerititid cyclostomes. Pp. 311–318 in Nielsen, C. & Larwood, G.P. (eds) *Bryozoa: Ordovician to Recent*. Olsen & Olsen, Fredensborg.

Taylor, P.D. (1987) Fenestrate colony-form in a new melicerititid bryozoan from the U. Cretaceous of Germany. *Mesozoic Research* **1**, 71–77.

Taylor, P.D. (1988a) Colony growth pattern and astogenetic gradients in the Cretaceous cheilostome bryozoan *Herpetopora*. *Palaeontology* **31**, 519–549.

Taylor, P.D. (1988b) Major radiation of cheilostome bryozoans: triggered by the evolution of a new larval type? *Historical Biology* **1**, 45–64.

Taylor, P.D. (1990a) Bioimmured ctenostomes from the Jurassic and the origin of the cheilostome Bryozoa. *Palaeontology* **33**, 19–34.

Taylor, P.D. (1990b) Preservation of soft-bodied and other organisms by bioimmuration – a review. *Palaeontology* **33**, 1–17.

Taylor, P.D. (1991) Observations on symbiotic associations of bryozoans and hermit crabs from the Otago Shelf of New Zealand. *Bulletin de la Société des Sciences Naturelle de l'Ouest de la France, Mémoire* **H.S. 1**, 487–495.

Taylor, P.D. (1994a) An early cheilostome bryozoan from the Upper Jurassic of Yemen. *Neues Jahrbuch für Geologie und Paläontologie Abhandlungen* **191**, 331–344.

Taylor, P.D. (1994b) Evolutionary palaeoecology of symbioses between bryozoans and hermit crabs. *Historical Biology* **9**, 157–205.

Taylor, P.D. (1994c) Systematics of the melicerititid cyclostome bryozoans; introduction and the genera *Elea*, *Semielea* and *Reptomultelea*.

Bulletin of The Natural History Museum, London, Geology Series **50**, 1–103.

Taylor, P.D. (1995) Late Campanian–Maastrichtian Bryozoa from the United Arab Emirates–Oman border region. *Bulletin of The Natural History Museum, London, Geology Series* **51**, 267–273.

Taylor, P.D. (1999) Bryozoa. Pp. 623–646 in E. Savazzi (ed.) *Functional morphology of the invertebrate skeleton*. Wiley, Chichester.

Taylor, P.D. (2000) Cyclostome systematics: phylogeny, suborders and the problem of skeletal organization. Pp. 87–103 in Herrera Cubilla, A. & Jackson, J.B.C. (eds) *Proceedings of the 11th International Bryozoology Association Conference*. Smithsonian Tropical Research Institute, Balboa.

Taylor, P.D. (2001) Preliminary systematics and diversity patterns of cyclostome bryozoans from the Neogene of the Central American Isthmus. *Journal of Paleontology* **75**, 578–589.

Taylor, P.D. (2005) Bryozoans and palaeoenvironmental interpretation. *Journal of The Palaeontological Society of India* **50**(2), 1–11.

Taylor, P.D. (2012) A new bryozoan genus from the Jurassic of Switzerland, with a review of the cribrate colony-form in bryozoans. *Swiss Journal of Palaeontology* **131**, 201–210.

Taylor, P.D. (2014) Possible serpulid worm affinities of the supposed bryozoan *Corynotrypoides* from the Triassic Cassian Formation of the Italian Dolomites. *Batalleria* **20**, 11–16.

Taylor, P.D. (2015) Differentiating parasitism and other interactions in fossilized colonial organisms. *Advances in Parasitology* **90**, 329–347.

Taylor, P.D. (2016) Competition between encrusters on marine hard substrates and its fossil record. *Palaeontology* **59**, 481–497.

Taylor, P.D. (2019) A brief review of the scanty fossil record of Cretaceous bryozoans from Gondwana. *Australasian Palaeontological Memoirs* **52**, 147–154.

Taylor, P.D. & Allison, P.A. (1998) Bryozoan carbonates in space and time. *Geology* **26**, 459–462.

Taylor, P.D. & Badve, R. (1995) A new cheilostome bryozoan from the Cretaceous of India and Europe: a cyclostome homeomorph. *Palaeontology* **38**, 627–657.

Taylor, P.D., Berning, B. & Wilson, M.A. (2013) Reinterpretation of the Cambrian 'bryozoan' *Pywackia* as an octocoral. *Journal of Paleontology* **87**, 984–990.

Taylor, P.D. & Curry, G.B. (1985) The earliest known fenestrate bryozoan, with a short review of Lower Ordovician Bryozoa. *Palaeontology* **28**, 147–158.

Taylor, P.D., Dick, M.H., Clements, D. & Mawatari, S.F. (2012) A diverse bryozoan fauna from Pleistocene marine gravels at Kuromatsunai, Hokkaido, Japan. Pp. 367–383 in Ernst, A., Schäfer, P. & Scholz, J. (eds) *Bryozoan studies 2010*. Springer, Berlin.

Taylor, P.D. & Di Martino, E. (2014) Why is the tropical Cenozoic fossil record so poor for bryozoans? *Studi Trentini di Scienze Naturali* **94**, 249–257.

Taylor, P.D., Di Martino, E. & Martha, S.O. (2018) Colony growth strategies, dormancy and repair in some Late Cretaceous encrusting bryozoans: insights into the ecology of the Chalk seabed. *Palaeobiodiversity and Palaeoenvironments* **99**, 425–446.

Taylor, P.D. & Ernst, A. (2004) Bryozoan diversification during the Ordovician. Pp. 147–156 in Webby, B.D., Droser, M.L. & Paris, F. (eds) *The Great Ordovician Biodiversification Event*. Columbia University Press, New York.

Taylor, P.D. & Ernst, A. (2008) Bryozoans in transition: the depauperate and patchy Jurassic biota. *Palaeogeography, Palaeoclimatology, Palaeoecology* **263**, 9–23.

Taylor, P.D. & Foster, T.S. (1998) Bryozoans from the Pliocene Bowden Shell Bed of Jamaica. *Contributions to Tertiary and Quaternary Geology* **35**, 63–83.

Taylor, P.D. & Furness, R.W. (1978) Astogenetic and environmental variation of zooid size within colonies of Jurassic *Stomatopora* (Bryozoa, Cyclostomata). *Journal of Paleontology* **52**, 1093–1102.

Taylor, P.D. & Gordon, D.P. (1997) *Fenestulipora*, gen. nov., an unusual cyclostome bryozoan from New Zealand and Indonesia. *Invertebrate Taxonomy* **11**, 689–703.

Taylor, P.D. & Gordon, D.P. (2003) Endemic new cyclostome bryozoans from Spirits Bay, a New Zealand marine-biodiversity 'hotspot'. *New Zealand Journal of Marine and Freshwater Research* **37**, 653–669.

Taylor, P.D. & Gordon, D.P. (2007) Bryozoans from the Late Cretaceous Kahuitara Tuff of the Chatham Islands, New Zealand. *Alcheringa* **31**, 339–363.

Taylor, P.D., Gordon, D.P. & Batson, P.B. (2004) Bathymetric distributions of modern populations of some common Cenozoic Bryozoa from New Zealand, and paleodepth estimation. *New Zealand Journal of Geology and Geophysics* **47**, 57–69.

Taylor, P.D. & Grischenko, A.V. (1999) *Rodinopora* gen. nov. and the taxonomy of fungiform cyclostome bryozoans. *Species Diversity* **4**, 9–33.

Taylor, P.D., Hara, U. & Jasionowski, M. (2006) Unusual early development in a cyclostome bryozoan from the Ukrainian Miocene. *Linzer biologische Beiträge* **38**, 55–64.

Taylor, P.D. & James, N.P. (2013) Secular changes in colony-forms and bryozoan carbonate sediments through geological time. *Sedimentology* **60**, 1184–1212.

Taylor, P.D., James, N.P., Bone, Y., Kuklinski, P. & Kyser, T.K. (2009) Evolving mineralogy of cheilostome bryozoans. *Palaios* **24**, 440–452.

Taylor, P.D., James, N.P. & Phillips, G. (2014) Mineralogy of cheilostome bryozoans across the Eocene–Oligocene boundary in Mississippi, USA. *Palaeobiodiversity and Palaeoenvironments* **94**, 425–438.

Taylor, P.D. & Jenkins, H.L. (2017) Evolution of larval size in cyclostome bryozoans. *Historical Biology* **30**, 535–545.

Taylor, P.D. & Jones, C.G. (1996) Use of the environmental chamber in uncoated SEM of Recent and fossil bryozoans. *Microscopy and Analysis* **1996** (March), 27–29.

Taylor, P.D. & Kuklinski, P. (2011) Seawater chemistry and biomineralization: did trepostome bryozoans become hypercalcified in the 'calcite sea' of the Ordovician? *Palaeobiodiversity and Palaeoenvironments* **91**, 185–195.

Taylor, P.D., Kuklinski, P. & Gordon, D.P. (2007) Branch diameter and depositional depth in cyclostome bryozoans: testing a potential paleobathymetric tool. *Palaios* **22**, 220–224.

Taylor, P.D., Kudryavtsev, A.B. & Schopf, J.W. (2008) Calcite and aragonite distributions in the skeletons of bimineralic bryozoans as revealed by Raman spectroscopy. *Invertebrate Biology* **127**, 87–97.

Taylor, P.D. & Larwood, G.P. (1988) Mass extinctions and the pattern of bryozoan evolution. Pp. 99–119 in Larwood, G.P. (ed.) *Extinction and survival in the fossil record.* Systematics Association Special Publication, London.

Taylor, P.D. & Larwood, G.P. (1990) Major evolutionary radiations in the Bryozoa. Pp. 209–233 in Taylor, P.D. & Larwood, G.P. (eds) *Major evolutionary radiations.* Systematics Association Special Volume, London.

Taylor, P.D., Lazo, D.G. & Aguirre-Urreta, M.B. (2009) Lower Cretaceous bryozoans from Argentina: a 'by-catch' fauna from the Agrio Formation (Neuquén Basin). *Cretaceous Research* **30**, 193–203.

Taylor, P.D. & Lewis, J.E. (2003) A new skeletal structure in a cyclostome bryozoan from Taiwan. *Journal of Natural History* **37**, 2959–2965.

Taylor, P.D., Lombardi, C. & Cocito, S. (2015) Biomineralization in bryozoans: present, past and future. *Biological Reviews* **90**, 1118–1150.

Taylor, P.D., Martha, S.O. & Gordon, D.P. (2018) Synopsis of 'onychocellid cheilostome bryozoan genera. *Journal of Natural History* **52**, 1657–1721.

Taylor, P.D. & Mawatari, S.F. (2005) Preliminary overview of the cheilostome bryozoan *Microporella.* Pp. 329–339 in Moyano G., H.I., Cancino, J.M. & Wyse Jackson, P.N. (eds) *Bryozoan studies 2004.* Balkema, Leiden.

Taylor, P.D. & McKinney, F.K. (1996) An *Archimedes*-like cyclostome bryozoan from the Eocene of North Carolina. *Journal of Paleontology* **70**, 218–229.

Taylor, P.D. & McKinney, F.K. (2006) Cretaceous Bryozoa from the Campanian and Maastrichtian of the Atlantic and Gulf Coastal Plains, United States. *Scripta Geologica* **132**, 1–346.

Taylor, P.D. & Michalik, J. (1991) Cyclostome bryozoans from the late Triassic (Rhaetian) of the West Carpathians, Czechoslovakia. *Neues Jahrbuch für Geologie und Paläontologie* **182**, 285–302.

Taylor, P.D. & Monks, N. (1997) A new cheilostome bryozoan genus pseudoplanktonic on molluscs and algae. *Invertebrate Biology* **116**, 39–51.

Taylor, P.D. & Palmer, T.J. (1994) Submarine caves in a Jurassic reef (La Rochelle, France) and the evolution of cave biotas. *Naturwissenschaften* **81**, 357–360.

Taylor, P.D., Schembri, P.J. & Cook, P.L. (1989) Symbiotic associations between hermit crabs and bryozoans from the Otago region, southeastern New Zealand. *Journal of Natural History* **23**, 1059–1085.

Taylor, P.D. & Schindler, K.S. (2004) A new Eocene species of the hermit-crab symbiont *Hippoporidra* (Bryozoa) from the Ocala Limestone of Florida. *Journal of Paleontology* **78**, 790–794.

Taylor, P.D. & Sendino, C. (2010) Latitudinal distribution of bryozoan-rich sediments in the Ordovician. *Bulletin of Geosciences* **85**, 565–572.

Taylor, P.D. & Sendino, C. (2013) Chirality in the Late Palaeozoic fenestrate bryozoan *Archimedes.* *Batalleria* **19**, 41–46.

Taylor, P.D., Tan, S.-H.A., Kudryavtsev, A.B. & Schopf, J.W. (2016) Carbonate mineralogy of a tropical bryozoan biota and its vulnerability to ocean acidification. *Marine Biology Research* **12**, 776–780.

Taylor, P.D. & Taylor, A.B. (2012) Bryozoans from the Pliocene Coralline Crag of Suffolk: a brief review. Pp. 163–173 in Dixon, R. (ed.) *A celebration of Suffolk geology.* GeoSuffolk, Ipswich.

Taylor, P.D., Vinn, O. & Wilson, M.A. (2010) Evolution of biomineralisation in 'lophophorates'. *Special Papers in Palaeontology* **84**, 317–333.

Taylor, P.D. & Voigt, E. (1999) An unusually large cyclostome bryozoan (*Pennipora anomalopora*) from the Upper Cretaceous of Maastricht. *Bulletin de l'Institut Royal des Sciences Naturelles de Belgique, Sciences de la Terre* **69**, 165–171.

Taylor, P.D. & Voigt, E. (2006) Symbiont bioclaustrations in Cretaceous cyclostome bryozoans. *Courier Forschungsinstitut Senckenberg* **257**, 131–136.

Taylor, P.D. & Waeschenbach, A. (2015) Phylogeny and diversification of bryozoans. *Palaeontology* **58**, 585–599.

Taylor, P.D. & Waeschenbach, A. (2019) Phylogenetic affinities of *Crisulipora* and the dual origin of branch articulation in cyclostome bryozoans. *Australasian Palaeontological Memoirs* **52**, 155–161.

Taylor, P.D., Waeschenbach, A. & Florence, W. (2011) Phylogenetic position and systematics of the bryozoan *Tennysonia*: further evidence for convergence and plasticity in skeletal morphology among cyclostome bryozoans. *Zootaxa* **3010**, 58–68.

Taylor, P.D., Waeschenbach, A., Smith, A.M. & Gordon, D.P. (2015) In search of phylogenetic congruence between molecular and morphological data in bryozoans with extreme adult heteromorphy. *Systematics and Biodiversity* **13**, 525–544.

Taylor, P.D. & Weedon, M.J. (1996) Skeletal ultrastructure and affinities of eleid (meliceritid) cyclostomate bryozoans. Pp. 341–350 in Gordon, D.P., Smith, A.M. and Grant-Mackie, J.A. (eds) *Bryozoans in space and time*. NIWA, Wellington, 442 pp.

Taylor, P.D. & Weedon, M.J. (2000) Skeletal ultrastructure and phylogeny of cyclostome bryozoans. *Zoological Journal of the Linnean Society* **128**, 337–399.

Taylor, P.D. & Wilson, M.A. (1994) *Corynotrypa* from the Ordovician of North America: colony form in a primitive stenolaemate bryozoan. *Journal of Paleontology* **68**, 241–257.

Taylor, P.D. & Wilson, M.A. (1996) *Cuffeyella*, a new bryozoan genus from the Late Ordovician of North America, and its bearing on the origin of the post-Paleozoic cyclostomes. Pp. 351–360 in Gordon, D.P., Smith, A.M. & Grant-Mackie, J.A. (eds) *Bryozoans in space and time*. NIWA, Wellington.

Taylor, P.D. & Wilson, M.A. (1999a) *Dianulites*: an unusual Ordovician bryozoan with a high-magnesium calcite skeleton. *Journal of Paleontology* **73**, 38–48.

Taylor, P.D. & Wilson, M.A. (1999b) Middle Jurassic bryozoans from the Carmel Formation of southwestern Utah. *Journal of Paleontology* **73**, 816–830.

Taylor, P.D. & Wilson, M.A. (2003) Palaeoecology and evolution of marine hard substrate communities. *Earth-Science Reviews* **62**, 1–103.

Taylor, P.D., Wilson, M.A. & Bromley, R.G. (1999) A new ichnogenus for etchings made by cheilostome bryozoans into calcareous substrates. *Palaeontology* **42**, 595–604.

Taylor, P.D. & Zaborski, P.M. (2002) A Late Cenomanian bryozoan biostrome from north-eastern Nigeria. *Cretaceous Research* **23**, 241–253.

Temereva, E.N. (2017) Innervation of the lophophore suggests that the phoronid *Phoronis ovalis* is a link between phoronids and bryozoans. *Scientific Reports* **7**, 14440, 16 pp.

Temkin, M.H. (1996) Comparative fertilization biology of gymnolaemate bryozoans. *Marine Biology* **127**, 329–339.

Thiel, D.L., Cuffey, R.J. & Kowalczyk, F.J. (1996) Fossil 'Rolling Stones': bryozoan nodules in the Keyser Limestone (Latest Silurian) at the Mexico Railroad Cut, Central Pennsylvania. *Pennsylvania Geology* **27**, 2–7.

Thiel, M. & Gutow, L. (2005) The ecology of rafting in the marine environment. II. The rafting organisms and community. *Oceanography and Marine Biology: An Annual Review* **43**, 279–418.

Thomas, F.C., Hardy, I.A. & Rashid, H. (2003) Bryozoan-rich layers in surficial Labrador Slope sediments, eastern Canadian Arctic. *Canadian Journal of Earth Science* **40**, 337–350.

Thompson, T.E. (1958) The natural history, embryology, larval biology and post-larval development of *Adalaria proxima* (Alder and Hancock) (Gastropoda Opithobranchia). *Proceedings of the Royal Society, Series B* **686**, 1–57.

Thomsen, E. (1976) Depositional environment and development of Danian bryozoan biomicrite mounds (Karlby Klint, Denmark). *Sedimentology* **23**, 485–509.

Thomsen, E. (1977) Relations between encrusting bryozoans and substrate: an example from the Danian of Denmark. *Bulletin of the Geological Society of Denmark* **26**, 133–145.

Thomsen, E. & Håkansson, E. (1995) Sexual versus asexual dispersal in clonal animals: examples from cheilostome bryozoans. *Paleobiology* **21**, 496–508.

Thorpe, J.P., Shelton, G.A.B. & Laverack, M.S. (1975) Electrophysiology and co-ordinated behavioural responses in the colonial bryozoan *Membranipora membranacea* (L.). *Journal of Experimental Biology* **62**, 389–404.

Tilbrook, K.J. (1997) Barnacle and bivalve associates of a bryozoan-coral symbiosis from the Coralline Crag (Pliocene) of England. *Tertiary Research* **18**, 7–22.

Tilbrook, K.J. & Cook, P.L. (2005) Petraliellidae Harmer, 1957 (Bryozoa: Cheilostomata) from Queensland, Australia. *Systematics and Biodiversity* **2**, 319–339.

Tilbrook, K.J. & De Grave, S. (2005) A biogeographical analysis of Indo-West Pacific cheilostome bryozoans. Pp. 341–349 in Moyano G., H.I., Cancino, J.M. & Wyse Jackson, P.N. (eds) *Bryozoan studies 2004*. Balkema, Leiden.

Tilbrook, K.J. & Gordon, D.P. (2015) Bryozoa from the Straits of Johor, Singapore, with the description of a new species. *Raffles Bulletin of Zoology Supplement* **31**, 255–263.

Todd, C.D. & Turner, S.J. (1988) Ecology of intertidal and sublittoral cryptic epifaunal assemblages. 2. Nonlethal overgrowth of encrusting bryozoans by colonial ascidians.

Journal of Experimental Marine Biology and Ecology **115**, 113–126.

Todd, J.A. (1994) The role of bioimmuration in the exceptional preservation of fossil ctenostomates, including a Jurassic species of *Buskia*. Pp. 187–192 in Hayward, P.J., Ryland, J.S. & Taylor, P.D. (eds) *Biology and palaeobiology of bryozoans*. Olsen & Olsen, Fredensborg.

Todd, J.A. (2000) The central role of ctenostomes in bryozoan phylogeny. Pp. 104–135 in Herrera Cubilla, A. & Jackson, J.B.C. (eds) *Proceedings of the 11th International Bryozoology Association Conference*. Smithsonian Tropical Research Institute, Balboa.

Todd, J.A. & Hagdorn, H. (1993) First record of Muschelkalk Bryozoa: the earliest ctenostome body fossils. *Sonderbände der Gesellschaft für Naturkunde in Württemberg* **2**, 285–286.

Todd, J.A., Taylor, P.D. & Favorskaya, T.A. (1997) A bioimmured ctenostome bryozoan from the Early Cretaceous of the Crimea and the new genus *Simplicidium*. *Geobios* **30**, 205–213.

Tolokonnikova, Z. & Ernst, A. (2010) Palaeobiogeography of Famennian (Late Devonian) bryozoans. *Palaeogeography, Palaeoclimatology, Palaeoecology* **298**, 360–369.

Tolokonnikova, Z.A., Ernst, A. & Wyse Jackson, P.N. (2014) Palaeobiogeography and diversification of Tournaisian-Viséan bryozoans (lower-middle Mississippian, Carboniferous) from Eurasia. *Palaeogeography, Palaeoclimatology, Palaeoecology* **414**, 200–211.

Tompsett, S., Porter, J.S. & Taylor, P.D. (2009) Taxonomy of the fouling cheilostome bryozoans *Schizoporella unicornis* (Johnston) and *Schizoporella errata* (Waters). *Journal of Natural History* **43**, 2227–2243.

Toscano, F. & Raspini, A. (2005) Epilithozoan fauna associated with ferromanganese crustgrounds on the continental slope segment between Capri and Li Galli Islands (Bay of Salerno, Northern Tyrrhenian Sea, Italy). *Facies* **50**, 427–441.

Tsyganov-Bodounov, A., Hayward, P.J., Porter, J.S. & Skibinskic, D.O.F. (2009) Bayesian phylogenetics of Bryozoa. *Molecular Phylogenetics and Evolution* **52**, 904–910.

Tuckey, M.E. (1990a) Biogeography of Ordovician bryozoans. *Palaeogeography, Palaeoclimatology, Palaeoecology* **77**, 91–126.

Tuckey, M.E. (1990b) Distributions and extinctions of Silurian Bryozoa. *Geological Society, London, Memoirs* **12**, 197–206.

Tuckey, M.E. & Anstey, R L. (1992) Late Ordovician extinctions of bryozoans. *Lethaia* **25**, 111–117.

Ulrich, E.O. (1890) Palaeozoic Bryozoa. *Report of the Geological Survey of Illinois* **8**, 283–688.

Ulrich, E.O. & Bassler, R.S. (1904) A revision of the Paleozoic Bryozoa: part I – on genera and species of Ctenostomata. *Smithsonian Miscellaneous Collections* **45**, 256–294.

Urbanek, A. (2004) Morphogenetic gradients in graptolites and bryozoans. *Acta Palaeontologica Polonica* **49**, 485–504.

Utgaard, J. (1973) Mode of colony growth, autozooids and polymorphism in the bryozoan order Cystoporata. Pp. 317–360 in Boardman, R.S., Cheetham, A.H. and Oliver, W.J. (eds) *Animal colonies*. Dowden, Hutchinson and Ross, Stroudsburg.

Valentine, J.W. (1973) Coelomate superphyla. *Systematic Biology* **22**, 97–102.

Vávra, N.R. (1987) Bryozoa from the Early Miocene of the Central Paratethys: biogeographical and biostratigraphical aspects. Pp. 285–292 in Ross, J.R.P. (ed.) *Bryozoa: present and past*. Western Washington University, Bellingham.

Vávra, N.R. (2000) Biogeographical aspects of bryozoan faunas of the Central Paratethys from the Miocene. Pp. 392–399 in Herrera Cubilla, A. & Jackson, J.B.C. (eds) *Proceedings of the 11th International Bryozoology Association Conference*. Smithsonian Tropical Research Institute, Balboa.

Vávra, N.R. (2012) The use of Early Miocene bryozoan faunal affinities in the Central Paratethys for inferring climatic change and seaway connections. Pp. 401–418 in Ernst, A., Schäfer, P. & Scholz, J. (eds) *Bryozoan studies 2010*. Lecture Notes in Earth System Sciences **143**. Springer, Berlin.

Vennin, E. (2007) Coelobiontic communities in Neptunian fissures of synsedimentary tectonic origin in Permian reef, southern Urals, Russia. *Geological Society, London, Special Publications* **275**, 211–227.

Vermeij, G.J. (1977) The Mesozoic marine revolution: evidence from snails, predators and grazers. *Paleobiology* **3**, 245–258.

Vermeij, G.J. (1987) *Evolution and escalation*. Princeton University Press, Princeton, 527 pp.

Vieira, L.M., Migotto, A.E. & Winston, J.E. (2014) Ctenostomatous Bryozoa from São Paulo, Brazil, with descriptions of twelve new species. *Zootaxa* **3889**, 485–524.

Vieira, L.M., Spencer Jones, M.E., Winston, J.E., Migotto, Alvaro E. & Marques, A.C. (2014) Evidence for polyphyly of the genus *Scrupocellaria* (Bryozoa: Candidae) based on a phylogenetic analysis of morphological characters. *PLoS ONE* **9**(4), e95296.

Vieira, L.M. & Stampar, S.N. (2014) A new *Fenestrulina* (Bryozoa, Cheilostomata) commensal with tube-dwelling anemones (Cnidaria, Ceriantharia) in the tropical southwestern Atlantic. *Zootaxa* **3780**, 365–374.

Vignols, R.M., Valentine, A.M., Finlayson, A.G., Harper, E.M., Schöne, B.R., Leng, M.J., Sloane, H.J. & Johnson, A.L.A. (2018) Marine climate and hydrography of the Coralline Crag (early Pliocene, UK): isotopic evidence from 16 benthic invertebrate taxa. *Chemical Geology* **526**, 62–83.

Vinn, O. (2012) Palaeobiology of cryptic fauna beneath early Sheinwoodian (Silurian) stromatoporoids from Saaremaa, Estonia. *GFF* **134**, 335–337.

Vinn, O., Ernst, A., Wilson, M.A. & Toom, U. (2019) Symbiosis of conulariids with trepostome bryozoans in the Upper Ordovician of Estonia (Baltica). *Palaeogeography, Palaeoclimatology, Palaeoecology* **518**, 89–96.

Vinn, O. & Mutvei, H. (2009) Calcareous tube-worms of the Phanerozoic. *Estonian Journal of Earth Sciences* **58**, 286–296.

Vinn, O. & Toom, U. (2016a) Rugosan epibionts on vertical stems from the Ludlow and Pridoli of Saaremaa, Estonia (Baltica). *Palaios* **31**, 35–40.

Vinn, O. & Toom, U. (2016b) A sparsely encrusted hardground with abundant *Trypanites* borings from the Llandovery of the Velise River, western Estonia (Baltica). *Estonian Journal of Earth Sciences* **65**, 19–26.

Vinn, O., Wilson, M.A. & Motus M.-A. (2014) The earliest giant *Osprioneides* borings from the Sandbian (Late Ordovician) of Estonia. *PLoS ONE* **9**(6), e99455

Vinn, O. & Zaton, M. (2012) Phenetic phylogenetics of tentaculitoids – extinct problematic calcareous tube-forming organisms. *GFF* **134**, 145–156.

Vinogradov, A.V. (1996) New fossil freshwater bryozoans from the Asiatic part of Russia and Kazakhstan. *Paleontological Journal* **30**, 284–292.

Viskova, L.A. (1991) Patterns of organization of bryozoan articulated colonies. *Paleontological Journal* **25**, 1–12.

Viskova, L.A. (1992) Morskie Postpaleozoyskie Mshanki. *Trudy Paleontologicheskogo Instituta. Akademiya Nauk SSSR, Moscow* **250**, 1–187. [in Russian]

Viskova, L.A. (2009) New data on the colonial morphology of the Jurassic bryozoans of the Class Stenolaemata. *Paleontological Journal* **43**, 543–549.

Viskova, L.A. (2011) Rare bryozoans (Stenolaemata) with bilateral colonies from the Jurassic and Cretaceous of the East European Platform. *Paleontological Journal* **45**, 40–51.

Viskova, L.A. (2016) Bryozoans of the Order Melicerititida: morphological features and position of the order in the taxonomic structure of the class Stenolaemata. *Paleontological Journal* **50**, 153–162.

Viskova, L.A. & Ivantsov, A. Yu (1999) The earliest uncalcified bryozoan from the Ordovician from the Ordovician in the vicinity of Saint Petersburg. *Paleontological Journal* **33**, 26–29.

Viskova, L.A. & Koromyslova, A.V. (2012) *Tamanicella* gen. nov., a new genus of bryozoans forming the Late Miocene bioherms of Cape Panagia in the Taman Peninsula (Russia). *Paleontological Journal* **46**, 29–43.

Viskova, L.A. & Pakhnevich, A.V. (2009) A new boring bryozoan from the Middle Jurassic of the Moscow Region and its micro-CT research. *Paleontological Journal* **44**, 157–167.

Voigt, E. (1979) The preservation of slightly or non-calcified fossil Bryozoa (Ctenostomata and Cheilostomata) by bioimmuration. Pp. 541–564 in Larwood, G.P. & Abbott, M.B. (eds) *Advances in bryozoology*. Academic Press, London.

Voigt, E. (1981a) Repartition et Utilisation Stratigraphique des Bryozoaires due Crétacé Moyen (Aptien-Coniacien). *Cretaceous Research* **2**, 439–462.

Voigt, E. (1981b) Upper Cretaceous bryozoan-seagrass association in the Maastrichtian of the Netherlands. Pp. 281–298 in Larwood, G.P. & Nielsen, C. (eds) *Recent and fossil Bryozoa*. Olsen & Olsen, Fredensborg.

Voigt, E. (1982) Über die wahrscheinliche Funktion der Frontalwand-Tuberkeln als Distanzhalter bei cheilostomen Bryozoen (fossil und rezent). *Verhandlungen des Naturwissenschaftlichen Vereins in Hamburg* **33**, 131–154.

Voigt, E. (1983) Zur Biogeographie der europäischen Oberkreide-bryozoenfauna. *Zitteliana* **10**, 317–347.

Voigt, E. (1987) Thalassinoid burrows in the Maastrichtian Chalk Tuff near Maastricht (The Netherlands) as a fossil hardground microcavern biotope of Cretaceous bryozoans. Pp. 293–300 in Ross, J.R.P. (ed.) *Bryozoa: present and past*. Western Washington University, Bellingham.

Voigt, E. (1991) Mono- or polyphyletic evolution of cheilostomatous bryozoan divisions? *Bulletin de la Societe des Sciences Naturelles de l'Ouest de la France Mémoire* **H.S. 1**, 505–522.

Voigt, E. (1992) Stütz-, Anker- und Haftorgane bei rezenten und fossilen Bryozoen (Cyclostomata und Cheilostomata). *Verhandlungen des Naturwissenschaftlichen Vereins in Hamburg, NF* **33**, 155–189.

Voigt, E. & Flor, F.D. (1970) Homöomorphien bei fossilen cyclostomen Bryozoen, dargestellt am Beispiel der Gattung *Spiropora* Lamouroux 1821. *Mitteilungen aus dem Geologisch-Paläontologischen Institut der Universität Hamburg* **39**, 7–96.

Voigt, E. & Vávra, N. (2006) *Stylodefranciopora turris* nov. gen. nov. sp. – eine neue cyclostome Bryozoe aus der Oberkreide von Schleswig-Holstein (Deutschland). *Courier Forschungsinstitut Senckenberg* **257**, 137–149.

Von Dassow, M. (2005) Effects of ambient flow and injury on the morphology of a fluid transport system in a bryozoan. *Biological Bulletin* **208**, 47–59.

Waeschenbach, A., Cox, C.C., Littlewood, D.T.J., Porter, J.S. & Taylor, P.D. (2009) First molecular estimate of cyclostome bryozoan phylogeny confirms extensive homoplasy among skeletal characters used in traditional taxonomy. *Molecular Phylogenetics and Evolution* **52**, 241–251.

Waeschenbach, A., Porter, J.S. & Hughes, R.N. (2012) Molecular variability in the *Celleporella hyalina* (Bryozoa; Cheilostomata) species complex: evidence for cryptic speciation from complete mitochondrial genomes. *Molecular Biology Reports* **39**, 8601–8614.

Waeschenbach, A., Taylor, P.D. & Littlewood, D.T.J. (2012) A molecular phylogeny of bryozoans. *Molecular Phylogenetics and Evolution* **62**, 718–735.

Wahl, M. (1989) Marine epibiosis. I. Fouling and antifouling: some basic aspects. *Marine Ecology Progress Series* **58**, 175–189.

Walker, S.E. (1988) Taphonomic significance of hermit crabs (Anomura: Paguridea): epifaunal hermit crab–infaunal gastropod example. *Palaeogeography, Palaeoclimatology, Palaeoecology* **63**, 45–71.

Walker, S.E. (1992) Criteria for recognizing marine hermit crabs in the fossil record using gastropod shells. *Journal of Paleontology* **66**, 535–558.

Walter, B. (1970) Les Bryozoaires Jurassiques en France. *Documents des Laboratoires de Géologie de la Faculté des Sciences de Lyon*, **35** [for 1969], 1–328.

Walter, B. (1977) Un gisement de bryozoaires Aptiens dans le Gard. *Geobios* **10**, 325–336.

Walter, B. (1996) La faune de bryozoaires de la transgression hauterivienne dans le Bassin de Paris. *Geobios* **29**, 5–11.

Ward, M.A. & Thorpe, J.P. (1989) Assessment of space utilisation in a subtidal temperate bryozoan community. *Marine Biology* **103**, 215–224.

Wass, R. (1977) Branching patterns and phylogeny of the family Vittaticellidae (Bryozoa: Cheilostomata). *Australian Journal of Zoology* **25**, 103–119.

Wass, R.E., Conolly, J.R. & MacIntyre, R.J. (1970) Bryozoan carbonate sand continuous along southern Australia. *Marine Geology* **9**, 63–73.

Waters, A.W. (1891) On chilostomatous characters in Melicertitidae and other fossil Bryozoa. *Annals and Magazine of Natural History, Series* 6 **8**, 48–53.

Watts, P.C. & Thorpe, J.P. (2006) Influence of contrasting larval developmental types upon the population-genetic structure of cheilostome bryozoans. *Marine Biology* **149**, 1093–1101.

Watts, P.C., Thorpe, J.P. & Taylor, P.D. (1998) Natural and anthropogenic dispersal mechanisms in the marine environment: a study using cheilostome Bryozoa. *Philosophical Transactions of the Royal Society, Series B* **353**, 453–464.

Weedon, M.J. (1997) Mural 'hoods' in the basal walls of cyclostome bryozoans and their taxonomic distribution. *Species Diversity* **2**, 105–119.

Wender, P.A., Hardman, C.T., Ho, S., Jeffreys, M.S., Maclaren, J.K. and 5 others (2017) Scalable synthesis of bryostatin 1 and analogs, adjuvant leads against latent HIV. *Science* **358**, 218–223.

Wendt, D.E. (1998) Effect of larval swimming duration on growth and reproduction of *Bugula neritina* (Bryozoa) under field conditions. *Biological Bulletin* **195**, 126–135.

Whitehead, J.W., Seed, R. & Hughes, R.N. (1996) Factors controlling spinosity in the epialgal bryozoan *Flustrellidra hispida* (Fabricius). Pp. 367–375 in Gordon, D.P., Smith, A.M. & Grant-Mackie, J.A. (eds) *Bryozoans in space and time*. NIWA, Wellington.

Wieczorek, S.K. & Todd, C.D. (1997) Inhibition and facilitation of bryozoan and ascidian settlement by natural multi-species biofilms: effects of film age and the roles of active and passive larval attachment. *Marine Biology* **128**, 463–473.

Williams, M., Haywood, A.M., Harper, E.M., Johnson, A.L.A., Knowles, T., Leng, M.J., Lunt, D.J., Okamura, B., Taylor, P.D. & Zalasiewicz, J. (2009) Pliocene climate and seasonality in North Atlantic shelf seas. *Philosophical Transactions of the Royal Society, Series A* **367**, 85–108.

Wilson, M.A. (1985) Disturbance and ecologic succession in an Upper Ordovician cobble-dwelling hardground fauna. *Science* **228**, 575–577.

Wilson, M.A. (1986) Coelobites and spatial refuges in a Lower Cretaceous cobble-dwelling hardground fauna. *Palaeontology* **29**, 691–703.

Wilson, M.A., Bosch, S. & Taylor, P.D. (2014) Middle Jurassic (Callovian) cyclostome bryozoans from the Tethyan tropics (Matmor Formation, southern Israel). *Bulletin of Geosciences* **90**, 51–63.

Wilson, M.A., Buttler, C.J. & Taylor, P.D. (2019) Bryozoans as taphonomic engineers, with examples from the Upper Ordovician (Katian) of Midwestern North America. *Lethaia* **52**, 403–409.

Wilson, M.A., Ozanne, C.R. & Palmer, T.J. (1998) Origin and paleoecology of free-rolling oyster accumulations (ostreoliths). *Palaios* **13**, 70–78.

Wilson, M.A. & Palmer, T.J. (1992) Hardgrounds and hardground faunas. *University of Wales, Aberystwyth, Institute of Earth Sciences* **9**, 1–131.

Wilson, M.A. & Taylor, P.D. (2001) Palaeoecology of hard substrate faunas from the Cretaceous Qahlah Formation of the Oman Mountains. *Palaeontology* **44**, 21–41.

Wilson, M.A. & Taylor, P.D. (2006) Predatory drillholes and partial mortality in Devonian colonial metazoans. *Geology* **34**, 565–568.

Wilson, M.A. & Taylor, P.D. (2012) Palaeoecology, preservation and taxonomy of encrusting ctenostome bryozoans inhabiting ammonite body chambers in the Late Cretaceous Pierre Shale of Wyoming and South Dakota, USA. Pp. 419–433 in Ernst, A., Schäfer, P. & Scholz, J. (eds) *Bryozoan studies 2010*. Springer, Berlin.

Wilson, M.A. & Taylor, P.D. (2014) The morphology and affinities of *Allonema* and *Ascodictyon*, two abundant Palaeozoic encrusters commonly misattributed to the ctenostome bryozoans. *Studi Trentini di Scienze Naturali* **94**, 259–266.

Wilson, M.A. & Taylor, P.D. (2016) A new runner-like cyclostome bryozoan from the Bromide Formation (Sandbian, Upper Ordovician) of Oklahoma and its phylogenetic affinities. *Journal of Paleontology* **90**, 413–417.

Wilson, M.A. & Taylor, P.D. (2017) Exceptional pyritized cyanobacterial mats encrusting brachiopod shells from the Upper Ordovician (Katian) of the Cincinnati, Ohio, region. *Palaios* **32**, 673–677.

Winston, J.E. (1977) Distribution and ecology of estuarine ectoprocts: a critical review. *Chesapeake Science* **18**, 34–57.

Winston, J.E. (1978) Polypide morphology and feeding behavior in marine ectoprocts. *Bulletin of Marine Science* **28**, 1–31.

Winston, J.E. (1981) Feeding behavior of modern bryozoans. *University of Tennessee Department of Geological Sciences Studies in Geology* **5**, 1–21.

Winston, J.E. (1982) Marine bryozoans (Ectoprocta) of the Indian River area, Florida. *Bulletin of the American Museum of Natural History* **173**, 99–176.

Winston, J.E. (1984) Why bryozoans have avicularia – a review of the evidence. *American Museum Novitates* **2789**, 1–26.

Winston, J.E. (1986) Victims of avicularia. *PSZNI: Marine Ecology* **7**, 193–199.

Winston, J.E. (1991) Avicularian behaviour – a progress report. *Bulletin de la Société des Sciences Naturelle de l'Ouest de la France, Mémoire* **H.S. 1**, 531–540.

Winston, J.E. (2009) Cold comfort: systematics and biology of Antarctic bryozoans. Pp. 205–221 in Krupnik, M., Lang, A. & Miller, S.E. (eds) *Smithsonian at the poles: contribution to International Polar Year science, I.* Smithsonian Institution Scholarly Press, Washington DC.

Winston, J.E. (2010) Life in the colonies: learning the alien ways of colonial organisms. *Integrative and Comparative Biology* **50**, 919–933.

Winston, J.E. (2016) Bryozoa of Floridan *Oculina* reefs. *Zootaxa* **4071**, 1–81.

Winston, J.E. & Cheetham, A.H. (1984) The bryozoan *Nellia tenella* as a living fossil. Pp. 257–265 in Eldredge, N. & Stanley, S. (eds) *Living fossils.* Springer Verlag, New York.

Winston, J.E. & Håkansson, E. (1986) The interstitial fauna of the Capron Shoals, Florida. *American Museum Novitates* **2865**, 1–98.

Winston, J.E. & Håkansson, E. (1989) Molting by *Cupuladria doma*, a free-living bryozoan. *Bulletin of Marine Science* **44**, 1152–1158.

Winston, J.E. & Hayward, P.J. (1994) Bryozoa of the US Antarctic Research Program: preliminary report. Pp. 205–210 in Hayward, P.J., Ryland, J.S. & Taylor, P.D. (eds) *Biology and palaeobiology of bryozoans.* Olsen & Olsen, Fredensborg.

Winston, J.E. & Migotto, A.E. (2005) A new encrusting interstitial marine fauna from Brazil. *Invertebrate Biology* **124**, 79–87.

Winston, J.E. & Vieira, L.M. (2013) Systematics of interstitial encrusting bryozoans from southeastern Brazil. *Zootaxa* **3710**, 101–146.

Wisshak, M., Berning, B., Jakobsen, J. & Freiwald, A. (2015) Temperate carbonate production: biodiversity of calcareous epiliths from intertidal to bathyal depths (Azores). *Marine Biodiversity* **45**, 87–112.

Wood, A.C.L. & Probert, P.K. (2013) Bryozoan-dominated benthos of Otago shelf, New Zealand: its associated fauna, environmental setting and anthropogenic threats. *Journal of the Royal Society of New Zealand* **43**, 231–249.

Wood, A.C.L., Probert, P.K., Rowden, A.A. & Smith, A.M. (2012) Complex habitat generated by marine bryozoans: a review of its distribution, structure, diversity, threats and conservation. *Aquatic Conservation: Marine and Freshwater Ecosystems* **22**, 547–563.

Wood, T.S. (1983) General features of the Class Phylactolaemata. Pp. 287–303 in Boardman, R.S., Cheetham, A.H., Blake, D.B., Utgaard, J., Karklins, O.L., Cook, P.L., Sandberg, P.A., Lutaud, G. & Wood, T.S. *Treatise on invertebrate paleontology, part G, Bryozoa, revised.* Geological Society of America and University of Kansas, Boulder and Lawrence.

Wood, T.S. (2002) Freshwater bryozoans: a zoogeographical assessment. Pp. 339–345 in Wyse Jackson, P.N., Buttler, C.J. & Spencer Jones, M.E. (eds) *Bryozoan studies 2001: proceedings of the 12th International Bryozoology Association Symposium.* Balkema, Lisse.

Wood, T.S. (2019) What phylactolaemate bryozoans actually eat. P. 83 in *Abstracts of the 18th International Bryozoology Association Conference, Liberec,* http://18iba.tul.cz/images/Liberec_16_to_22_June_2019-abstract_book_final2.pdf.

Wood, T.S. & Lore, M.B. (2005) The higher phylogeny of phylactolaemate bryozoans inferred from 18S ribosomal DNA sequences. Pp. 361–367 in Moyano, H.I., Cancino, J.M. & Wyse Jackson, P.N. (eds) *Bryozoan studies 2004.* Taylor & Francis, London.

Wood, T., Anurakpongsatorn, P. & Mahujchariyawong, J. (2006) Swimming zooids: an unusual dispersal strategy in the ctenostome bryozoan, *Hislopia. Linzer biologische Beiträge* **38**, 71–75.

Woollacott, R.M. & Zimmer, R.L. (1972) Origin and structure of the brood chamber in *Bugula neritina* (Bryozoa). *Marine Biology* **16**, 165–170.

Wulff, J.I. (1990) Biostratinomic utility of *Archimedes* in environmental interpretation. *Palaios* **5**, 160–166.

Wyer, D.W. & King, P.E. (1973) Relationships between some British littoral and sublittoral pycnogonids. Pp. 199–208 in Larwood, G.P. (ed.) *Living and fossil Bryozoa*. Academic Press, London.

Wyse Jackson, P.N. (2006) Bryozoa from Waulsortian buildups and their lateral facies (Mississippian, Carboniferous) in Belgium and Ireland. *Courier Forschungsinstitut Senckenberg* **257**, 149–159.

Wyse Jackson, P.N. & Bancroft, A.J. (1994) Possible opercular structures in the fenestrate bryozoan *Thamniscus* from the Upper Carboniferous of northern England. Pp. 215–218 in Hayward, P.J., Ryland, J.S. & Taylor, P.D. (eds) *Biology and palaeobiology of bryozoans*. Olsen & Olsen, Fredensborg.

Wyse Jackson, P.N. & Bancroft, A.J. (1995) Generic revision of the cryptostome bryozoan *Rhabdomeson* Young and Young, 1874, with descriptions of two species from the Lower Carboniferous of the British Isles. *Journal of Paleontology* **69**, 28–45.

Wyse Jackson, P.N. & Buttler, C.J. (2015) Part G, revised, volume 2, chapter 3: preparation, imaging, and conservation of Paleozoic bryozoans for study. *Treatise Online* **63**, 1–15.

Wyse Jackson, P.N., Ernst, A. & Suárez Andrés, J.L. (2017) Articulation in the Family Rhabdomesidae (Cryptostomata: Bryozoa) from the Mississippian of Ireland. *Irish Journal of Earth Sciences* **35**, 35–44.

Wyse Jackson, P.N. & Key, M.M., Jr (2007) Borings in trepostome bryozoans from the Ordovician of Estonia: two ichnogenera produced by a single maker, a case of host morphology. *Lethaia* **40**, 237–252.

Wyse Jackson, P.N. & Key, M.M., Jr (2014) Epizoic bryozoans on cephalopods through the Phanerozoic: a review. *Studi Trentini di Scienze Naturali* **94**, 283–291.

Wyse Jackson, P.N., Key, M.M., Jr & Coakley, S.P. (2014) Epizoozoan trepostome bryozoans on nautiloids from the Upper Ordovician (Katian) of the Cincinnati Arch region, USA: an assessment of growth, form, and water flow dynamics. *Journal of Paleontology* **88**, 475–487.

Wyse Jackson, P.N., Taylor, P.D. & Tilsley, J.W. (1999) The 'Balladoole Coral' from the Lower Carboniferous of the British Isles, reinterpreted as the unusual cystoporate bryozoan *Meekoporella* Moore & Dudley, 1944. *Proceedings of the Yorkshire Geological Society* **52**, 257–268.

Xia, F.-s. (2002) Fenestrate Bryozoa with avicularia-like structures from the Middle Jurassic of north Tibet and the origin of cheilostome bryozoans. *Acta Micropalaeontologica Sinica* **19**, 237–255.

Xia, F.-s., Zhang, S.-g. & Wang, Z.-Z. (2007) The oldest bryozoans: new evidence from the Late Tremadocian (Early Ordovician) of East Yangtze Gorges in China. *Journal of Paleontology* **81**, 1308–1326.

Xing, J. & Qian, P. (1999) Tower cells of the marine bryozoan *Membranipora membranacea*. *Journal of Morphology* **239**, 121–130.

Yagunova, E.B. & Ostrovsky, A.N. (2008) Encrusting bryozoan colonies on stones and algae: variability of zooidal size and its possible causes. *Journal of the Marine Biological Association of the United Kingdom* **88**, 901–908.

Yagunova, E.B. & Ostrovsky, A.N. (2010) The influence of substrate type on sexual reproduction of the bryozoan *Cribrilina annulata* (Gymnolaemata, Cheilostomata): a case study from Arctic seas. *Marine Biology Research* **6**, 263–270.

Yoshioka, P.M. (1982a) Predator-induced polymorphism in the bryozoan *Membranipora membranacea* (L.). *Journal of Experimental Marine Biology and Ecology* **61**, 233–242.

Yoshioka, P.M. (1982b) Role of planktonic and benthic factors in the population dynamics of the bryozoan *Membranipora membranacea*. *Ecology* **63**, 457–468.

Yoshioka, P.M. (1986) Competitive coexistence of the dorid nudibranchs *Dordella steinbergae* and *Corame pacifica*. *Marine Ecology Progress Series* **33**, 81–88.

Zabala, M. & Maluquer, P. (1988) Illustrated keys for the classification of Mediterranean Bryozoa. *Treballs del Museu de Zoologia, Barcelona* **4**, 1–284.

Zabala, M., Maluquer, P. & Harmelin, J.-G. (1993) Epibiotic bryozoans on deep-water scleractinian corals from the Catalonia Slope (western Mediterranean, Spain, France). *Scientia Marina* **57**, 65–78.

Zabin, C.J., Obernolte, R., Mackie, J.A., Gentry, J., Harris, L. & Geller, J. (2010) A non-native bryozoan creates novel substrate on the mud-flats in San Francisco Bay. *Marine Ecology Progress Series* **412**, 129–139.

Zágoršek, K. (1996) Paleoecology of the Eocene bryozoan marl in the Alpine-Carpathian region. Pp. 413–422 in Gordon, D.P., Smith, A.M. & Grant-Mackie, J.A. (eds) *Bryozoans in space and time*. NIWA, Wellington.

Zágoršek, K., Ramalho, L.V., Berning, B. & Araújo Távora, V. de (2014) A new genus of the family Jaculinidae (Cheilostomata, Bryozoa) from the Miocene of the tropical western Atlantic. *Zootaxa* **3838**, 98–112.

Zágoršek, K., Vávra, N. & Holcová, K. (2007) New and unusual Bryozoa from the Badenian (Middle Miocene) of the Moravian part of the Vienna Basin (Central Paratethys, Czech Republic). *Neues Jahrbuch für Geologie und Paläontologie, Abhandlungen* **243**, 201–215.

Zahl, P.A. & McLaughlin, J.J.A. (1957) Isolation and cultivation of zooxanthellae. *Nature* **180**, 199–200.

Zamora, S., Mayoral, E., Gámez Vintaned, J.A., Bajo, S. & Espílez, E. (2008) The infaunal echinoid *Micraster*: Taphonomic pathways indicated by sclerozoan trace and body fossils from the Upper Cretaceous of northern Spain. *Geobios* **41**, 15–29.

Zaton, M. & Borszcz, T. (2013) Encrustation patterns on post-extinction early Famennian (Late Devonian) brachiopods from Russia. *Historical Biology* **25**, 1–12.

Zaton, M., Machocka, S., Wilson, M.A., Marynowski, L. & Taylor, P.D. (2011) Origin and paleoecology of Middle Jurassic hiatus concretions from Poland. *Facies* **57**, 275–300.

Zaton, M., Wilson, M.A. & Zavar, E. (2011) Diverse sclerozoan assemblages encrusting large bivalve shells from the Callovian (Middle Jurassic) of southern Poland. *Palaeogeography, Palaeoclimatology, Palaeoecology* **307**, 232–244.

Zimmerman, L.S. & Cuffey, R.J. (1987) Species involved on Permian bryozoan bioherms, West Texas. Pp. 309–316 in Ross, J.R.P. (ed.) *Bryozoa: present and past*. Western Washington University, Bellingham.

Zuchsin, M. & Baal, C. (2007) Large gryphaeid oysters as habitats for numerous sclerobionts: a case study from the northern Red Sea. *Facies* **53**, 319–327.

Zuchsin, M. & Stachowitsch, M. (2009) Epifauna-dominated benthic shelf assemblages: lessons from the modern Adriatic Sea. *Palaios* **24**, 211–221.

Index

Note: page numbers in *italics* refer to figures.

Bryozoan Paleobiology, First Edition. Paul D. Taylor.
© 2020 Natural History Museum. Published 2020 by John Wiley & Sons Ltd.